Van Nostrand Reinhold Soil Science Series

Editor: Charles W. Finkl, Jnr., Florida Atlantic University

SOIL CLASSIFICATION / *Charles W. Finkl, Jnr.*
CHEMISTRY OF IRRIGATED SOILS / *Rachel Levy*
SOIL SALINITY: Two Decades of Research in Irrigated Agriculture /
    H. Frenkel and A. Meiri
ANDOSOLS / *Kim H. Tan*
PODZOLS / *Peter Buurman*
SOIL NUTRIENT AVAILABILITY: Chemistry and Concepts / *Y. K. Soon*
ADSORPTION PHENOMENA / *Robert D. Harter*
SOIL EROSION AND ITS CONTROL / *R. P. C. Morgan*
SOIL MICROMORPHOLOGY / *Georges Stoops and Hari Eswaran*
SOIL MINERAL WEATHERING / *J. A. Kittrick*
CHEMISTRY OF SOIL SOLUTIONS / *Adel M. Elprince*

Related Titles
ENCYCLOPEDIA OF SOIL SCIENCE, PART 1 / *Rhodes W. Fairbridge and
    Charles W. Finkl, Jnr.*
ORE FIELDS AND CONTINENTAL WEATHERING / *Jean-Claude Samama*
SOIL MECHANICS, 3rd ed. / *R. F. Craig*
SURFICIAL DEPOSITS OF THE UNITED STATES / *Charles B. Hunt*

SOIL MINERAL WEATHERING

Edited by
J. A. KITTRICK
Washington State University
Pullman, Washington

A Hutchinson Ross Publication

VNR VAN NOSTRAND REINHOLD COMPANY
New York

Copyright © 1986 by **Van Nostrand Reinhold Company Inc.**
Van Nostrand Reinhold Soil Science Series
Library of Congress Catalog Card Number: 85-26359
ISBN: 0-442-24642-0

All rights reserved. No part of this work covered by the copyrights hereon may be reproduced or used in any form or by any means— graphic, electronic, or mechanical, including photocopying, recording, taping, or information storage and retrieval systems— without permission of the publisher.

Manufactured in the United States of America.

Published by Van Nostrand Reinhold Company Inc.
115 Fifth Avenue
New York, New York 10003

Van Nostrand Reinhold Company Limited
Molly Millars Lane
Wokingham, Berkshire RG11 2PY, England

Van Nostrand Reinhold
480 Latrobe Street
Melbourne, Victoria 3000, Australia

Macmillan of Canada
Division of Gage Publishing Limited
164 Commander Boulevard
Agincourt, Ontario MIS 3C7, Canada

15 14 13 12 11 10 9 8 7 6 5 4 3 2 1

Library of Congress Cataloging-in-Publication Data
Main entry under title:
Soil mineral weathering.
   (Van Nostrand Reinhold soil science series)
   Reprints of articles from various periodicals.
   "A Hutchinson Ross publication."
   Includes index.
   1. Soil mineralogy—Addresses, essays, lectures.   2. Weathering— Addresses, essays, lectures.   I. Kittrick, J. A.   II. Series.
S592.55.S65   1986   552'.5   85-26359
ISBN 0-442-24642-0

CONTENTS

Series Editor's Foreword	vii
Preface	xi
Contents by Author	xiii
Introduction	1

PART I: MINERALS AND OCCURRENCE

Editor's Comments on Papers 1 Through 7   4

1   HENDRICKS, S. B., and W. H. FRY: The Results of X-ray and Microscopical Examinations of Soil Colloids
Soil Sci. **29**:457-471, 474-479 (1930)   9

2   JACKSON, M. L., S. A. TYLER, A. L. WILLIS, G. A. BOURBEAU, and, R. P. PENNINGTON: Weathering Sequence of Clay-Size Minerals in Soils and Sediments. I: Fundamental Generalizations
J. Phys. and Coll. Chem. **52**:1237-1260 (1948)   30

3   RICH, C. I., and S. S. OBENSHAIN: Chemical and Clay Mineral Properties of a Red-Yellow Podzolic Soil Derived from Muscovite Schist
Soil Sci. Soc. Am. Proc. **19**:334-339 (1955)   54

4   YOSHINAGA, N., and S. AOMINE: Imoglolite in Some Ando Soils
Soil Sci. Plant Nutr. (Tokyo) **8**:22-29 (1962)   60

5   JONES, R. C., and G. UEHARA: Amorphous Coatings on Mineral Surfaces
Soil Sci. Soc. Am. Proc. **37**:792-798 (1973)   68

6   SCHWERTMANN, U., D. G. SCHULZE, and E. MURAD: Indentification of Ferrihydrite in Soils by Dissolution Kinetics, Differential X-ray Diffraction, and Mossbauer Spectroscopy
Soil Sci. Soc. Am. J. **46**:869-875 (1982)   75

7   ESWARAN H., and B. C. WONG: A Study of a Deep Weathering Profile on Granite in Peninsular Malaysia: III. Alteration of Feldspars
Soil Sci. Soc. Am. J. **42**:154-158 (1978)   82

Contents

PART II: DISSOLUTION AND PRECIPITATION

Editor's Comments on Papers 8 Through 12   88

8   RAUSELL-COLOM, J. A., T. R. SWEATMAN, C. B. WELLS, and K. NORRISH: Studies in the Artificial Weathering of Mica
Experimental Pedology, E. G. Hallsworth and D. V. Crawford, ed.,
Butterworths, London, 1965, pp. 40–72   91

9   MORTLAND, M. M., K. LAWTON, and G. UEHARA: Alteration of Biotite to Vermiculite by Plant Growth
Soil Sci. **82**:477–481 (1956)   124

10   KITTRICK, J. A.: Mica-Derived Vermiculites as Unstable Intermediates
Clays Clay Miner. **21**:479–488 (1973)   129

11   VAN BREEMEN, N., J. MULDER, and C. T. DRISCOLL: Acidification and Alkalinization of Soils
Plant and Soil **75**:283–308 (1983)   139

12   EGGLETON, R. A., and P. R. BUSECK: High Resolution Electron Microscopy of Feldspar Weathering
Clays Clay Miner. **28**:173–178 (1980)   165

PART III: CHEMICAL PRINCIPLES

Editor's Comments on Papers 13 Through 18   172

13   GARRELS, R. M.: Some Free Energy Values from Geologic Relations
Amer. Miner. **42**:780–791 (1957)   176

14   KITTRICK, J. A.: Soil Minerals in the Al_2O_3-SiO_2-H_2O System and a Theory of Their Formation
Clays Clay Miner. **17**:157–167 (1969)   188

15   KITTRICK, J. A.: Solubility of Two High-Mg and Two High-Fe Chlorites Using Multiple Equilibria
Clays Clay Miner. **30**:167–179 (1982)   199

16   TARDY, Y., and R. M. GARRELS: A Method of Estimating the Gibbs Energies of Formation of Layer Silicates
Geochim. Cosmochim. Acta **38**:1101–1116 (1974)   212

17   HELGESON, H. C., R. M. GARRELS, and F. T. MACKENZIE: Evaluation of Irreversible Reactions in Geochemical Processes Involving Minerals and Aqueous Solutions—II: Applications
Geochim. Cosmochim. Acta **33**:455–481 (1969)   228

18   TSUZUKI, Y.: Solubility Diagrams for Explaining Zone Sequences in Bauxite, Kaolin, and Pyrophyllite-Diaspore Deposits
Clays Clay Miner. **24**:297–302 (1976)   255

Author Citation Index   261
Subject Index   267
About the Editor   271

SERIES EDITOR'S FOREWORD

The Van Nostrand Reinhold Soil Science Series attempts to provide cogent summaries of the field by reproducing classical and modern papers, ones that provide keys to understanding of critical turning points in the development of the discipline. Scientific literature today is so vast and widely dispersed, especially in a multifaceted discipline like soil science, that much valuable information becomes ignored by default. Many pioneering works are now coveted by libraries, and retrieval from the archives is not easy. In fact, many important papers published in the ephemeral literature are no longer available to serious or committed researchers through interlibrary loan. Other professionals devoted to teaching or burdened with administrative duties must be hard pressed to keep up with comprehensive arrays of technical literature spread through scores of journals. Most of us can, at best, skim only a few select journals to make copies of tables of contents, abstracts and summaries, and reviews in order to remain abreast of specialized and often limited aspects of the robust field of soil science as a whole.

This series in soil science, developed as a practical solution to this problem, reprints key papers and investigative landmarks that relate to a common theme. The papers are reproduced in facsimile, either in their entirety or in significant part, so readers can follow major original events in the field, not peruse paraphrased or abbreviated versions of others. Some foreign works have been especially translated for use in the series. Occasionally short, foreign language articles are reproduced from French or German journals.

Essays by the volume editor provide running commentaries that introduce readers to highlights in the field, provide critical evaluation of the significance of the various papers, and discuss the development of selected topics or subject areas. It is hoped that the volume editor's comments will ease the transition for the seasoned investigator who wishes to step into a new field of research as well as provide students and professors with a compact working library of most important scientific advances in soil science.

Areas of specialization in soil science are divided by the International Society of Soil Science into seven divisions or "commissions." The first six commissions cover soil physics, chemistry, mineralogy, biology, fertility, and technology. Because the scope of the field is so great, we concentrate initially on topics traditionally devoted to the seventh commission: soil morphology, genesis, classification, and geography. The series thus begins

Series Editor's Foreword

with volumes dealing with the major soils of the world: their recognition, characteristics, formation, distribution, and classification. Other volumes concentrate on topics in agronomy, soil-plant relationships, soil engineering topics, or melds of pure science with soil systems. The Van Nostrand Reinhold Soil Science Series plows deeply through the field, picking significant but timely topics on an eclectic basis.

Each volume in the series is edited by a specialist or authority in the area covered by the book. The volume editor's efforts reflect a concerted worldwide search, review, selection, and distillation of the primary literature contained in journals and monographs and in industrial and governmental reports. Individual volumes thus represent an information-selection and repackaging program of value to libraries, students, and professionals.

The books contain a preface, introduction, and highlight commentaries by the volume editor. Many volumes contain rare papers that are hard to locate and obtain, as well as landmark papers published in English for the first time. All volumes contain author citation and subject indexes of the contained papers, usually twenty to fifty key papers in a given subject area.

The present volume deals with soil mineral weathering, a complex subject that embraces aspects of epidiagenesis and pedogenesis. According to generally accepted definitions, weathering is the broad group of processes—such as the chemical action of air and rainwater and of plants and bacteria and the mechanical action of changes of temperature—whereby rocks on exposure are changed or altered, that is, they decay and finally crumble into soil. A somewhat narrower point of view is traditionally elicited in the soil sciences where three distinct phenomena in the weathering phase are specifically distinguished: (1) the disintegration and progressive alteration of rocks or other (preweathered) initial materials, (2) the displacement or migration of soluble or very fine-grained fractions of the weathering complex under the influence of percolating solutions, and (3) the role of biological activity in the weathering sequence as it involves plants and microbes as well as organic acids.

Generally speaking, most of the reactions studied under the topic of weathering involve the unmixing of complex molecules, but sometimes the reverse is true, as when new complex compounds are produced by authigenesis and pedogenesis, namely, the neoformation of clay minerals. Some of the more important chemical reactions include, for example: hydration-dehydration, hydrolysis-dehydrolysis, ion adsorption, cation or base exchange, oxidation-reduction, carbonization, and halmyrolysis. Weathering reactions that take place in the soil solum constitute the basis of pedochemical weathering and involve wide ranging processes of soil mineral weathering that include, for example: alkalization (solonization)-dealkalization (solodization), calcification-decalcification, desilication (ferrallitization, ferritization, allitization), ferrugination, melanization-leucanization, salination-desalinization, and silicification-deslicification.

A much-studied aspect of soil mineral weathering is the formation of clay minerals, including authigenic differentiation and subsequent transformation (neoformation). Although a very large number of clay minerals are known, seven principal types are important in soils: kaolinite, halloysite, montmorillonite, hydrous mica, vermiculite, chlorite, and allophane. The first six contain crystalline structures composed of silicon tetrahedral, aluminum hydroxide, and magnesium hydroxide sheets in various combinations. Allophane is a noncrystalline to microcrystalline material that usually occurs as a brown isotropic mixture with ferric hydroxide in Andosols.

Grouped here between two covers are copies of papers, printed in full or *pro parte,* that contain some of the most crucial turning points in our understanding of soil mineral weathering. The conceptual development of ideas for clay mineral formation, and other processes associated with soil mineral weathering, are carefully explained in the editorial commentaries that accompany the papers of historical and scientific interest.

CHARLES W. FINKL, JR.

PREFACE

Almost all minerals that occur at the Earth's surface can also be expected to occur in at least minor amounts in soils somewhere. Soil mineral weathering could thus include all the chemical and physical reactions undergone by thousands of minerals. The papers reprinted in this volume involve some of the major soil minerals and a few important chemical reactions and principles.

Some of the papers demonstrate how information blossomed when new analysis equipment, such as X-ray diffraction, became available. Yet small-budget researchers need not be discouraged entirely. The theory and analysis techniques for the application of chemical equilibria to soil minerals at room temperature were available for at least fifty years before they were applied to soil minerals. Geologists had been applying more complicated equilibrium theory at higher temperatures years before soil scientists thought to tackle the simpler case at 25°C.

Many of the papers included in this book bear on rate considerations in a qualitative way. Because actual reaction rate constants from laboratory experiments have not yet been applied in nature, papers on that topic do not appear in this volume. Something must also be said about other papers that do not appear in this volume. I polled colleagues on their selections of benchmark papers in soil mineral weathering. The response was enthusiastic, but unfocused. Not any one paper was mentioned by a majority of respondents. Mentioned most often were a few of my own papers, indicating that I had mostly polled my own friends. Further, one of the papers that shaped my thinking the most was not mentioned by any of the respondents. So much for my poll! Be advised that the papers selected for this volume are a very personal choice, but they are good ones that you should enjoy reading.

J. A. KITTRICK

CONTENTS BY AUTHOR

Aomine, S., 60
Bourbeau, G. A., 30
Buseck, P. R., 165
Driscoll, C. T., 139
Eggleton, R. A., 165
Eswaran, H., 82
Fry, W. H., 9
Garrels, R. M., 176, 212, 228
Helgeson, H. C., 228
Hendricks, S. B., 9
Jackson, M. L., 30
Jones, R. C., 68
Kittrick, J. A., 129, 188, 199
Lawton, K., 124
Mackenzie, F. T., 228
Mortland, M. M., 124
Mulder, J., 139
Murad, E., 75

Norrish, K., 91
Obershain, S. S., 54
Pennington, R. P., 30
Rausell-Colom, J. A., 91
Rich, C. I., 54
Schulze, D. G., 75
Schwertmann, U., 75
Sweatman, T. R., 91
Tardy, Y., 212
Tsuzuki, Y., 255
Tyler, S. A., 30
Uehara, G., 68, 124
Van Breemen, N., 139
Wells, C. B., 91
Willis, A. L., 30
Wong, B. C., 82
Yoshinaga, N., 60

SOIL MINERAL WEATHERING

INTRODUCTION

Long before clay minerals could be identified, pedologists and geologists understood that soil materials change with time. Total chemical analyses and petrographic examination of the sand fraction of soil horizons, and depth profiles of weathered rocks and sediments often showed systematic bulk composition changes. The decrease in SiO_2 relative to Al_2O_3, and the decrease in feldspars relative to quartz, are examples of broadscale weathering changes that were found to be taking place everywhere in the world. It was evident that these weathering changes were producing new clay-size minerals, but until the advent of X-ray diffraction in the 1930s it was not possible to reliably identify them. The development of the X-ray diffractometer and sample preparation techniques has made identification of common clay minerals easy. Complicated interstratifications are now the challenge. With the advent of high resolution analytical electron microscopes and electron microprobes in the late 1970s, it is now possible to examine soil mineral weathering on a microscale, where reactions are relatively rapid. Because it takes a long time for microscale changes to produce broadscale changes in bulk composition, the initial broadscale concept of soil mineral weathering was that of a slowly changing, relatively inert system. Now the concept is one of complex activity at mineral surfaces.

Most soil systems, when examined in detail, are found to contain dozens of major minerals and perhaps hundreds of minor ones. The system is so complex, and the solubility of most of the minerals is so low, that the soil system bears little resemblance to textbook chemical systems. Thus, in the early 1950s it was still common to consider that regular chemical principles, such as the solubility product, did not apply to clay minerals. Clay minerals were thought to be "different" from ordinary chemical compounds. It has now become evident that clay minerals are different from ordinary textbook compounds only in that they are more complicated, and when they occur in soils they are in much more complicated systems. That regular chemical principles apply is now so obvious as to not be worthy of the statement. We have come a long way in a short time in understanding soil mineral weathering and the pace is accelerating.

Part I

MINERALS AND OCCURRENCE

Editor's Comments
on Papers 1 Through 7

1 **HENDRICKS and FRY**
   The Results of X-ray and Microscopical Examinations of Soil Colloids

2 **JACKSON et al.**
   Weathering Sequence of Clay-Size Minerals in Soils and Sediments. I: Fundamental Generalizations

3 **RICH and OBENSHAIN**
   Chemical and Clay Mineral Properties of a Red-Yellow Podzolic Soil Derived from Muscovite Schist

4 **YOSHINAGA and AOMINE**
   Imogolite in Some Ando Soils

5 **JONES and UEHARA**
   Amorphous Coatings on Mineral Surfaces

6 **SCHWERTMANN, SCHULZE, and MURAD**
   Identification of Ferrihydrite in Soils by Dissolution Kinetics, Differential X-ray Diffraction, and Mössbauer Spectroscopy

7 **ESWARAN and WONG**
   A Study of a Deep Weathering Profile on Granite in Peninsular Malaysia: III. Alteration of Feldspars

Weathering of minerals in soils is the transformation of constituents of minerals into forms that are more stable under current conditions. This process usually involves primary rock minerals weathering to secondary minerals, but may involve secondary minerals changing to other secondary minerals. Elements not incorporated into the new minerals are lost in drainage waters. Soil mineral weathering is so complex a process that many details will never be known. In terms of elements, the broad outlines of weathering changes were known in the early 1900s. However, it is essential to know at least which minerals are involved if weathering processes are to be even approximately understood.

Mineral changes in fine sand and larger particles were initially determined with the petrographic microscope. Goldich (1938) wrote one of the best papers, tracing mineral changes by chemical analyses and the use of the petrographic microscope. Unfortunately for soil scientists, clays are too small for petrographic analysis and it is this fraction that contains the greatest surface area and is, therefore, the source of the most important physical and chemical properties of soils. Mineral changes in the soil clay fraction were a mystery in the early 1900s. However, by the time Goldich published in 1938, a new technique had already come upon the scene that radically changed the understanding of soil mineral weathering: the identification of clay-size minerals by X-ray diffraction (XRD). The papers in Part I of this book generally follow the progress of XRD at approximately 10 year intervals.

Hendricks and Fry (Paper 1) still depended upon the petrographic microscope and chemical analyses, but they were also among the first to apply XRD to soil clays. Their powder patterns were as indistinct in the original publication as they are in this book. Notice in their Table 5 that the longest smectite (montmorillonite) XRD spacing is 4.44 Å. Although they supposedly identified smectite in some of their soil clays, the large central beam hole obscured any diagnostic (00l) peaks (Of course, use of oriented samples and diagnostic polar liquids had not yet been discovered). This XRD work also showed that soil clays were indeed crystalline, a theory previously unaccepted by the majority of soil scientists.

During the next 18 years considerable information accumulated on the relative abundance of clay-size minerals in soils and sediments, mainly through the application of XRD. In Paper 2, Jackson, Tyler, Willis, Bourbeau, and Pennington observed that certain soil clay minerals were generally restricted to specific occurrences. For reasons of time, temperature, and rainfall, these occurrences could be characterized as early, intermediate, and late stages of weathering. If these initial assumptions regarding weathering stages were correct, then the clay minerals could be used to characterize the weathering stage of the soils and sediments in which they were found. This concept is similar to the one used by metamorphic geologists (of which S. A. Tyler was one) to indicate the degree of metamorphism.

Jackson et al. (Paper 2) assembled a series of 13 weathering stages, and by examining the available clay-mineral information, formulated a series of "fundamental generalizations." Two of the most useful generalizations are (1) that the stage of weathering increases with the fineness of the particles and (2) the stage of weathering increases with increasing proximity to the soil surface. The authors were influenced by the erroneous ideas of their time about clays being stable end products of weathering, and by "binary transformations," wherein a given primary mineral can only weather to a certain secondary mineral. Still, they adopted the view that soils may not be at equilibrium, but may be slowly advancing to a more weathered stage. No one, before or since, has interpreted so much soil mineral weathering information so usefully.

By 1955, numerous mineralogical analyses were being performed on soil clays. Rich and Obenshain (Paper 3) give one of the best examples of these analyses. They applied XRD, along with centrifuge particle size separation, X-ray diffraction of random and oriented samples (with polar liquid), differential thermal analysis (DTA), total chemical analyses, and surface determinations. Most of these methods had been developed or improved (and continually refined) by M. L. Jackson and co-workers. Rich and Obenshain (Paper 3) also found interesting random and regular interstratifications resulting from muscovite weathering to dioctahedral vermiculite in an acid soil. They could have applied the common operational definition of chlorite to their 14.7 Å material that refused to contract to 10 Å with 1NKCl treatment at 300°C. Instead, they applied enough techniques and treatments to alert them to the fact that they were not dealing with typical chlorite behavior. Using 1NNH$_4$F and other treatments, they discovered nonexchangeable Al in the interlayer positions of vermiculite. This discovery touched off extensive research on that subject, which has only recently subsided.

Although a clay mineral may occur in soils and sediments, it has to be identified there for the first time by someone. Once others know what characteristics to look for, that mineral may henceforth be frequently identified in soils. For example, one looks in vain for vermiculite in the 13 weathering stages in Paper 2 by Jackson et al. because vermiculite was identified in a soil for the first time by MacEwan in that same year, 1948. Prior to 1948, investigators must have been ignoring vermiculite XRD lines or were probably attributing them to smectite. As more major minerals are found in soils, it becomes increasingly difficult to identify new ones. However, Yoshinaga and Aomine (Paper 4) managed to identify imogolite. They showed distinctive XRD, DTA, and transmission electron microscope (TEM) characteristics for imogolite by starting with volcanic-ash derived soils high in this previously "unknown mineral colloid." They further concentrated the imogolite by dispersing and removing the allophane at a high pH where imogolite does not disperse. The distinctive threadlike morphology of imogolite by TEM in their Figure 2 is particularly impressive. Imogolite is a transition product in the rapid transformation of primary to secondary minerals and does not occur in older sediments. Because of its restricted occurrence, it is possible that soil scientists did not actually overlook imogolite as often as they once overlooked vermiculite.

Prior to Hendricks and Fry (Paper 1) and others, it was commonly believed that soil clays were amorphous. Mattson (1938) wrote a series of papers on the laws of soil colloidal behavior based upon this concept. When it became overwhelmingly evident that soil clays contained crystalline material, Mattson concluded that only the core of clay minerals was crystalline, and that this core was surrounded by a coating of amorphous material that determined its major physical and chemical properties. At this time, no one was listening and investigators generally ignored Mattson's work. However, in 1973 with the use of a good electron microscope (good for its time, that is) and careful specimen preparation work, Jones and Uehara (Paper 5) did find gel-like coatings on soil mineral grains. This work has not been much advanced by others and the chemical and physical significance of the coatings is still speculative.

In recent years many soil mineralogists shifted their interest from the major minerals to subsystems containing minerals that are present in small amounts or are poorly crystalline. One example is ferrihydrite, a fast-forming intermediate in the formation of hematite. It is poorly crystalline and is soluble in oxalate extraction. We cannot help wondering if we may be looking at some ferrihydrite in Paper 5. Ferrihydrite could not be directly identified in soils until Schwertmann, Schulze, and Murad (Paper 6) in 1982 used XRD and Mössbauer spectroscopy before and after oxalate extraction to show, by difference, the spectra of the ferrihydrite that had been there originally.

Part I closes with Paper 7, the final paper from a series of three by Eswaran and Wong (1978a, 1978b). This series set high standards for identifying soil minerals, making it possible to accurately assess what happens during mineral weathering. Paper 7 is also a good example of how the application of sophisticated instrumental techniques to soil mineral weathering has escalated since the publication of Paper 3 in 1955. Further, Eswaran and Wong have shown how advantageous it is to examine mineralogically not just a bulk sample, but also the micromorphology of the soil profile and even the weathering surface of mineral grains. This more detailed examination of the soil has led to the significant conclusion that the secondary minerals formed are a function of the chemical microenvironment. The primary weatherable minerals merely furnish the constituents (a fundamental assumption of the work reported in Part III).

REFERENCES

Eswaran, H., and C. B. Wong, 1978a, A Study of a Deep Weathering Profile on Granite in Peninsular Malaysia: I. Physico-Chemical and Micromorphological Properties, *Soil Sci. Soc. Am. J.* **42**:144–149.

Eswaran, H., and C. B. Wong, 1978b, A Study of Deep Weathering Profile on Granite in Peninsular Malaysia: II. Mineralogy of the Clay, Silt, and Sand Fractions, *Soil Sci. Soc. Am. J.* **42:**149-153.

Goldich, S. S., 1938, A Study of Rock Weathering. *J. Geol.* **46:**17-58.

MacEwan, D. M. C., 1948, Les mineraux argileux de quelques sols ecossals, *Verres Silic. Ind.* **13:**41-46.

Mattson, S., 1938, The Constitution of the Pedosphere, *Ann. R. Agric. Coll. Sweden* **5:**261-276.

Copyright © 1930 by the Williams & Wilkins Company
Reprinted by permission from pages 457-471 and 474-479 of *Soil Sci.*
29:457-479 (1930)

THE RESULTS OF X-RAY AND MICROSCOPICAL EXAMINATIONS OF SOIL COLLOIDS

STERLING B. HENDRICKS AND WILLIAM H. FRY[1]

Bureau of Chemistry and Soils, U. S. Department of Agriculture

Received for publication November 22, 1929

In 1924, W. O. Robinson and R. S. Holmes published the results of an investigation on the chemical composition of soil coilloids (7), in which they showed that a partial classification of soils could be made on the basis of their silica-sesquioxide ratios. R. S. Holmes (5) later showed that a specific soil type, Leonardtown, is characterized by constancy of composition. It was suggested by them, though they lacked definite proof, that the greater part of the silica, alumina, iron oxide, and water might be present as constituents of some mineral such as kaolinite, nontronite, halloysite, or pyrophyllite. Robinson and Holmes remarked: "With a further development of X-ray methods, it may be possible to decide definitely whether such complex mixtures as soil colloids are composed ultimately of crystals, and what compounds, if any, are present." The present work is essentially a consummation of this expressed hope.

MICROSCOPICAL EXAMINATION OF SOIL COLLOIDS

The colloids used in the following studies were extracted from soils by the method described in an earlier publication (4). The method limits the maximum size of the particles extracted to about 1μ; the greater part of the material is far smaller. Particles of this size are too small to admit of much observation in visible light and are far too small for the determination of the optical crystallographic properties in such light. These particles, however, have a tendency to orient themselves during the evaporation and drying of the suspension into the semblance of a regular cyrstallographic, or at least optical, arrangement as is shown from the petrographic examination. The mechanics of such a tendency, manifested even when the suspension was occasionally stirred, are not obvious. It might be suggested that if the colloidal particles are platy in

---

[1] The successful completion of this investigation was markedly aided by the assistance and advice of a number of people. Mr. R. S. Holmes, Mr. W. O. Robinson, and Dr. M. S. Anderson supplied the majority of the soil colloid samples examined. We are especially indebted to them and to Mr. Glen Edgington for analyses of the materials used. Dr. E. T. Wherry of the Bureau of Chemistry and Soils, and Dr. Clarence S. Ross of the Geological Survey of the Department of Interior kindly supplied us with fundamental information concerning the clay minerals. The success of the work, in a great measure, was due to the fact that pure samples of clay minerals were given to us by Dr. Ross. Mr. M. E. Jefferson did much of the X-ray photographic work.

shape (11) there would be a tendency for them to lie flat. The presence of predominantly platy minerals in the samples (see in the following), supports this hypothesis.

When the colloidal suspension was drying the particles agglomerated into masses large enough for petrographic study. The formation of aggregates and the orientation of the particles were perhaps subject to disturbing influences such as convection currents in the suspension, interference among the particles while settling, and cohesion. A specific soil colloid, as it will appear later, is not necessarily a single pure substance. For these reasons perfect optical reactions from the materials could not be expected, although such phenomena as interference figures and their modifications produced by the gypsum plate were surprisingly distinct.

Colloidal aggregates from a few soils were found to be opaque. Opacity itself is, of course, an optical property, but one that prevents the determination of other characteristics usually observed or measured in petrographic microscopic work. This property is always associated with a high iron or organic matter content. In mounts of even the most opaque of these colloids occasional doubly refracting aggregates were found. These, however, were deeply colored, translucent rather than transparent, and either failed to give interference figures at all or else gave figures of so vague an appearance as to be unrecognizable. Apparently the iron and organic matter is present in sufficient quantity to mask any doubly refracting material.

The double refraction of the large majority of the colloidal aggregates was surprisingly distinct. The interference colors ranged from the first to about the second order of Newton's scale and in some cases colors of an apparently higher order were noted.

The extinction was occasionally sharp, but in general it was wavy and sometimes absent, with the aggregate remaining bright during a complete rotation between crossed nicols. No spheroidal extinction was noted. The general wavy character of the extinction strongly suggested strain structure, but the study of the other optical properties of the aggregates made this belief untenable.

The interference figures were never perfect. Although the general shape of the hyperbolas and bars was easily recognizable and sometimes distinct, they were broad and blurred. The figures were, nevertheless, at least as good as and often better than those given by some minerals carrying a recognizable coating of colloidal material.

The blurred appearance of the interference figures made any measurement of the optic axial angle uncertain. In general this angle varied from zero up to about $2V = 30°$ or higher. Whether the figures that appeared to have a zero axial angle were actually uniaxial or were in reality biaxial with a small axial angle could not be decided with the imperfect figures obtainable. For this reason the listing of certain aggregates as giving unixial figures is purely tentative. The same sample of colloidal aggregates usually contained individual aggregates showing quite variable separations of the hyperbolas.

The optical character, determined with the gypsum plate, was perfectly definite except as noted in table 1. The refractive indexes of different aggregates in the same mount varied slightly. The values given are about the mean for the sample in question. It should be stated, however, that the colloidal aggregates were mostly more or less colored. For this reason perfect matching against the immersion oil was not possible. For these two reasons—variation of material and inaccuracy of determination—it is believed that a third decimal in the index figures would have been unjustified. Because of the imperfectly crystallized state of the aggregates and of their probable impurity, no effort was made to determine their indexes in different directions. The figures as given in table 1 represent simply the mean indexes.

The optical properties of the crystalline aggregates of the colloids examined are listed in table 1. Both decomposed organic matter and iron hydroxide are opaque in relatively thick sections. When present in sufficient quantity they effectively mask the presence of other substances. The observed scarcity of doubly refracting material in the Beckett, 0–6 inch, and the Superior, 3–8 inch, samples is probably due to the imperfectly crystallized state of the aggregates. The high iron content of the Nipe and Cecil samples accounts for the masking of the double refracting material in them.

Interference figures could not, of course, appear for samples free from doubly refracting material. Although there is but a small amount of doubly refracting material in the samples mentioned in the foregoing, some few particles could be observed. These particles were apparently quite impure. The results obtained from other samples (table 1) suggest that the Beckett, Superior, Cecil, and Nipe materials would also have given interference figures had there been less admixture with opaque substances.

The optical character of the aggregates, as determined on the interference figures by the gypsum plate, is predominantly negative. This is in accord with the known character of the clay minerals. Several samples, however, contained optically positive aggregates. The explanation for this is not known, but it might be suggested that quartz is positive. Quartz, as shown by the X-ray examination, is the only substance other than the clay minerals that is known to be present.

The refractive indexes given in table 1 are in general somewhat higher than those of the clay minerals shown to be present by the X-ray examination. The presence of free iron compounds in the aggregates is sufficient to explain this increase in index of refraction.

There is a partial correlation between the refractive indexes and the silica-sesquioxide ratio (table 2). In general, an increase in the silica-sesquioxide ratio is accompanied by a decrease in the refractive index. This change might be due to the concomitant variation in ferric oxide and alumina content. It can be said more truly that the refractive indexes fall into two large classes depending upon the silica-sesquioxide ratio. Numerous factors such as induration, chemical combination, and impurities probably combine to cause the observed irregularities.

X-RAY EXAMINATION OF SOIL COLLOIDS

X-ray technic

X-ray powder photographs were made from samples mounted in celluloid dissolved in amyl acetate, or from samples packed into very thin-walled glass

TABLE 1
A summary of the optical properties of soil colloids

COLLOID	DEPTH	DOUBLY REFRACTING MATERIAL	INTERFERENCE FIGURES	UNIAXIAL OPTICAL CHARACTER	BIAXIAL OPTICAL CHARACTER	REFRACTIVE INDEX
	inches					
Beckett	0–6	Scarce	None found	....	....	1.58
Beckett	6–11	Plentiful	Poor	....	−(?)	1.57–1.58
Beckett	24–36	Plentiful	Poor	....	−	1.60
Carrington	0–12	Moderate	Poor	....	Indeterminate	1.57
Carrington	15–36	Plentiful	Poor	....	−	1.57–1.58
Cecil	0–9	Scarce	None found	....	....	1.62
Cecil	9–18	Scarce	None found	....	....	1.62
Chester	0–8	Plentiful	Fair	−	....	1.60
Chester	8–32	Plentiful	Poor	−	−	1.60
Clarksville	0–10	Plentiful	Fair	−	....	1.59
Clarksville	10–36	Plentiful	Fair	....	−	1.60
Fallon	0–12	Plentiful	Fair	−	....	1.57
Huntington	0–8	Plentiful	Poor	−	....	1.61
Huntington	8–30	Plentiful	Poor	−	....	1.61
Manor	0–7	Plentiful	Poor	....	....	1.59
Manor	7–20	Plentiful	Fair	−	+ and −	1.59
Marshall	0–14	Plentiful	Fair	−	−	1.58
Marshall	14–36	Plentiful	Fair	−	....	1.58
Miami	0–10	Plentiful	Fair	−	−	1.59
Miami	10–24	Plentiful	Fair	−	−	1.59–1.60
Nipe	0–12	Scarce	None found	....	....	1.70
Norfolk	0–8	Plentiful	Fair	Indeterminate	+ and −	1.62
Norfolk	12–36	Plentiful	Fair	Indeterminate	+ and −	1.62
Ontario	12–22	Plentiful	Fair	....	−	1.61
Orangeburg	0–10	Plentiful	Fair	+ and −	+ and −	1.60
Orangeburg	10–36	Plentiful	Poor	....	+ and −	1.62
Sassafras	0–8	Plentiful	Fair	−	−	1.59
Sassafras	8–22	Plentiful	Fair	....	−	1.59
Sharkey	0–4	Plentiful	Poor	....	−	1.57
Stockton	0–38	Plentiful	Fair	−	−	1.57
Stockton	38–50	Plentiful	Fair	−	−	1.57–1.58
Superior	0–3	Scarce	None found	....	....	
Superior	3–8	Scarce	Poor	−(?)	−(?)	1.57–1.58
Superior	12–30	Plentiful	Poor	....	....	1.60–1.61
Superior	30–40	Plentiful	Poor	....	....	1.60–1.61
Wabash	15–36	Plentiful	Fair	−	....	1.57–1.58

TABLE 2
Comparison of refractive index, silica, alumina, and iron oxide contents of soil colloids

SOURCE OF SAMPLE	DEPTH	SiO₂ RECALCULATED*	Al₂O₃ RECALCULATED*	Fe₂O₃ RECALCULATED*	$\dfrac{\text{MOLS SiO}_2}{\text{MOLS Al}_2\text{O}_3 + \text{Fe}_2\text{O}_3}$	REFRACTIVE INDEX
	inches					
Fallon	0–12	65.91	20.67	13.42	3.82	....
Sharkey	0–4	62.05	27.33	10.62	3.07	1.57
Stockton	0–38	59.94	27.26	12.80	2.87	....
Carrington	0–12	59.67	30.03	10.30	2.75	....
Superior	3–8	68.52	24.72	6.75	4.01	....
Wabash	15–36	62.42	25.77	11.82	3.33	....
Stockton	38–50	60.03	27.45	12.51	2.85	1.575
Beckett	6–11	68.52	24.72	6.75	2.85	....
Carrington	15–36	58.56	30.71	10.73	....	....
Marshall	14–36	60.07	27.43	12.50	2.87	....
Marshall	0–14	59.97	28.36	11.67	2.82	1.58
Beckett	0–6	....	....	....	2.20	....
Miami	0–10	56.69	29.29	14.02	2.50	....
Clarksville	0–10	53.71	33.56	12.73	2.18	....
Sassafras	0–8	50.26	36.69	13.05	1.85	1.59
Manor	7–20	49.31	38.12	12.57	1.81	....
Manor	0–7	48.39	38.74	12.86	1.74	....
Miami	10–24	58.38	29.04	12.57	2.66	1.595
Clarksville	10–36	52.70	33.42	13.89	2.10	....
Orangeburg	0–10	49.51	38.09	12.40	1.83	....
Chester	8–32	48.85	37.32	13.84	1.79	1.60
Chester	0–8	48.60	37.36	14.04	1.77	....
Beckett	24–36	....	....	....	1.67	....
Superior	30–40	55.14	28.77	16.09	1.79	....
Superior	12–30	45.01	35.81	19.18	1.59	1.605
Ontario	12–22	51.47	30.00	18.54	2.08	....
Sassafras	8–22	49.49	35.20	15.31	1.89	....
Huntington	8–30	49.65	33.61	16.74	1.89	1.61
Huntington	0–8	49.57	35.63	14.80	1.86	....
Norfolk	12–36	49.54	37.03	13.43	1.84	....
Orangeburg	10–36	48.21	39.25	12.04	1.71	....
Norfolk	0–8	47.39	38.67	13.94	1.67	1.62
Cecil	0–9	41.90	44.50	13.60	1.34	....
Cecil	9–18	39.72	47.75	12.52	1.20	....
Nipe	0–12	11.51	17.89	70.60	0.31	>1.70

* Recalculated on basis of Fe₂O₃ + Al₂O₃ + SiO₂ = 100 per cent.

tubes. Photographs were made with copper, iron, and molybdenum characteristic radiation. The casettes used with molybdenum radiation have radiuses of approximately 10 cm. and are hemicylindrical. Zirconium oxide filters were used for removing the MoK_β radiation. The X-ray tube used was operated at a peak voltage of 33 K. V. and a current of 15 M. A. A small cylindrical casette of 3.583 cm. radius was used with the copper and iron radiation. To produce these radiations the X-ray tube, an electron tube of laboratory construction, was operated under a tension of 20 to 30 K. V. and a current of 10 to 40 M. A. An oxide-coated platinum filament was used to prevent contamination of the anticathode, which, nevertheless, was cleaned at frequent intervals. A filter was not used with the copper and iron radiation.

Each pure substance examined has a characteristic powder diffraction pattern. It should be remarked that a mineral, of the type found, present as 10 per cent or less of the crystalline material of a sample would not be indicated by the diffraction pattern obtained. Truly amorphous material does not give an X-ray diffraction pattern. Mixtures of crystalline minerals with amorphous material give the diffraction pattern of the mineral. However, the contrast between the diffraction lines and the general photographic background is changed.

X-ray diffraction characteristics of pure clay minerals

A cursory examination of the X-ray powder diffraction patterns obtained from a few soil colloid samples indicated that the mineral present in each is closely related to kaolinite ($Al_2O_3 \cdot 2SiO_2 \cdot 2H_2O$). We were fortunate at this juncture in obtaining analyzed samples of carefully purified clay mineral.[2] The analyses together with the specific locality of the clay minerals used in this work are listed in table 3. Six distinct types of X-ray powder diffraction patterns were obtained from these minerals. The samples of montmorillonite and beidellite gave identical powder diffraction patterns. In the case of sample 4 (table 3), which is an iron-rich beidellite from Spokane, Washington, there is a definite change in the values of some of the interplaner spacings compared with those of the type of montmorillonite from Otay, California. Ross and Shannon (10) have assigned the formula $(Mg \cdot Ca)O \cdot Al_2O_3 \cdot 5SiO_2 \cdot nH_2O$ to montmorillonite. However, Wherry, Ross, and Kerr (12) have indicated that the ratio $SiO_2:Al_2O_3$ might vary from 2/1 to 5/1, without change in structure; they commented that the compound

[2] From Dr. Clarence S. Ross of the U. S. Geological Survey, who with Prof. Paul F. Kerr of Columbia University is carrying out an extended investigation on the mineralogical, petrographic, and diffraction pattern characteristics of these minerals. Their work entirely supersedes the earlier descriptions given by Dana (3); it is, perforce, the basis for our investigation. We were kindly allowed to examine in manuscript a partial presentation of their results. The mineralogical and analytical material presented in this section is largely derived from that source; the X-ray diffraction results independently obtained are in accord with those of Professor Kerr.

probably does not have a definite silica to alumina ratio. Beidellite (10) has been shown to be essentially $Al_2O_3 \cdot 3SiO_2 \cdot nH_2O$. Here again wide variation in the silica-alumina ratio does not produce a noticeable change in physical properties including the X-ray diffraction patterns.

If these two suggested formulas, neglecting water, are written as $Ca_3Al_6Si_{12}O_{36}$ and $Al_8Si_{12}O_{36}$, it is seen that they differ chiefly in replacement of aluminum

TABLE 3

Analyses of pure clay minerals

	MONTMORILL-ONITE		BEIDELLITE		ORDOVICIAN BENTONITE	DICKITE	KAOLINITE		HALLO-YSITE
	1	2	3	4	5	6	7	8	9
SiO_2	50.06	49.56	47.28	46.06	55.27	45.00	44.92	44.70	44.08
Al_2O_3	21.32	15.08	20.27	12.22	20.73	40.70	40.22	38.64	39.20
Fe_2O_3	0.22	3.44	8.68	18.54	2.18	tr.	0.54	0.96	0.10
FeO	tr.	....	....	0.28	0.41	....	....	....	None
MgO	4.42	7.84	0.70	1.62	3.82	tr.	0.14	0.08	0.05
CaO	1.26	1.08	2.75	1.66	tr.	0.22	0.08	0.24	None
Na_2O	0.33	....	0.97	....	tr.	tr.	....	0.62	0.20
K_2O	0.19	....	tr.	....	6.21	....	....	0.14	....
H_2O-	14.06	22.96	19.72	17.26	4.87	0.00	0.08	0.64	1.44
H_2O+	7.56				6.50	14.08	14.22	13.88	14.74
TiO_2	tr.	0.40	....	0.84	0.30	tr.	....	0.22	tr.
SrO	....	....	....	....	N.F.	....	....	....	....
MnO	0.13	0.01	....	....	N.D.	....	....	....	....
SiO_2/Al_2O_3	....	....	....	....	....	188/100	189/100	....	....

1. Pink, from a pegmatite, Pala, San Diego County, Calif., J. G. Fairchild, Analyst (Unpublished).
2. Bentonite, crude, selected pink material, near Otay, San Diego County, Calif., Earl V. Shannon, Analyst.
3. Beidellite gouge clay from Beidell, Sagauche County, Colo., E. S. Larsen and E. T. Wherry (6).
4. Iron rich beidellite; Spokane, Washington.
5. Ordivician bentonite, Evansville, Tenn.; J. G. Fairchild, U. S. Geol. Survey (unpublished).
6. Dickite, Gusihuiriachic, Mexico; J. G. Fairchild, Analyst (9).
7. Kaolinite, massive, from U. V. X. mine, Jerome, Ariz. Collected by F. L. Ransome; F. A. Gonyer, Analyst.
8. Kaolinite from one mile south of Ione, Amador County, Calif. F. A. Gonyer, Analyst.
9. Halloysite, Hickory, N. C., J. G. Fairchild, Analyst. (Unpublished).

by calcium, the ratio of silicon to oxygen remaining constant within the limit of error of analysis and the purity of the product. As is shown by sample 4, iron can partially replace aluminum with a resulting slight modification of the lattice.

The results of the investigations of Prof. W. L. Bragg and his co-workers on the structure of some silicate minerals show that quite often such compounds are structurally similar. Some structures are, for instance, based upon a hexagonal or cubic close packing of oxygen atoms, with the smaller atoms of higher valence occupying interstitial positions (1). An examination of the X-ray diffraction data shows that the spacings and intensities of the lines cannot be explained on the basis of a simple close packing of oxygen atoms. The similarity of diffraction patterns, nevertheless, points to some common structural characteristics rather than to actual identity obscured by analysis of impure material. These minerals are far too complicated in structure to allow structure determinations without the use of very good crystals.

Crystalline kaolin minerals give diffraction patterns markedly different from those of montmorillonite-beidellite. The patterns of kaolinite, dickite, and nacrite can be easily differentiated from one another (see pl. 1, fig. 1 and table 4). The composition of these minerals can be represented by the formula $Al_2O_3 \cdot 2SiO_2 \cdot 2H_2O$; here again there is a wide variation in the observed silica-alumina ratio. Halloysite, from Hickory, N. C., gives a pattern closely similar to that of kaolinite, but different from it in relative intensities of reflection, in particular, of the planes of small spacings. The halloysite patterns were markedly poorer than those of dickite, kaolinite, or nacrite.

Potassium-bearing Ordovician bentonite, from Evansville, Tenn., shows a pattern similar to that of montmorillonite, but differing from it by the presence of a number of extra lines. As shown in plate 1, figure 1, a mixture of montmorillonite and quartz gives a diffraction pattern very similar to that of the material from Evansville, Tenn. We hesitate at this time to describe the material as a separate type of mineral.

The six general types of photographs: montmorillonite, Ordovician bentonite, halloysite, kaolinite, dickite, and nacrite, together with that of quartz and a mixture of montmorillonite with quartz for comparison with ordovician bentonite are reproduced as figure 1. These photographs were made with copper radiation (K_α, = 1.53730 Å) unfiltered, in a cylindrical camera with a radius of 3.583 cm. The photograph of quartz is indicative of the quality of photographs obtained from materials composed of not too fine particles. Accurate determinations, from specific samples, of the spacings and intensities of reflection of the principal reflecting planes are listed in tables 5 to 7. Sodium chloride was used as a reference substance.

X-ray diffraction characteristics of soil colloids

Descriptions of the colloid samples, together with designations of the types of diffraction patterns given by them are listed in table 8. The analyses of the corresponding samples are given in table 9. Many of these are the same as were studied by Robinson and Holmes (7).

Some typical photographs representative of each type of pattern obtained from these colloid materials are shown in plate 1, figure 2. It is to be noted that in general the pattern is more diffuse and the background darkening, due to scattered radiation, is more pronounced than is the case for the pure clay

TABLE 4
A summary of x-ray powder diffraction data from Kaolinite and Dickite

KAOLINITE CLAY MINERAL				DICKITE CLAY MINERAL			
Angle	Line	Spacing Å	Intensity	Angle	Line	Spacing Å	Intensity
6° 11'	CuKα_1	7.15	s.	8° 53'	CuKβ	4.500	v.w.
10° 5'	α_1	4.39	m.w.	9° 51'	α_1	4.494	m.
10° 36'	α_1	4.18	v.w.	10° 38.5'	α_1	4.162	m.
11° 20'	α_1	3.912	v.w.	12° 25	β	3.232	v.w.
12° 22.5'	α_1	3.585	v.s.	11° 37.5'	α_1	3.815	m.w.
15° 42'	β	2.568	v.w.	12° 19'	α_1	3.605	s.
16° 10'	β	2.491	v.w.	12° 53.5'	α_1	3.446	v.w.
17° 26.5'	α_1	2.565	s.	13° 34.5'	α_1	3.276	v.w.
17° 58'	α_1	2.492	m.s.	14° 27'	α_1	3.08	v.w.
19° 9'	α_1	2.345	s.	15° 2'	α_1	2.96	v.w.
19° 37'	α_1	2.290	m.	15° 55'	α_1	2.802	w.
20° 31'	α_1	2.19	v.w.	17° 22'	α_1	2.574	m.s.
22° 47'	α_1	1.985	m.w.	17° 51'	α_1	2.509	m.
23° 31'	α_1	1.927	v.w.	19° 19'	α_1	2.329	v.s.
24° 40'	α_1	1.84	v.w.	20° 11'	α_1	2.221	w.
25° 37.5'	α_1	1.78	v.w.	20° 32'	β	2.19	v.w.
27° 38.5'	α_1	1.657	w.	22° 53.5'	α_1	1.976	m.s.
28° 21'	α_1	1.618	w.	24° 10'	α_1	1.88	v.w.
30° 3'	α_1	1.535	v.w.	24° 35'	β	1.67	v.w.
31° 10'	α_1	1.485	s.	26° 29'	α_1	1.725	v.w.
32° 7'	α_1	1.446	v.w.	27° 46'	α_1	1.651	s.
35° 18'	α_1	1.330	v.w.-w.	29° 0'	α_1	1.58	v.w.
36° 9'	α_1	1.303	m.w.	29° 30'	α_1	1.56	v.w.
37° 1'	α_1	1.277	m.w.	30° 6'	α_1	1.533	m.w.
38° 45'	α_1	1.228	v.w.	31° 6'	α_1	1.489	m.-m.s.
39° 34'	α_1	1.207	v.w.	31° 54'	β	1.316	m.w.
				32° 37'	α_1	1.426	m.w.-w.
				33° 31'	α_1	1.392	w.
				34° 7.5'	α_1	1.371	w.
				35° 46'	α_1	1.315	m.s.
				36° 53'	α_1	1.281	m.w.
				37° 51'	α_1	1.253	m.w.
				38° 33'	α_1	1.231	m.w.
				39° 5'	α_1	1.219	v.w.
				40° 22'	α_1	1.187	m.w.
				42° 13.5'	β	1.033	v.w.
				44° 12.5'	α_1	1.103	m.-m.w.
				45° 45'	α_1	1.073	m.w.
				47° 5'	β	0.948	v.w.
				48° 15'	α_1	1.031	m.
				48° 22'	α_1	1.028	v.w.
				51° 14'	α_1	0.985	m.w.
				52° 28'	α_1	0.970	w.
				53° 32'	α_1	0.956	w.-m.w.
				54° 10'	α_1	0.948	w.
				55° 8'	α_1	0.937	v.w.

TABLE 4—Concluded

KAOLINITE CLAY MINERAL				DICKITE CLAY MINERAL			
Angle	Line	Spacing	Intensity	Angle	Line	Spacing	Intensity
		Å				Å	
....	..	.....	......	55° 54′	α_1	0.928	m.w.
....	..	.....	......	57° 29′	α_1	0.912	v.w.
....	..	.....	......	59° 32′	α_1	0.892	v.w.
....	..	.....	......	60° 18′	α_1	0.885	m.
....	..	.....	......	61° 56′	α_1	0.872	w.
....	..	.....	......	63° 32′	α_1	0.859	m.w.

minerals (pl. 1, fig. 1). These facts are accounted for by smallness of particle size and by the possible presence of an amorphous constituent in the samples. Some of the photographs taken with copper radiation show a general photographic darkening due to the secondary iron radiation produced by the iron in the sample. Many of the samples were photographed with iron radiation without great improvement in the contrast except in those cases in which iron was present in large amounts.

The diffraction patterns obtained from soil colloids could be divided into three general classes; namely, montmorillonite, halloysite, and Ordovician bentonite. The close similarity of the pattern given by ordovician bentonite to that given by a mixture of montmorillonite and quartz (the optimum amount of quartz being 10 to 15 per cent) is particularly confusing in the case of the soil colloids. A differentiation between these two possible patterns could sometimes be made on the basis of variation in relative intensities of the lines obtained from the mixture due to changes in the amount of quartz present.

Accurate interplaner spacing and intensity measurements for three types of soil colloids—montmorillonite, Ordovician bentonite, and halloysite—are listed in tables 5 through 7. Data obtained from the pure clay minerals are listed for comparison. These measurements were made on photographs from samples of the soil colloid having a low amount of iron present. Sodium chloride was used as a reference substance. It is to be noted that within the limit of error the spacings are the same for the two groups of materials; especially when the poor character of the patterns obtained from the soil colloid samples is considered.

The diffraction patterns obtained from a few of the soil colloid samples were dissimilar to those of the previously described clay minerals. A very fine material separated from the C horizon of a Durham soil gave the same type of pattern as a sample of a corresponding Chester soil, namely, that of bauxite. It is to be noted that these samples have a very high alumina content. A Nipe colloid, characterized by a very high index of refraction, gave a distinctly different type of pattern but since it was not representative, no great effort was made to identify it. Neither these samples nor others examined showed the presence of limonite, hematite, magnetite, the common micas, or

TABLE 5
A summary of x-ray diffraction data from Montmorillonite type patterns

\multicolumn{4}{c}{MONTMORILLONITE CLAY MINERAL}	\multicolumn{4}{c}{MONTMORILLONITE SOIL COLLOID}
Angle	Line	Spacing	Intensity	Angle	Line	Spacing	Intensity
		Å				Å	
8° 53′	CuKβ	4.499	m.w.	......	..	.....	....
9° 51′	α₁	4.493	s.	9° 57′	α₁	4.44	s.
14° 32′	α₁	3.063	v.w.	......	..	.....	....
15° 48′	β	2.550	v.w.	......	..	.....	....
17° 32′	α₁	2.551	v.s.	17° 27′	α₁	2.56	s.
18° 4′	α₁	2.478	v.w.	......	..	.....	....
24° 17′	α₁	1.869	v.w.	......	..	.....	....
27° 9′	α₁	1.685	w.	26° 56′	α₁	1.697	w.
27° 47′	β	1.490	m.w.	27° 33′	β	1.501	m.w.
31° 3′	α₁	1.491	v.s.	30° 59′	α₁	1.493	v.s.
32° 32′	β	1.291	v.w.	......	..	.....	....
36° 42′	α₁	1.286	m.w.	36° 32′	α₁	1.290	w.
38° 23′	α₁	1.238	m.w.	38° 12′	α₁	1.243	w.

TABLE 6
A summary of x-ray diffraction data from Ordovician bentonite (Montmorillonite-quartz ?) type patterns

\multicolumn{4}{c}{ORDOVICIAN BENTONITE CLAY MINERAL}	\multicolumn{4}{c}{ORDOVICIAN BENTONITE SOIL COLLOID}
Angle	Line	Spacing	Intensity	Angle	Line	Spacing	Intensity
		Å				Å	
9° 56′	CuKα₁	4.45	s.	9° 53′	α₁	4.48	s.
12° 4′	β	3.321	w.-m.w.	11° 58′	β	3.35	m.w.
13° 23.5′	α₁	3.320	s.	13° 16′	α₁	3.35	s.
14° 28′	α₁	3.076	v.w.	....	..	.....	....
15° 34.5′	β	2.587	w.-m.w.	....	..	.....	....
17° 27′	α₁	2.565	s.	17° 27′	α₁	2.56	s.
18° 52′	α₁	2.376	v.w.	....	..	.....	....
21° 4′	α₁	2.138	w.	....	..	.....	....
22° 48′	α₁	1.984	w.	22° 42′	α₁	1.993	w.
25° 11′	α₁	1.807	v.w.	24° 53.5′	α₁	1.827	v.w.
26° 55′	α₁	1.698	m.w.	26° 53.5′	α₁	1.700	m.
27° 29′	β	1.505	w.	27° 41.5′	β	1.495	v.w.
28° 7′	α₁	1.631	w.	....	..	.....	....
30° 51′	α₁	1.498	v.s.	31° 8′	α₁	1.487	v.s.
32° 19′	β	1.300	v.w.	....	..	.....	....
34° 2′	α₁	1.372	m.	34° 2′	α₁	1.371	m.
36° 28′	α₁	1.293	m.s.	36° 23′	α₁	1.296	m.s.
38° 18′	α₁	1.230	w.	....	..	.....	....
40° 2′	α₁	1.195	v.w.	39° 34′	α₁	1.207	v.w.
40° 39′	α₁	1.180	v.w.	40° 29′	α₁	1.184	v.w.

the common feldspars. Quartz was definitely shown to be present in several samples, notably the Sharkey colloids.

It is to be remarked that samples with high iron content often gave the halloysite type pattern. One might expect that replacement of aluminum by iron would produce a change in the interplaner distances. A slight change is produced by increased iron content in the case of the pure clay mineral beidellite, but there is no detectable change in the case of the soil colloids. Similarly the presence of excess magnesium oxide might lead to displacement of the diffraction lines; but close examination of the Fallon loam colloid showed no measurable effect. In both of the cases cited, change of spacing was tested

TABLE 7

A summary of x-ray diffraction data from Halloysite type patterns

HALLOYSITE PURE CLAY MINERAL				HALLOYSITE SOIL COLLOID			
Angle	Line	Spacing	Intensity	Angle	Line	Spacing	Intensity
		Å				Å	
10° 2.5'	α_1	4.41	....	10° 2.5'	$CuK\alpha_1$	4.41	....
....	..	.....	....	12° 23'	α_1	3.584	....
....	..	.....	....	15° 37'	β	2.58	v.w.
16° 8'	β	2.50	v.w.	16° 8'	β	2.50	v.w.
17° 29'	α_1	2.560	s.	17° 26'	α_1	2.567	s.
17° 50.5'	α_1	2.508	m.s.	17° 49.5'	α_1	2.510	m.s.
19° 13'	α_1	2.337	m.s.	19° 16'	α_1	2.328	s.
....	..	.....	....	19° 40'	α_1	2.284	m.
....	..	.....	....	20° 38.5'	α_1	2.18	v.w.
....	..	.....	....	22° 45'	α_1	1.988	m.w.
....	α_1	1.680	m.w.	27° 20'	α_1	1.674	m.w.
28° 00'	α_1	1.637	m.w.	27° 58'	α_1	1.639	m.w.
....	..	.....	....	30° 12'	α_1	1.528	m.w.-w.
31° 17'	α_1	1.480	v.s.	31° 17'	α_1	1.480	v.s.
....	..	.....	....	32° 7'	α_1	1.45	w.
....	..	.....	....	34° 19'	α_1	1.36	v.w.
37° 00'	α_1	1.278	m.w.	37° 5'	α_1	1.275	m.w.
38° 43'	α_1	1.229	m.	38° 39'	α_1	1.231	m.
....	..	.....	....	40° 26'	α_1	1.185	......

for by observing the line width in patterns obtained from mixtures of the specific material with standard samples of the type mineral.

Isomorphous replacement of one element by another must undoubtedly play a prominent rôle in the composition of these minerals. The X-ray photographs indicate that in such a replacement there is no great deformation of the lattice. The absence of pattern variation does not deny the possibility of variable composition.

It was desired to see whether or not colloid samples obtained from as widely different localities as possible and from varying depths in a specific locality but still belonging to the same soil type gave similar patterns. For this reason a study was made of a series of Leonardtown, Miami, Chester, and

Cecil colloids from localities as indicated and analysis as listed in table 10. With the exception of the Leonardtown colloids, the photographs from a particular series were the same in all but a few cases in which quartz was present as an impurity. The series of photographs obtained from the Cecil colloids with molybdenum radiation is reproduced in plate 2, figure 1. These photo-

TABLE 8
Type of x-ray diffraction pattern and refractive indices of soil colloids

SOURCE OF SAMPLE	DEPTH	TYPE PATTERN	RATIO SiO_2/R_2O_3	REFRACTIVE INDEX
	inches			
Sharkey Clay, Miss.	0–4	Montmorillonite-Beidellite (some quartz present)	3.07	1.57
Fallon loam, surface, Nev.	0–12	Montmorillonite-Beidellite	3.82	1.57
Houston black clay, surface, Texas	0–12	Montmorillonite-Beidellite	3.56	.....
Marshall silt loam, surface, Nebr.	0–14	Montmorillonite-Beidellite	2.82	1.58
Huntington loam subsoil, Md.	8–30	Montmorillonite-Beidellite	1.89	1.61
Sassafras silt loam soil, Md.	8–22	Montmorillonite-Beidellite	1.89	1.61
Iredell Clay subsoil, N. C.	6–24	Montmorillonite-Beidellite	1.93	.....
Beckett	6–11	Ordovician bentonite or Montmorillonite and quartz	2.85	1.575
Superior	3–8	Ordovician bentonite or Montmorillonite and quartz	4.01	1.575
Miami silty clay loam, surface, Ind.	0–10	Ordovician bentonite or Montmorillonite and quartz	2.50	1.59
Miami silty clay loam, subsoil, Ind.	10–24	Ordovician bentonite or Montmorillonite and quartz	2.66	1.595
Leonardtown, Md.	7–17	Ordovician bentonite or Montmorillonite and quartz	2.07	1.60
Leonardtown, hardpan, Md.	18–28	Halloysite	2.13	1.60
Cecil Clay loam, Ga.	0–9	Halloysite	1.34	1.62
Cecil Clay loam, Ga.	9–18	Halloysite	1.20	1.62
Cecil Clay loam, N. C.	5–36	Halloysite	1.18	1.62
Ontario loam, subsoil, N. Y.	12–22	Halloysite	2.08	1.61
Vega Baja clay loam, P. R.	0–12	Halloysite	1.34	.....
Chester loam, Md.	0–8	Halloysite	1.77	1.60
Chester loam, Md.	8–32	Halloysite	1.79	1.60
Chester	.....	Bauxite	0.68	1.60
Sub-colloid from Durham	.....	Bauxite with some clay mineral	....	.....
Nipe colloid	.....	Unidentified	0.31	>1.70

graphs are typical halloysite type patterns as produced with molybdenum radiation.

In the case of the Miami, Chester, and Cecil colloids, it will be noted that there are marked variations in gross analysis accompanied by a constancy of diffraction pattern. Colloids having a halloysite type pattern sometimes show

[*Editor's Note:* Table 10 has been omitted.]

TABLE 9
Chemical analyses of soil colloids

NUMBER	SOIL TYPE FROM WHICH COLLOID WAS EXTRACTED	DEPTH	SiO$_2$	TiO$_2$	Al$_2$O$_3$	Fe$_2$O$_3$	MnO	CaO	MgO	K$_2$O	Na$_2$O	P$_2$O$_5$	SO$_3$	Cl	CO$_2$	IGNITION LOSS	COMBINED WATER	H$_2$O 110°	ORGANIC MATTER	N	MOLS SiO$_2$ / MOLS R$_2$O$_3$
		inches	per cent	per cent	per cent	per cent	per cent	per cent	per cent	per cent	per cent	per cent	per cent	per cent	per cent	per cent	per cent	per cent	per cent	per cent	
4618	Amarillo, Texas†	0–5	50.51	0.56	22.04	8.80	0.14	1.48	2.08	2.68	0.06	0.20	0.18		0.00	12.08	8.06		4.02	0.33	3.10
4619	Amarillo, Texas†	10–20	51.51	0.57	22.71	8.61	0.09	1.59	2.66	2.54	0.01	0.12	0.14		0.00	9.50	7.36		2.14	0.25	3.09
4620	Amarillo, Texas†	30–40	51.32	0.58	22.43	8.46	0.05	2.27	2.80	2.50	0.06	0.14	0.14		0.35	10.19	7.38		1.96	0.13	3.13
4621	Amarillo, Texas†	54–64	51.23	0.55	24.08	8.19	0.07	1.73	2.83	2.42	0.10	0.11	0.13		0.00	8.72	7.76		0.96	0.08	2.97
4622	Amarillo, Texas†	70–75	38.42	0.37	17.64	5.71	0.06	16.38	2.53	1.83	0.06	0.09	0.10		11.22	17.85	6.21		0.42	0.07	3.21
4623	Amarillo, Texas†	96–100	45.88	0.52	20.09	6.89	0.05	9.05	2.87	2.26	0.01	0.11	0.13		5.87	12.41	6.41		0.13	0.05	3.18
4624	Beckett, Mass.†	0–6	10.58	0.49	5.99	3.64	0.02	0.65	0.31	0.49	0.25	0.31	0.99		0.00	75.87	5.89		69.98		2.16
4625	Beckett, Mass.†	6–11	38.39	1.48	20.26	5.55	0.02	0.40	1.03	2.51	0.24	0.25	0.23		0.00	29.25	2.47		26.78		2.74
4628	Beckett, Mass.†	24–36	34.31	0.72	27.55	11.60	0.01	0.17	2.10	4.10	0.25	0.18	0.19		0.00	18.82	9.34		9.48		1.67
.....	Carrington, Iowa*	0–12	44.89	0.47	22.59	7.75	0.057	1.48	1.44	1.36	0.22	0.28	0.19	0.065	0.00	20.15	8.56	11.22	11.59	0.61	2.75
.....	Carrington, Iowa*	15–36	48.04	0.65	25.19	8.80	0.032	1.29	1.53	0.89	0.38	0.14	0.08	0.037	0.00	13.75	9.23	10.98	4.52	0.20	2.64
.....	Clarksville, Ky.*	0–10	42.40	0.81	26.49	10.05	0.372	0.62	1.22	1.76	0.25	0.65	0.21	0.045	0.00	15.60	9.89	10.21	5.71	0.38	2.18
.....	Clarksville, Ky.*	10–36	43.68	0.63	27.70	11.51	0.150	0.48	1.39	1.65	0.34	0.38	0.09	0.031	0.00	12.79	10.51	8.00	2.28	0.20	2.10
.....	Fallon, Nev.*	0–12	52.57	0.30	16.49	10.70	0.210	2.20	5.67	1.97	1.02	0.26	0.11			8.87	7.04	11.78	1.83	0.09	3.82
.....	Hagerstown, Md.*	0–8	39.93	0.55	29.02	9.45	0.108	1.25	1.53	1.61	0.50	0.30	0.11	0.033	0.00	16.91	9.94	3.69	6.97	0.24	1.91
.....	Hagerstown, Md.*	8–30	41.63	0.57	29.86	11.31	0.034	0.74	1.42	1.84	0.11	0.20	0.07	0.032	0.00	13.20	10.51	4.16	2.69	0.10	1.89
.....	Huntington, Md.*	0–8	37.79	0.40	27.16	11.28	0.263	0.70	1.30	2.67	0.54	0.64	0.27	0.033	0.00	17.96	8.14	5.33	9.82	0.55	1.86
.....	Huntington, Md.*	8–30	39.54	0.47	26.77	13.33	0.083	0.54	1.32	2.67	0.45	0.35	0.08	0.032	0.00	14.35	8.24	4.89	6.11	0.20	1.89
.....	Manor, Md.*	0–7	38.86	0.72	31.11	10.33	0.112	0.64	1.23	1.37	0.63	0.28	0.09	0.029	0.00	15.63	11.94	4.63	3.69	0.21	1.74
.....	Manor, Md.*	7–20	40.36	0.70	31.20	10.29	0.077	0.38	1.35	1.28	0.45	0.46	0.11	0.020	0.00	13.91	11.35	5.34	2.56	0.15	1.81
.....	Marshall, Neb.*	0–14	45.93	0.48	21.72	8.94	0.066	1.19	2.01	2.28	0.21	0.46	0.21	0.060	0.00	16.47	7.92	7.75	8.55	0.51	2.82
.....	Marshall, Neb.*	14–36	48.18	0.50	22.00	10.03	0.111	1.36	1.62	2.07	0.15	0.21	0.06	0.032	0.00	13.28	9.32	11.00	3.96	0.22	2.87
.....	Nipe, Cuba†	0–12	10.19		15.84	62.51		0.23	0.05	Trace	Trace					9.86					0.31
.....	Norfolk, N. C.*	0–8	38.25	0.79	31.21	11.25	0.033	0.54	0.53	0.26	0.09	0.23	0.08	0.047	0.00	16.51	11.92	4.45	4.69	0.22	1.67

....	Norfolk, N. C.*	12–36	41.60	0.71	31.10	11.28	0.005	0.34	0.49	0.46	0.22	0.26	0.05	0.028	0.00	13.81	11.50	3.25	1.60	0.25	1.84
....	Ontario, N. Y.*	12–22	42.40	0.56	24.71	15.27	0.138	1.18	2.59	2.39	0.51	0.25	0.06	0.028	0.00	10.71	7.22	9.50	3.49	0.23	2.08
....	Orangeburg, Miss.*	0–10	40.35	0.54	31.04	10.11	0.185	0.51	0.73	0.81	0.24	0.42	0.10	0.026	0.00	14.89	10.63	4.34	4.26	0.26	1.83
....	Orangeburg, Miss.*	10–36	40.35	0.44	33.27	10.08	0.595	0.45	0.67	0.81	0.29	0.17	0.09	0.024	0.00	13.60	11.92	5.44	1.68	0.25	1.71
....	Sassafras, Md.*	0–8	39.24	0.63	28.64	10.19	0.123	0.75	1.16	1.17	0.38	0.47	0.20	0.035	0.00	17.29	10.97	5.44	6.32	0.43	1.85
....	Sassafras, Md.*	8–22	41.14	0.70	29.26	12.73	0.031	0.53	1.07	1.35	0.42	0.08	0.07	0.020	0.00	14.08	12.49	5.32	1.59	0.32	1.89
....	Sharkey, Miss.*	0–4	51.34	0.51	22.61	8.79	....	1.41	2.48	....	....	....	....	....	....	10.93	7.45	9.30	3.48	0.57	3.07
....	Stockton, Calif.*	0–38	49.50	1.14	22.51	10.57	0.023	1.96	2.68	0.26	0.66	0.06	0.03	0.020	0.00	11.95	10.75	10.70	1.20	0.05	2.87
....	Stockton, Calif.*	38–50	48.61	0.94	22.23	10.13	0.063	1.70	3.37	0.78	0.50	0.20	0.04	0.019	0.00	12.51	9.69	8.76	2.82	0.07	2.85
4630	Superior, Wisc.†	3–8	46.20	0.97	16.67	4.55	0.08	1.01	1.25	2.14	0.34	0.39	0.35	....	0.00	25.58	5.88	....	19.70	....	4.01
4631	Superior, Wisc.†	12–30	31.60	0.88	25.14	13.47	0.08	0.66	1.71	1.37	0.15	0.96	0.18	....	0.00	23.35	13.39	....	9.96	....	1.59
4632	Superior, Wisc.†	30–40	42.20	1.57	22.02	12.31	0.12	0.61	3.41	1.23	0.06	0.31	0.17	....	0.00	15.59	10.24	....	5.35	....	2.40
....	Wabash, Neb.*	15–36	50.55	0.56	20.87	9.57	0.033	1.59	2.21	1.97	0.14	0.24	0.14	0.034	0.00	12.34	6.74	10.24	5.60	0.27	3.33

*Analyses taken from W. O. Robinson and R. S. Holmes (7).
†Analyses by Glen Edgington.

a silica-sesquioxide ratio as low as 1.0. This low ratio might be explained by the presence of a variable amount of an amorphous impurity, or by a variable composition of the clay mineral, halloysite, as previously indicated. This variable composition of the pure mineral might be due to the formation of solid solutions with other hydrous aluminum silicates.

A series of photographs of Leonardtown colloids shows that with but one exception the pattern is that of Ordovician bentonite or a mixture of montmorillonite with quartz. The hardpan from beneath a sample giving an Ordovician bentonite type pattern clearly gives a halloysite type pattern. Such a change is not entirely unexpected because of the obviously close structural similarity of the clay minerals. This one exception does not vitiate the general rule that a colloid of a specific soil type is characterized by the presence of an essential mineral constituent that does not vary from locality to locality (5).

A series of photographs, taken under identical conditions, of quartz mixed with amorphous aluminum silicate prepared by precipitation is shown in plate 2, figure 2. These patterns illustrate the increase in general scattering due to the presence of non-crystalline material. They are to be compared with those of the clay minerals, as is also the pattern of the homogeneous nickel oxide sample. In the case of the nickel oxide diffraction pattern it will be seen that the lines are very broad. This is due to the finely divided nature of the material, 10^{-5} to 10^{-6} cm. on the edge. Although the clay minerals and soil colloids gave markedly poor diffraction patterns, nevertheless it will be observed that the individual lines are rather well defined. The extra line width must sometimes be due to near superposition of two diffraction lines. There is no marked graduation of particle size over a large range, as would be suggested by a fuzziness of the lines. It is thus suggested that the crystalline clay minerals are not degenerated beyond a limit of about 10^{-5} to 10^{-6} cm. on an edge. The poor quality of the patterns is perhaps due to distortion of the small individual crystals.

DISCUSSION

It was impossible to answer the question whether or not the constituent giving the powder diffraction pattern was the only mineral present in the colloid. It should be remarked that with the exception of the presence of quartz in some samples the diffraction patterns showed the presence of only one crystalline component in each soil. This indicates that either the colloid sample is of uniform mineral composition or that only those minerals of the clay type retain their crystallinity sufficiently well to give an X-ray diffraction pattern. The primary soil minerals, such as micas, feldspars, and quartz, are not shown as crystalline components of the colloid fraction. Since these minerals can be microscopically identified in the parent soil, this rather indicates that they are not reduced, without chemical alteration, to small "colloidal" dimensions by natural processes, at least in the colloids examined.

Robinson and Holmes calculated the theoretical percentage of kaolinite and nontronite in a number of soil colloids. These calculations are based on the assumption that all of the silica is combined first with alumina to form kaolinite, then with the iron to form nontronite, given the formula $Fe_2O_3 \cdot 2SiO_2 \cdot 2H_2O$ by Clark (2). With but one exception (table 9), that of Huntington loam, these colloids having excess silica, as indicated by Robinson and Holmes, have the montmorillonite type pattern. This indicates that the calculation based upon the assumed presence of kaolinite leads to the evident presence of free silica since in reality the mineral present, montmorillonite, has a higher silica-sesquioxide ratio.

Robinson and Holmes (7) point out that as the silica-sesquioxide ratio increases in the soil colloid there is a concomitant change in the lime soda-sesquioxide ratio $(CaO+Na_2O:R_2O_3)$. In particular in the region near a silica-sesquioxide ratio of 2 to 1, the point of change from halloysite to montmorillonite, there is an increase in the lime soda-sesquioxide ratio. Ross and Shannon have concluded that montmorillonite contains $(Mg \cdot Ca)O$ as an essential constituent, the formula being $(Mg \cdot Ca)O \cdot Al_2O_3 \cdot 4\text{-}5SiO_2 \cdot nH_2O$. As previously pointed out there must be wide variations in the compositions of type montmorillonites. Here again there is indication that a constituent of the soil colloid, as shown by analysis, varies in a manner to be expected from the X-ray diffraction patterns.

Ordovician bentonite, as found in the United States, is a mineral of quite specific origin, namely, volcanic ash (8, 10). The analysis of Shannon shows that potassium is a prominent constituent of the type mineral. Considering these factors, its possible occurrence as a constituent of soil colloids seems rather surprising, especially since in the colloid samples there is no evidence of the presence of an unusually large amount of potassium. This, too, rather suggests that the material present in the soil colloids is a mixture of montmorillonite and quartz or else that potassium is not an essential constituent of the mineral.

In conclusion it can be said that the chemical correlations observed by Robinson and Holmes can be fully explained by the presence of montmorillonite, ordovician bentonite, and halloysite, as the characteristic minerals of the soil colloids.

Ross and Shannon (10) commented on the great variation in the amount of water present in samples of montmorillonite. From the fact that the water could be easily lost and regained and that during the process the physical properties of the minerals were not changed they suggested that the water is probably adsorbed. "This tendency to adsorb such large amounts of water is undoubtedly a function of the physical structure of the material which, *being extremely finely micaceous*, gives relatively enormous surface areas." The presence of the micaceous clay minerals as components of the soil colloids might explain some of the adsorption phenomena that characterize such materials. The absorption phenomena are perhaps also to be explained by the presence of montmorillonite as a mineral that has a great amount of water of crystallization or constitution.

SUMMARY

Microscopical examination and X-ray powder diffraction photographs of the finely divided materials separated from soils by suspension methods show that these fractions contain crystalline substances.

A specific sample gives a characteristic powder diffraction pattern that can be identified as arising from one of the clay minerals.

By comparison with the powder diffraction patterns of known clay minerals, it has been shown that montmorillonite-beidellite, Ordovician bentonite (or a mixture of montmorillonite and quartz), and halloysite are the common mineral constituents of soil colloids. Bauxite was found to be present in two samples.

Fine fractions from a specific type of soil obtained from widely different localities give the same type of diffraction pattern.

REFERENCES

(1) BRAGG, W. L. 1929 Atomic arrangement in the silicates. *Trans. Faraday Soc.* 25 (pt. 6): 291–314.
(2) CLARKE, F. W. 1920 The data of geochemistry, ed. 4. *U. S. Geol. Survey Bul.* 695: 486.
(3) DANA, E. S. 1914 System of Mineralogy, ed. 6, p. 684 et seq. New York.
(4) GILE, P. L., ET AL. 1924 Estimation of colloidal materials in soils by adsorption. U. S. Dept. Agr. Bul. 1193.
(5) HOLMES, R. S. 1928 Variations of the colloidal materials in typical areas of the Leonardtown silt loam soil. *Jour. Agr. Res.* 36: 459–470. (Also unpublished data.)
(6) LARSEN, E. S., AND WHERRY, E. T. 1917 Leverrierite from Colorado. *Jour. Wash. Acad. Sci.* 7: 208–217.
(7) ROBINSON, W. O., AND HOLMES, R. S. 1924 The chemical composition of soil colloids. U. S. Dept. Agr. Dept. Bul. 1311.
(8) ROSS, C. S. 1928 Altered paleozoic volcanic materials and their recognition. *Bul. Amer. Assoc. Petroleum Geol.* 12: 143–164.
(9) ROSS, C. S., AND KERR, P. F. The kaolin minerals. (Manuscript.)
(10) ROSS, C. S., AND SHANNON, E. V. 1926. The minerals of bentonite and related clays and their physical properties. *Jour. Amer. Ceramic Soc.* 9: 77–96.
(11) WHERRY, E. T. 1925 Bentonite as a one-dimensional colloid. *Amer. Mineralogist* 10: 120–123.
(12) WHERRY, E. T., ROSS, C. S., AND KERR, P. F. 1929 From manuscript of Dr. E. T. Wherry, presented before the Colloid Symposium, June.

PLATE 1

TYPE CLAY MINERAL AND SOIL "COLLOID" DIFFRACTION PATTERNS

FIG. 1. Type Clay Mineral Diffraction Patterns. 1. Quartz; 2. Dickite; 3. Nacrite; 4. Kaolinite; 5. Halloysite; 6. Ordovician Bentonite; 7. Montmorillonite; 8. Montmorillonite-Quartz mixture; 9. Bauxite

FIG. 2. Diffraction Patterns of Soil "Colloids." 1. Bauxite; 2. Ordovician Bentonite (Montmorillonite Quartz); 3. Halloysite from Cecil Soil; 4. Halloysite; 5. Montmorillonite

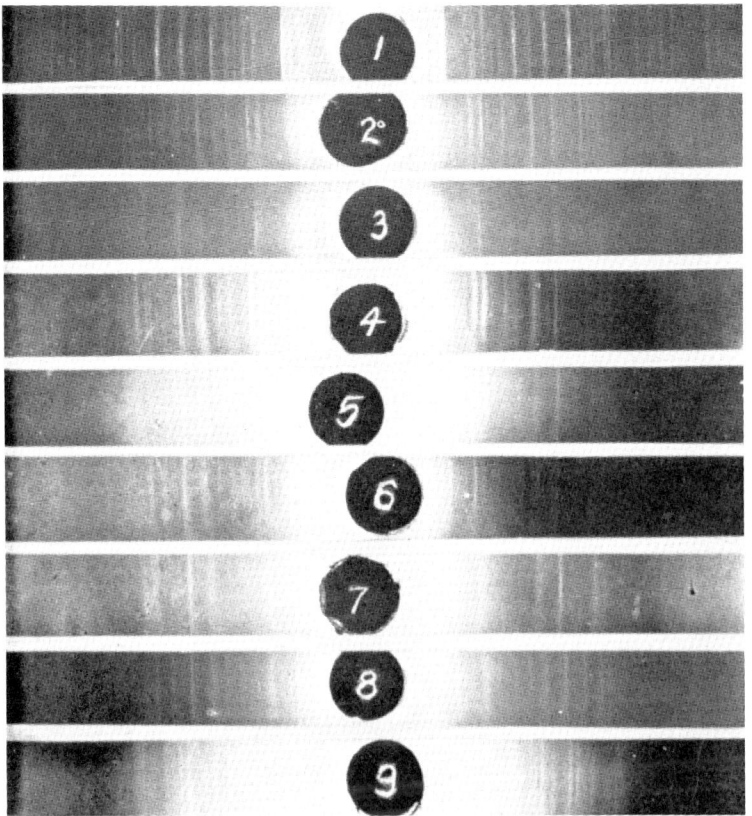

Fig. 1

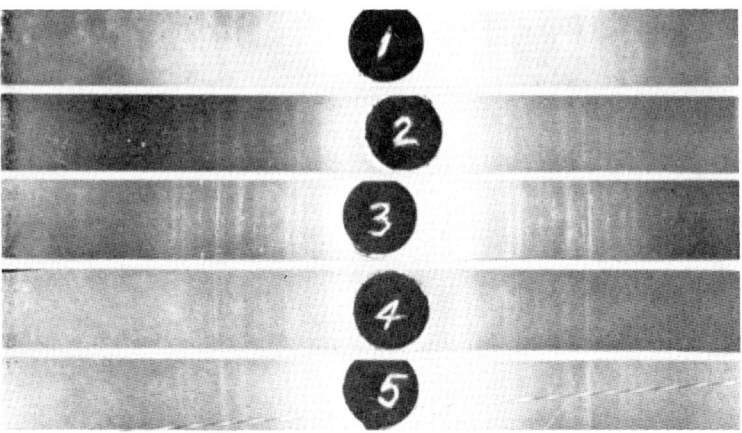

Fig. 2

PLATE 2

DIFFRACTION PATTERNS OF CECIL "COLLOIDS," AMORPHOUS ALUMINUM SILICATE, AND NICKEL OXIDE

FIG. 1. Series of Diffraction Patterns of Cecil "Colloids"; Mo K_α Radiation. 1. Cecil #5102 (see table 10); 2. #3922; 3. #3888; 4. #3889; 5. #3901.

FIG. 2. Quartz with Amorphous Aluminum Silicate and Nickel Oxide Showing Small Particle Size. 1. Nickel Oxide finely divided; 2. 25 per cent Quartz, 75 per cent Aluminum Silicate; 3. Quartz; 4. 50 per cent Quartz + 50 per cent Aluminum Silicate.

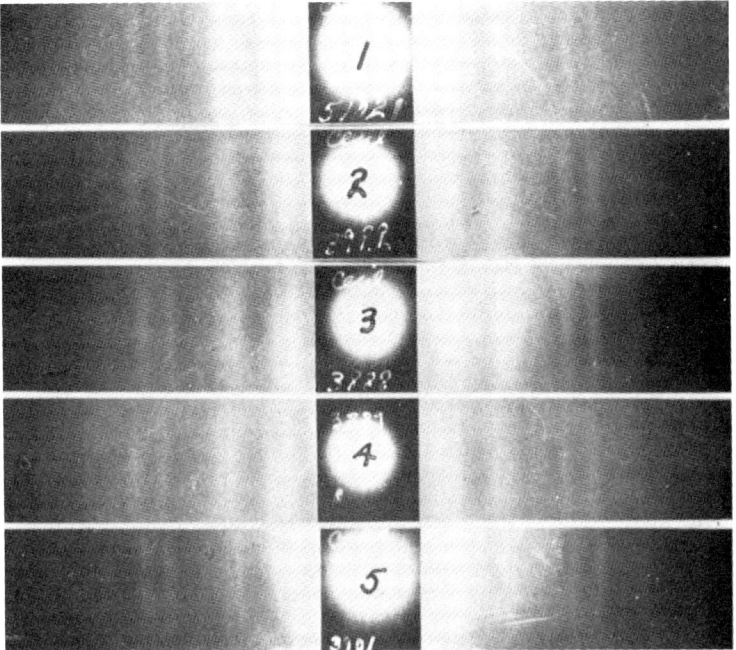

Fig. 1

Fig. 2

WEATHERING SEQUENCE OF CLAY-SIZE MINERALS IN SOILS AND SEDIMENTS. I

FUNDAMENTAL GENERALIZATIONS[1]

M. L. JACKSON, S. A. TYLER, A. L. WILLIS, G. A. BOURBEAU, AND R. P. PENNINGTON

University of Wisconsin, Madison, Wisconsin

Received February 20, 1948

Considerable progress has been made in the improvement of technics of preparation and identification of mineral species present in the colloidal fractions of soils and sediments, particularly of montmorillonite and the micas, through the x-ray diffraction method. As a result, considerable information has been accumulated on the relative abundance of the different specific minerals present in colloids of soils and sediments of diverse origins. Since the colloidal portions are the resultant products of weathering processes, they represent more or less stable end-products, which are more resistant to weathering than their parent materials.

It is the purpose of the present article to trace the course of weathering ("weathering sequence") of the finer mineral particles in soils and sediments, and to interpret the order of succession of minerals in the sequence on the basis of present concepts of crystal chemistry.

One view of weathering holds that the fundamental principle is embodied in the concept of direct weathering of a primary to a secondary mineral, each particle being a more or less closed system. Thus a primary mineral such as labradorite alters to a secondary mineral such as kaolinite (the "kaolinization" reaction), or "volcanic ash" alters to montmorillonite (the bentonite reaction), the calcium oxide, sodium oxide, and excess silica being carried away in solution. Accordingly, the source of each colloidal silicate clay mineral is sought as a specific parent mineral from which the colloidal mineral is a "primary weathering product." Observations of secondary colloidal products in pseudomorphic form of a parent crystal establish this binary transformation as a fundamentally sound view, in detail. The viewpoint herein developed recognizes this direct primary-secondary transformation, but seeks to extend the concept of weathering to include a summation, or integration, of a multiplicity of such binary transformations which may simultaneously be occurring in the soil or sediment, and responsible for its colloid composition.

In addition, the views are adopted that (*a*) one colloidal mineral may in some cases be a parent material of successive colloidal products as the weathering processes continue, (*b*) the weathering reactions are reversible, and (*c*) the entire mineral content of the clay-size fraction should be considered in the sequence,

---

[1] Joint contribution of the Wisconsin Agricultural Experiment Station and the Department of Geology, University of Wisconsin, Madison. Published by permission of the Director of the Agricultural Experiment Station. Supported in part by a grant from the Wisconsin Alumni Research Foundation.

without emphasis being placed on the older distinctions as to whether they are secondary or primary (unweathered) residuals. Interpretation of the mineralogy of youthful soils (important both to theoretical pedology and to agriculture), together with the recognition of the reversibility of the weathering reaction in both soils and sediments, requires this consideration of the entire mineral content of the clay-size fraction rather than merely the silicate "clay minerals." The unweathered residual minerals *truly enter the sequence* by virtue of their *relative resistance* to weathering, being residual *after more easily weathered minerals* of earlier stages have disappeared.

By "clay-size" is meant the finer portion of the soil or sediment, particularly the "fine clay" fraction of less than 0.2 micron equivalent spherical diameter, but also the "coarse clay" fraction, particles 0.2–2 microns in diameter, and to some extent the "fine silt" fraction, particles 2–5 microns in diameter. The inclusion of the coarser two fractions is justified partly because considerable of the clay minerals (illite and kaolinite particularly) occur in these fractions, but also because a functional consideration of the origin of clay-size particles requires *per se* a functional approach to the definition of clay size. Thus it is held that there is no definite "upper limit" of colloidal particle size of inorganic soil colloids, but rather a gradual change, depending in part on the mineral species being considered.[2]

THE MINERAL WEATHERING SEQUENCE: PRESENTATION

The minerals representing successive stages in the weathering sequence of clay-size minerals are listed in table 1. The more soluble or easily weathered substances appear in the first five stages. The first stages involving gypsum and other more or less freely soluble salts (stage 1) and calcite and related less soluble non-silicate minerals (stage 2) occur in well-developed soils only as secondary depositions in the lower horizons. When present, however, they usually dominate the important physical and chemical properties of the colloid, and are recognized therefore as a stage in the sequence for soils. They may constitute a small percentage of colloids of certain kinds of very young soils (e.g., reference 12, p. 43). Moreover, they are important constituents of the fine fractions of some sediments.

In stages 3 and 4, the silicates of the most easily weathered types (silicates groups I to V and certain members of VI) were first postulated on the basis of their occurrence in the fresh state in the finer sand and silt of young soils such as Rideau clay (16); later, amphibole (stage 3) and chlorite (stage 4) were actually found in the coarse clay of Abitibi soil, C horizon. This horizon is only slightly altered "rock flour," occurring as the subsoil of a soil developed from the glacial rock flour sediments of Lake Ojibway near James Bay in northern Quebec.[3]

[2] It is considered essential, however, that the clay-size minerals be studied in narrow particle-size ranges such as the three listed, for both theoretical and experimental reasons. Moreover, the limits of these arbitrary size ranges are also functional in character, rather than unique numerical limits.

[3] Collections obtained though the courtesy of Professor A. Scott, Laval University, St. Anne de la Pocatière, Quebec, Canada.

TABLE 1

Weathering sequence of clay-size minerals in soils and sedimentary deposits

WEATHERING STAGE AND SYMBOL		CLAY-SIZE MINERALS OCCURRING AT VARIOUS STAGES OF THE WEATHERING SEQUENCE	EXAMPLES OF OCCURRENCE OF MINERALS AT VARIOUS WEATHERING STAGES IN THE COLLOIDAL FRACTIONS	
			Of soils	Of sedimentary deposits
1	Gp	Gypsum (also halite, etc.)	Pierre clay, C horizon (South Dakota)	Polders clay (Holland)
2	Ct	Calcite (also dolomite, aragonite, etc.)	Minatare, B horizon (Nebraska)	Calcite limestones (Michigan)
3	Hr	Olivine-hornblende* (also diopside, etc.)	Abitibi, C horizon‡ (James Bay, Canada)	Fresh rock flour
4	Bt	Biotite† (also glauconite, chlorite, antigorite, nontronite, etc.	Abitibi, C horizon§ (James Bay, Canada)	Glauconite in Cambrian sandstone (Wisconsin)
5	Ab	Albite (also anorthite, microcline, stilbite, etc.)	Rideau clay, C_1 horizon (Ontario)	Authigenic feldspars (10)
6	Qr	Quartz (also cristobalite, etc.)	Rideau clay, B horizon (Ontario)	Authigenic quartz
7	Il	Illite (also muscovite, sericite, etc.)	Schomberg silt loam, B horizon (Ontario)	Pennsylvania underclays (Illinois)
8	X	Hydrous mica-intermediates ("X")	Dodgeville silt loam, A horizon¶ (Wisconsin)	Ordovician bentonite (Kentucky)
9	Mt	Montmorillonite (also beidellite, etc.)	Barnes silt loam, A horizon (South Dakota)	Bentonite (Wyoming)
10	Kl	Kaolinite (also halloysite, etc.)	Cecil clay, B horizon (Alabama)	China clay deposit (Georgia)
11	Gb	Gibbsite (also boehmite, etc.)	Fannin sandy loam, C horizon (North Carolina)	Bauxite (Arkansas)
12	Hm	Hematite (also goethite, limonite, etc.)	Nipe clay, B horizon (Puerto Rico)	Bog iron (Minnesota)
13	An	Anatase (also rutile, ilmenite, corundum, etc.)	Naiwa A_2 or B horizon (Kauai Island, Hawaii)	Metamorphosed bauxite

* Silicate groups I to V, with structures ranging from independent tetrahedra (I) to the line hexagonal structure of the amphiboles (V).

† Silicate group VI containing Fe^{+++} or Fe^{++} and Mg^{++} in the octahedral layer.

‡ Slightly altered "rock flour," bearing amphibole. Other members of stages 3 and 4 are postulated as occurring in soils on the basis of petrographic microscopic observations of occurrence in sand and silt of youthful soils.

§ Slightly altered "rock flour," bearing chlorite. Biotite may also occur in the clay of various young soils in association with illite, for example, in the mica of Rideau clay, C_1 horizon.

¶ In association with small amounts of illite (stage 7) and montmorillonite (stage 9).

This sediment has entered the weathering sequence, however, as evidenced, on the one hand, by extinction of the early stages 1 and 2 and near extinction of the early-intermediate stages 3 and 4, and on the other, by the occurrence of small amounts of illite and mica-intermediate and a slight amount of montmorillonite (stages 7, 8, and 9) in the fine clay. This case illustrates the essential meaning (i.e., persistence) of stages 3 and 4 in the weathering sequence. The sequence view fits the observed mineralogical content, and is more fundamental than the oversimplified characterization of this rock flour as being "unweathered" or exhibiting "absence of weathering," which could not be true in a surface material exposed for 10,000–20,000 years.

The group I to group IV silicates would be expected to occur in dominant amounts only in the clay fraction of certain kinds of youthful soils, for example, from fresh glacial rock flour derived from basic rocks. It is singular that five of the seven silicate structural groups occur in these scarcely represented early five stages of weathering. These primary silicates are, in most soils, largely decomposed from the fractions of particle-size range below 5 microns. However, they occur more commonly in sedimentary deposits, where protected from leaching.

The occurrence of stages 5 to 10 in the clay-size range of soils is well established. In the young soil Rideau clay, albite (stage 5) constitutes nearly one-half of the fine colloid of the C_1 horizon ($< 0.2 \mu$), and quartz (stage 6) the most of the remainder (16). Both albite and quartz occur in the coarse clay (0.2–2 μ) throughout the profile. Illite (clay mica with a 10 Å. basal spacing line, stage 7) has been reported extensively in soil clays, most commonly in the coarse clay, but also to some extent in the fine. Stages 5 to 7 are recognized as being analogous to the familiar sericitization reaction of microcline, or of plagioclase through the introduction of potassium to quartz and sericite. Heavy sericitization of plagioclase to mica, possibly in part deuteric (late magmatic), was observed in the very fine sand of Rideau clay, B and C horizons. Thus sericite may be considered one of the earliest of the hydrous mica series in soils. The hydrous mica-intermediates ("X") or degraded micas (basal spacing line at 12–13 Å. or absent) are recognized here as a separate and distinct stage (stage 8). The mica-intermediates occur very commonly in the fine clay of soils. Stages 3 to 7 correspond to the stability series of Goldich (cited in reference 14, p. 52) except that quartz and muscovite are reversed in the weathering sequence presented in accordance with their relative persistence in soil colloids.

The montmorillonite group appears at stage 9, and kaolinite at stage 10, representing advancements in weathering over the illites and mica-intermediates. Gibbsite appears in stage 11 as weathering (desilication) of kaolinite proceeds. Hematite occurs at stage 12 and anatase at stage 13 under conditions of extreme weathering, eluviation, and good oxidation. Anatase and, hypothetically, corundum should be more stable than hematite, especially since not affected by reduction, and therefore should conclude the series (stage 13). Corundum is not known to form in soils, but it may occur in metamorphosed sediments, and would be expected, for example, in metamorphosed gibbsite, boehmite, or diaspore deposits.

The application of the weathering sequence to colloids of soils and sediments may be summarized according to the following fundamental generalizations:

    1. From three to five minerals of the weathering sequence are usually pres-

ent in the colloid of any one soil horizon. There is a tendency for the composition of the colloid to be in the form of a distribution curve, being dominated (40–60 per cent) by one or two minerals with other adjacent minerals of the sequence decreasing in amounts with remoteness in the sequence.

2. The percentage of minerals of the early stages of the weathering sequence present in a soil clay fraction decreases, and the percentage of the successive members increases, with increasing intensity of weathering.

3. One to three intermediate stages may occasionally be absent from the normal sequence, particularly those following quartz, giving, for example,

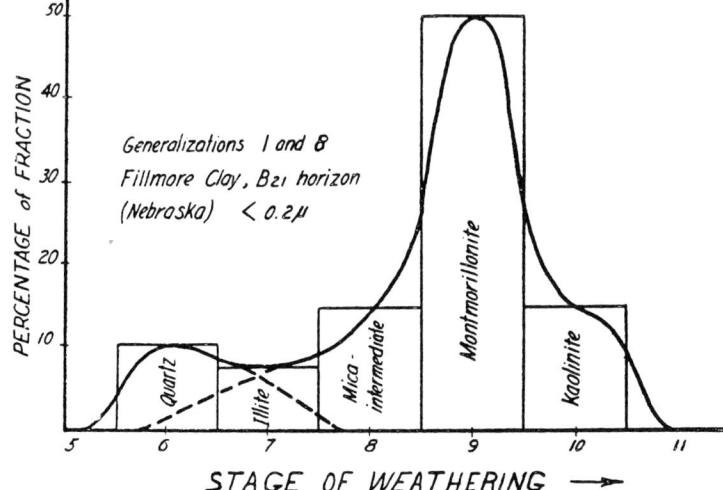

Fig. 1. Distribution curve of clay-size minerals in a Prairie planosol clay pan, showing asymmetry due to the quartz and illite being of a *coarser* fine-clay fraction.

a quartz–montmorillonite–kaolinite colloid, or a quartz–kaolinite–gibbsite colloid.

4. One or more stages may occasionally occur out of sequence as secondary depositions, particularly gypsum and calcite.

SOIL COLLOIDS AT VARIOUS STAGES OF WEATHERING: EXAMPLES

Application of the first generalization is illustrated in figure 1, wherein is illustrated the tendency for the occurrence of the minerals in the form of a distribution curve when plotted in order of their occurrence in the sequence. Important evidence[4] of the proper sequential order of quartz, illite, mica-intermediates, montmorillonite, and kaolinite is derived from the translation of this

[4] Confirmed by the particle-size function and soil-depth function, as discussed below.

distribution curve through the sequence. The almost continuous translation of the distribution curve is shown in figure 2, with a number of soil colloids, together with a summary of properties and implications.

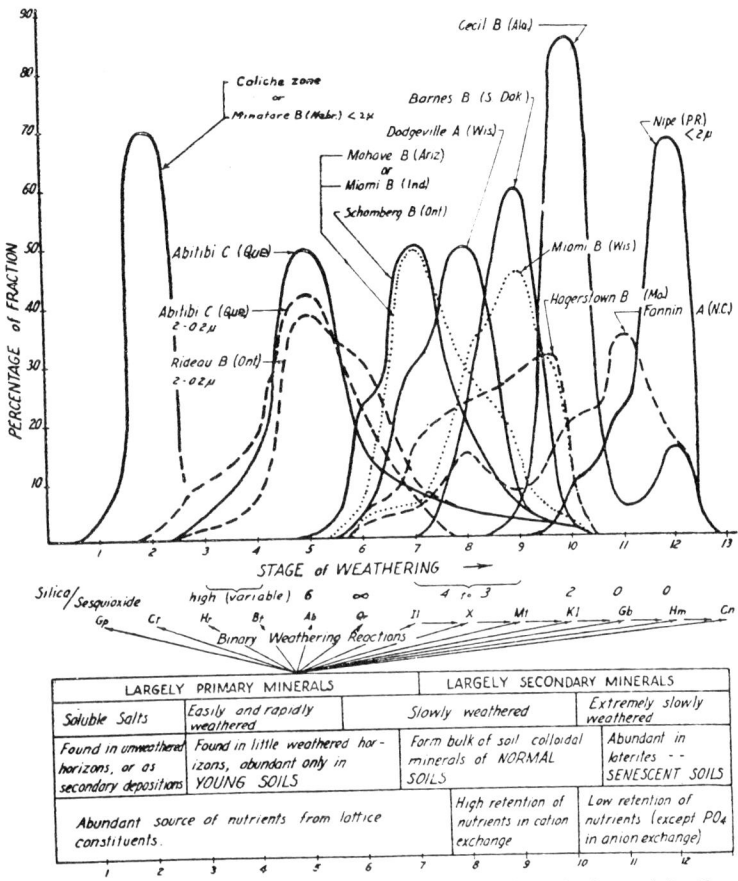

Fig. 2. Distribution curves for colloids of various stages of weathering and significance. (Particle sizes $< 0.2\,\mu$ unless otherwise stated.) "Cn" should read "An".

Early stages

Occurrence and disappearance of calcite (stage 2) in very young soils has been reviewed by Jenny (12, pp. 42–3). Under humid conditions, 5–10 per cent of calcium carbonate leached out of a dune sand during the course of 250 to 300 years of weathering, and of the Dutch polders reclaimed sea bottom in a similar period (12), i.e., during the embryonic stages of soil development.

The colloid of caliche zones (nearly pure calcite) or of Minatare silt loam B horizon (7), containing up to 75 per cent of calcite in the colloidal fraction, are examples taken from the semi-arid region (figure 2). Colloids high in calcite[5] also occur in the lower horizons of pedocals. The source of the calcite is generally secondary, through chemical precipitation, but in rare cases may involve sedimentation of colloids from eroded limestone.[5a] Preservation of calcite centers around the absence of sufficient leaching and the maintenance of high pH values. The dominating effect of such calcite on the properties of these colloids and of the soils themselves is well known.

In an analogous way, the presence of colloidal gypsum and other more soluble salts (stage 1) has a dominating influence on the nature of saline and solonized soils, particularly in the Pedocal region.[5b] These stage 1 and 2 colloids, representing secondary depositions, may occur out of sequence with respect to the silicate colloids present, as observed in the fourth generalization. It follows therefore, that the occurrence of these two early stages is not always associated with young soils, although it may be. Moreover, not all young soils contain minerals of the calcite or gypsum stages.

Early intermediate stages

The colloid from Abitibi silt loam, C horizon, is the youngest and least weathered soil material from the humid region which was studied (figure 2). The plagioclase of the albite-andesine end of the series is the dominant constituent, with lesser amounts of amphibole and chlorite of the two preceding stages, and quartz and mica of the two succeeding stages, thus further illustrating the distribution curve (first generalization). The colloid from Rideau clay, B horizon (16), is also centered on albite and quartz (stages 5 and 6) but lacks appreciable percentages of the earlier stages. Rideau clay is from southern Ontario and, while a young soil, is considerably advanced over the Abitibi sample of northern Quebec. The coarse clays of these two colloids are shown in figure 2 to illustrate the earlier stages of weathering. The fine clays also contained abundant feldspars and quartz (16), but were slightly further along in the weathering sequence, and only traces of minerals of stages 3 and 4 were present. The occurrence of amphibole and chlorite and of high amounts of feldspars in these colloids is believed to be the first instance reported for the fine fraction of soils. The occurrence of feldspars in the fine colloids of Rideau clay and other soils developed on the geologic Champlain sea was further verified by further collections from southern Quebec[6] of the St. Rosalie, A horizon, from two localities,

[5] The identity of the precipitated calcium carbonate of soils as being calcite has been amply verified by x-ray diffraction analyses in these and other laboratories.

[5a] Calcite and dolomite crystals of colloidal size have been observed in a podzol A_0 horizon, pedogenically formed within fragments of organic matter (Bourbeau, G. A.: Ph.D. Thesis, University of Wisconsin Library, Madison, 1948.

[5b] Since this manuscript was presented, Rodrigues and Hardy (Soil Sci. **64**, 127–42 (1947)) have reported up to 30 per cent gypsum in the colloid in certain horizons (p. 140) of a tropical soil derived from shale.

[6] Collections obtained through the courtesy of Mr. Roger Baril, Chief of Soil Survey, Laval University, St. Anne de la Pocatière, Quebec, Canada.

and the St. Damase, A horizon, in which 20–35 per cent of the fine colloid ($< 0.2\,\mu$ diameter) was plagioclase of the albite–andesine end of the series. It is interesting to compare the age of these soils with ferromagnesians disappearing in the early intermediate stages of weathering (9,000 to 20,000 years) to the very young soils listed by Jenny (12, p. 32), ranging from 250 to 1000 years of age with calcite disappearing.

Intermediate stages

The second generalization is exemplified in progressing through successive clays of figure 2. The fine colloids of Schomberg silt loam of Ontario (young soil of humid region), Miami silt loam of Indiana (retarded development for its region because of occurrence only on erosional slopes), and Mohave silt loam of Arizona (mature sierozem, desert region) are dominantly illite (stage 7). The colloid of Barnes silt loam of South Dakota (mature chernozem, subhumid region) is centered on montmorillonite (stage 9). The colloid of Miami silt loam of Wisconsin (mature gray-brown podzolic, cool humid region) is centered on montmorillonite with considerable kaolinite appearing. That of Hagerstown of Missouri (gray-brown podzolic, warmer humid region) is transitional, but with an increasing amount of kaolinite (stage 10) appearing. Both Miami and Hagerstown also contain appreciable mica (stages 7 and 8). A colloid of Hagerstown silt loam of Pennsylvania showed a similar composition. With the colloids from these gray-brown podzolic soils, the distribution curves are broader, extending from stage 7 (illite) to stage 10 (kaolinite), than in the chernozem (Barnes) centered on stage 9 (montmorillonite), with some mica-intermediate. The greater breadth of the curves for the gray-brown podzolic soils examined may be due to greater content of mica in the parent materials, coupled with greater intensity of weathering with some of the montmorillonite being carried on over to kaolinite. A postulate that the montmorillonite of the chernozem might represent an earlier stage of an alternative illite to mica-intermediate to kaolinite sequence is in conflict with the various lines of evidence to be presented. The illite percentage is higher in soils and horizons which have been subjected to less weathering. For example, the unweathered D horizon of the Miami soil (Wisconsin) is higher in illite than the A, B, or C horizons.

Advanced stages

Proceeding to soils developed under further increased weathering intensity, the Susquehanna sandy loam, B horizon, of Alabama (red podzolic soil, warm humid region) is approximately equal in stage 9 (montmorillonite) and stage 10 (kaolinite), further linking the direct succession of montmorillonite to kaolinite (curve not shown). Only a little mica is present in the fine colloid of this soil. The colloid of Cecil clay of Alabama (red podzolic soil, warm humid region) is predominantly kaolinite (stage 10), but contains appreciable ferric oxide (hematite, stage 12). That of another Cecil clay of North Carolina contains considerable gibbsite (stage 11) along with the kaolinite (curve not shown).

The colloid of Fannin sandy loam, C horizon, of North Carolina is centered on kaolinite and gibbsite (stages 10 and 11) with appreciable quantities of hematite present (figure 2). The colloid of Nipe clay (figure 2) of Puerto Rico is dominantly hematite and goethite (stage 12) but contains about 20 per cent of gibbsite (stage 11) and a small quantity of kaolinite (stage 10). The A_2 or B_{21} horizon of Naiwa soil of Kauai Island (the oldest Hawaiian island) consists of 25 per cent titanium dioxide, most of it anatase (stage 13), together with dominant amounts of hematite (stage 12).[6a] The translation of the distribution curves across the sequence of mineral weathering stages without a break in continuity emphasizes the procession of colloid composition as a continuous function of increasing weathering intensity (supporting evidence for generalization 2).

THE WEATHERING RATE FUNCTION: PROCESSION, ARREST, REVERSAL

Components of weathering rate function

The weathering rate may be viewed as a product of intensity and capacity factors. The intensity factors of weathering are temperature (T) and its complementary relationship to accumulation of humus; rate of water movement, or leaching provided by internal drainage (water); acidity of the solution (proton intensity, H^+) with particular reference to carbonic acid supply; and the degree of oxidation (electron intensity) and its fluctuation (oxidation–reduction, Δe^-). The capacity factors of weathering are the specific surface of the particles (s), and the specific nature of the mineral being weathered (k_m). The weathering per unit time (t) may be expressed in terms of these factors in the form of an equation:

Weathering rate = intensity factor × capacity factor
(time rate)
$$= f(\text{temperature, water, protons, electrons}) \times (\text{surface, nature of mineral})$$

Then the weathering stage of the clay-size minerals may be represented as a summation:

$$\text{Weathering stage} = \Sigma f(T, H_2O, H^+, \Delta e^-, s, k_m, t)$$

The five cardinal factors of soil formation (climate, vegetation, relief, parent material, and time) may be recognized as being expressed in the various intensity and capacity factors.

For a given mineral species, and a given particle-size range (constant capacity factor), the weathering stage of the soil colloid is a product of the intensity functions multiplied by the time in which weathering has been occurring, or simply as "intensity × time" product. Procession through various stages of the sequence follows increase of this product.

In general, increasing acidity and increasing oxidation must be considered as having a positive sense, favoring increased weathering intensity. Thus, pro-

[6a] From coöperative studies by Dr. G. D. Sherman, University of Hawaii, and the authors.

longed leaching under reducing conditions would not lead through the sequence; laterite forms under oxidizing conditions. Byers (3) postulated that weathering is entirely expressible as a hydrolysis reaction, but noted the influence of temperature, water movement, particle size, and specific nature of the mineral on the rate.

Arrest of weathering in the absence of leaching

The normal weathering processes may be interrupted by lack of sufficient intensity of one of the functional factors. As an example, the occurrence of large amounts of montmorillonite in bentonites, in spite of their great age (time of weathering), is attributed to lack of leaching. Under normal weathering, in the absence of dissipation by erosion, a surface deposit would have weathered to the final stages (hematite). Absence of leaching has prevented weathering of the feldspars and other silicates of the volcanic ash from proceeding beyond stage 9. Frequently quartz or cristobalite (stage 6) is found occurring in association with montmorillonite in bentonites, residual because of little or no leaching; leaching under normal weathering would have removed the excess silica and allowed montmorillonite to undergo further weathering. The sodium bentonite (Upton, Wyoming) and the calcium bentonite (Monroe County, Mississippi) may well represent the product of weathering of albite-rich and anorthite-rich volcanic ash, respectively. The high montmorillonite content of Lufkin clay (Mississippi) and Alamance clay (North Carolina) similarly occurs as a resultant of poor drainage and partial arrest of weathering at stage 9, in localities where a more advanced weathering stage is normal. Arrest has not been complete in the Alamance, since appreciable kaolinite (stage 10) is also present.

Reversal of weathering in sediments

The interruption or shift of one or more of the weathering processes may serve to reverse the weathering equation, and lead to a reverse traverse of the mineral weathering sequence. Much support for the weathering sequence and its reversible character is found in the hydrothermal reactions of minerals, including the reactions occurring under pressure. A second major line of thought on the reversal sequence in sediments is embodied in the geologists' term "diagenesis," referring to reversion by metamorphic processes.

The sediments are thought to represent a certain amount of reversal of the weathering sequence, particularly because of little leaching and a lowered oxidation potential. Iron of hematite is reduced, for example, and is available for recrystallization, perhaps in glauconite. In some instances, slowly percolating waters may illuviate a new supply of soluble components to assist in the reversion. Thus, ordovician bentonite is thought possibly to be an ancient montmorillonite which has picked up potassium and reverted, to a certain extent, to mica-intermediate. Continuation of this process to a greater extent in the presence of a more adequate potassium supply would explain the conversion of vast sediments to shales rich in illite. Exclusion of water and further supplies of solutes largely arrest the reversion at the illite stage.[7] To the extent that reversal is slowly continued either with or without influx of solutes from the outside of the sediment, authigenic quartz (stage 6), feldspars (stage 5) (10), zircon and tourmaline, etc. would be expected to occur *as found* in sedimentary deposits. Accessory chlorite and biotite of shales, glauconite of sandstones, and the calcite of limestones represent the ultimate reversal to stages 4 and 2 of the sequence. Or, in general:

> 5. The alteration sequence of the colloidal minerals of sedimentary deposits, under the impact of decreased or excluded leaching and oxidation, tends to be the reverse of that in the weathering of colloidal minerals of soils.

Thus montmorillonitic sediments tend to revert by solution or metamorphic processes to micas and earlier stages. A gibbsitic (bauxite) sediment would be expected to resilicate to kaolinite or even to montmorillonite or illite to the extent that silica and potassium sources were available in the mixture of minerals in the sediment.[8]

To the extent that recent sediments (alluvium or aeolian) are left exposed to continued surface weathering and soil formation, the weathering sequence continues scarcely interrupted. Or, in general:

> 6. Colloids of soils being developed on fresh or recent alluvial or aeolian sediments continue from the weathering stage occupied by the soils from which the sediment was derived.

Thus, the colloid of a young soil from such source would usually represent the particular weathering stage of its source, which would not necessarily be an early stage.

THREE ANALOGOUS RECAPITULATIONS OF THE WEATHERING SEQUENCE OCCURRING IN NATURE

Three analogous recapitulations of the weathering sequence of clay-size minerals are found in nature: first, in accordance with the familiar geographic pattern of weathering; second, as a particle-size function; and third, as a depth function in soil or sediments. These trends may be set forth as generalizations 7 to 9.

> 7. *Soil geography.*—The mineralogical composition of soil colloids follows the weathering sequence geographically, in accordance with the variation in weathering intensity factors which are controlled by the geographic distribution of climate, together with time of weathering.

Sediments are influenced by the same geographic factors, to the extent that they are derived from soils so controlled. The mineral composition tends to vary in

---

[7] Evidence of reverse traverse of the weathering sequence from kaolinite (stage 10) back to illite (stage 7) in sediments formed in the Gulf of Lower California from suspended solids of the Colorado River has been noted by Dr. R. E. Grim (personal communication, September, 1947).

[8] Goldman and Tracey (Econ. Geol. **41**, 567 (1946)) recently described resilication to kaolinite occurring in Arkansas bauxites.

the various Great Soil groups, being far advanced (stages 11 and 12) in the laterites, intermediate (stages 8 and 9) in the chernozems, less advanced (stages 7 to 9) in the sierozems, and least advanced (stages 3 to 6) in certain types of young soils, for example, those developed on the sediments of the Champlain and Ojibway glacial seas (figure 2).

8. *Particle surface or size function.*—The rate of weathering of clay-size mineral particles of soils and sediments varies according to the surface of the particles, being more rapid with increasing fineness of the particles. Thus, in stages 1 to 9, the mineralogical composition of colloids from soils, and to some extent of sediments, advances in the weathering sequence with increasing fineness of the fraction separated for identification. Decreasing size is the "capacity factor equivalent" to translation through greater intensity of weathering. In stages 10 to 12, kaolinite, gibbsite, and hematite may show the

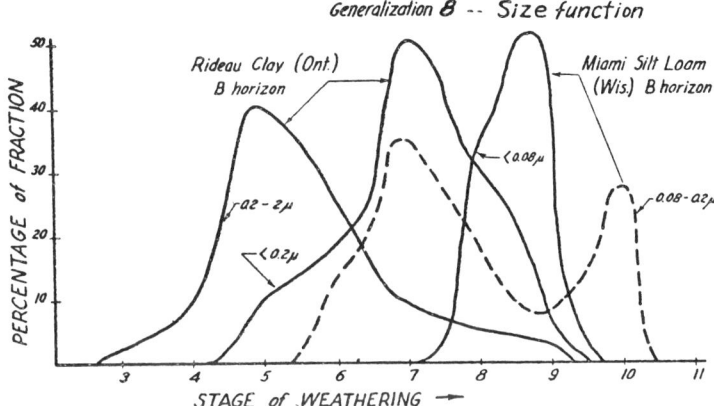

FIG. 3. Particle-size function in weathering stage of clay-size minerals

same trend; however, these minerals may undergo *crystal regrowth* and occur independently of the particle size.

With extreme fineness, weathering advances very rapidly with the result, for example, that hematite monolayers would be expected, and may be found, on almost any colloid developed under conditions of good oxidation. Likewise, the coarse clay fraction (0.2–2 μ diameter) almost invariably contains minerals of earlier stages of weathering than the fine clay ($< 0.2 \mu$ diameter). This is illustrated for these two fractions of Rideau clay, B horizon (Ontario), in figure 3. The coarser fraction is centered over feldspar (stage 5); the finer fraction over illite (stage 7). Both fractions contain quartz (stage 6); but the fine fraction contains small amounts of the more advanced stages (8 and 9), mica-intermediate and montmorillonite, while the coarser fraction contains small amounts of the earlier stages (3 and 4). One of the best lines of evidence that illite follows quartz in the sequence is that mica persists in increasing ratio to quartz as the particle size of the fraction decreases. The minerals which are more resistant to chemical weathering persist in the greater quantities in the finer size fractions. The asymmetry of the distribution curve of figure 1 can be explained on the basis of particle-size function. The quartz occurs in the *coarser* sizes present in the fine clay fraction, and thus persists overly long, simply because its particles are larger than the average for the fraction. The colloid of Miami silt loam (Wisconsin) is centered over mica-intermediate and montmorillonite (stages 8 and 9) in the fine particle size range of $< 0.08\ \mu$ diameter (figure 3), but the coarser 0.08–0.2 μ fraction contains predominantly quartz and illite of earlier stages 6 and 7. A second peak occurs over kaolinite (stage 10), as an illustration of the process of crystal regrowth. If the generalization held perfectly, kaolinite would appear in a finer fraction than montmorillonite. The fact that its crystals can grow larger, however, may account in part for its increased stability to weathering. It is interesting that the distribution curve of the entire colloidal fraction ($< 0.2\ \mu$) of Miami silt loam (Wisconsin) is a symmetrical monodistribution (figure 2) in contrast to the bimodal nature of the coarser subfraction in figure 3.

    9. *Horizon depth function.*—The weathering stage of the colloid of a soil horizon or of a sediment tends to advance with increasing proximity to the surface.

Thus, in soils, weathering stage *decreases* with depth down into the unweathered D horizon, but this change with depth usually is not great through the A, B, and C horizons, shifting only through one to three weathering stages. The change with depth is most pronounced in young soils in the humid region where weathering is in the early stages (Schomberg silt loam, Ontario, figure 4) (16). A strong line of evidence that montmorillonite follows illite and mica-intermediate in the sequence is the alteration of mica to montmorillonite in soils such as the Schomberg (figure 4). Bray (2) also noted a similar relationship for certain soils in Illinois. Weathering stages have been found to change little with depth through the A, B, and C horizons in many of the well-developed soils of the humid region. This is true of Miami silt loam (Wisconsin), although an analogous increase in illite with depth occurred in entering the D horizon of this soil.

In soils of the arid region, the weathering stage may increase with depth, in passing into horizons where the subsurface is kept more moist by protection from evaporation, while the surface mulch is dry over long periods. A somewhat analogous situation occurs in the planosol, wherein extra water supply and more sustained solution processes are provided by restricted rate of external drainage and consequent increased internal water supply. This situation of the accelerated weathering in the planosol, under restricted *rate* of drainage but ultimately *large total volume* of percolate, should not be confused with the arrest or reversal of weathering processes in sediments under conditions of virtually no internal drainage and percolation.

In sediments, the weathering stage would be expected in general also to advance with proximity to the surface, if the processes of change with time in the sediments are viewed as a reversal of the weathering equation. Thus the clay-size minerals of a very recent sediment might be abundant in kaolinite or montmorillonite, while in deeper lithology the same parent sediment would have reverted (diagenesis) to the earlier stage illite, as already discussed.

These polyfunctional processes of weathering are illustrated in figure 5, wherein weathering stage is represented on the vertical axis. Weathering stage advances with increase of the product "intensity × time" of weathering, and with decrease in particle diameter (or proportional increase of specific surface). Increments of weathering with decreased particle surface are indicated as Δw and Δs (figure 5). Weathering reaches completion in a short time with extreme fineness of particles, for example, the appearance of surface films of hematite early in weath-

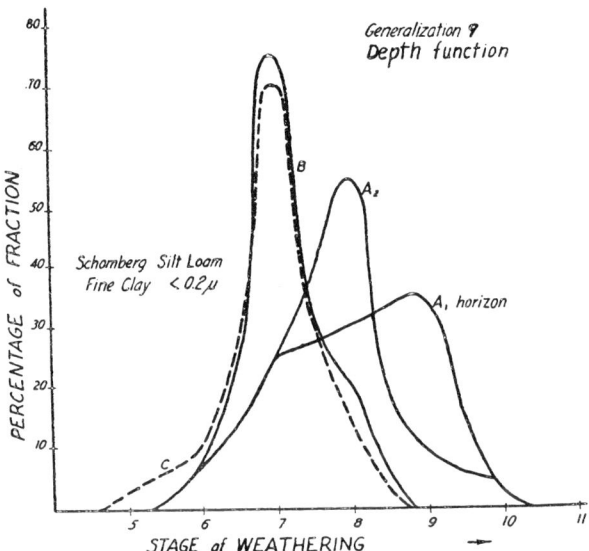

FIG. 4. Depth function in mineral weathering stage of colloids from a young soil

ering (figure 5). Specific properties of minerals (k_m) favoring ease of weathering have an influence similar to decrease in diameter in hastening weathering.

Stages 1 to 4 are traversed in relatively short "time × intensity" factors, while the sweep broadens (rate of change of weathering stage with "time × intensity" decreases) in the intermediate stages of 5 to 9. This trend has led to the general statements that basic rocks form laterites (5, 9), whereas acid rocks yield kaolinite and free silica (5, 8, 11). These statements indirectly state that the acidic rocks require a much larger "time × intensity" factor for complete weathering to stages 11 and 12, as compared to basic rocks.

The span of weathering stages found within the young Schomberg profile (figure 4) is represented by the increments A, B, C (figure 5) corresponding to these three horizons. This shortness of span with depth in the soil profile is considered typical—in fact, the span is usually much shorter in moderately well-developed soils. The increment A', B' represents the span for a desert soil, with weathering being somewhat further advanced in the B horizon.

CRYSTAL CHEMISTRY OF THE WEATHERING SEQUENCE

The discussion thus far has been concerned with relative persistence of the colloidal minerals as a measure of their relative stability. Emphasis has been placed on minerals "found" rather than the mechanism of their formation. The crystal chemistry of the weathering sequence may be divided into the questions of (a) the underlying reason for the relative stability as an expression of factors

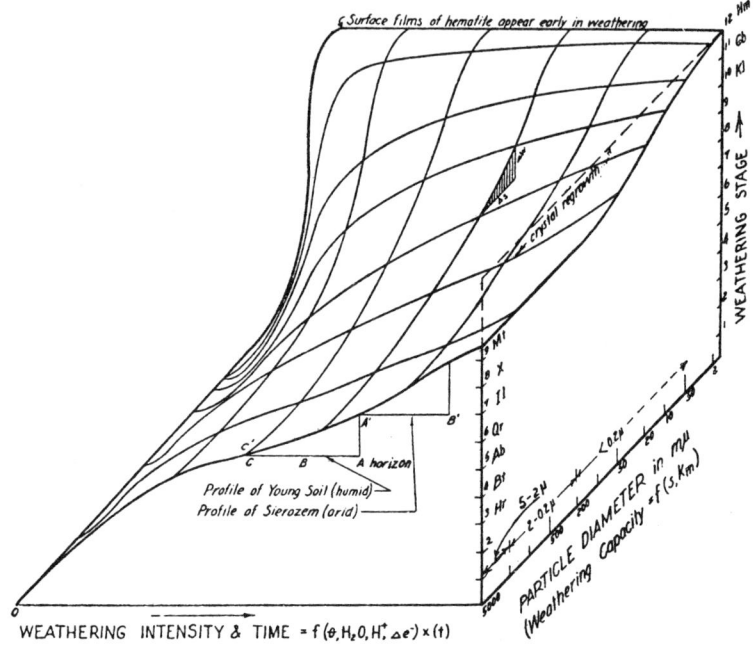

Fig. 5. Mineral weathering stage as a function of intensity × time of weathering, and the capacity factors of weathering. (The symbol θ should be T.)

of crystal structure, and (b) the chemical transformations involved in the formation of minerals of each stage. The second question resolves itself into an inquiry as to what extent each stage is the parent material for succeeding stages, or contrarily, to what extent each stage is a unique product of a primary parent material left either by eluviation of all else, or through a binary weathering reaction.

Crystal chemistry of relative stability

Chemical weathering takes place through simple solution, carbonation, hydrolysis, and oxidation and reduction. Simple solution and carbonation are the reactions chiefly concerned in weathering loss in (embryonic) stages 1 and 2. Discontinuity of the silica portion of the lattice, coupled with activity of the basic cations, appears to be the dominating factor in stages 3, 4, and 5. Independent or incompletely linked silica tetrahedra (silica groups I to V) permit ready hydrolysis of the strong bases in stage 3. Weakness along the octahedral planes caused by the presence of Fe^{++} and Fe^{+++} gives rise to the relative instability of stage 4 (biotite, etc.). The continuous silica sheets are a factor for stability of the other group VI silicates, and account for the stage 4 minerals succeeding the other ferromagnesian minerals of stage 3. The fourfold silica linkage of albite (feldspars in general, stage 5) brings their stability one level higher, but the high content of active bases introduces ease of hydrolysis. This places the feldspars one stage less stable than quartz (stage 6), with fourfold silica linkage but without the bases present. Quartz solubility is a linear function of specific surface, and therefore quartz is expected, and is found, to decrease tenfold in quantity for each tenfold decrease in particle size of fraction considered. The occurrence of the micas at stage 7, more stable than quartz, is contrary to Goldich's series for coarse particles (cited on p. 52 of reference 14), but has abundant experimental verification in the observed mineral content of colloids of various stages of weathering and in the particle-size function as discussed. The basis for the occurrence of mica after quartz is explained on the principle that a layer of aluminum ions beneath a layer of silica ions in the silica sheet (oxygen lattice) increases the stability of the silica layer. This finds support in the great stability of the alumina sheet, which persists through five stages of weathering (7 to 11).

The mica–montmorillonite order is analogous to the albite–quartz order. Of the less stable mineral of each pair, for example, muscovite contains 350 milliequiv. of non-exchangeable but hydrolyzable potassium per 100 g., while orthoclase contains 550 milliequiv. of potassium per 100 g. Of the more stable mineral of each pair, each has silica surfaces, and little or no hydrolyzable lattice bases. Montmorillonite has more stable (more quartz-like) surfaces of its crystal plates than mica.

It is noteworthy that, whereas the octahedral alumina layer stabilizes the silica sheet (stage: quartz < mica), the substitution of Al^{+++} for Si^{++++} within the silica sheet has the opposite effect (albite < quartz, or mica < montmorillonite), perhaps because of distortion in the silica sheet and the accompaniment of the hydrolyzable basic cation to balance the charge. The 60–100 milliequiev. of exchangeable bases per 100 g. of montmorillonite may be exchanged without disturbance of the lattice. Mica apparently cannot release its potassium without alteration of its structure (i.e., weathering). With regard to crystal lattice factors:

    10. The stability factor for silicates in the clay-size range is a resultant of the silicate structural groups in general, but is modified by considerable cross-over according to the ionic substitutions involved.

Beyond stage 7, the percentage occurrence no longer falls off with decreasing particle size, indicating that the lattice factors for stability (k_m) have made the increase in specific surface relatively much less important in determining weathering rate.

The octahedral alumina sheet persists as desilication occurs in forming kaolinite (stage 10) and gibbsite (stage 11), illustrated diagrammatically in figure 6. Hematite and anatase are the final endpoints of weathering under conditions of good aeration and high temperature. Their insolubility under these conditions slightly exceeds that of gibbsite and boehmite, and these two minerals eventually dissolve and eluviate, leaving hematite and anatase. Corundum, the isomorphous analogue of hematite, hypothetically might end the series, but higher than soil temperatures would be required for such alteration of gibbsite and boehmite. The requirement of oxidizing conditions is mandatory if hematite is to persist as an end-product; otherwise it is subject to the instability factor of the ferromagnesian minerals of stages 3 and 4. Transitory periods of reduction may account for the loss of iron and relative enrichment in anatase (and ilmenite) noted in the Naiwa soil.

Chemical transformations

Successive stages of the weathering sequence are invariably lacking in stoichiometric relationship to preceding stages. Thus, successive stages are not closed-system rearrangements of the chemical content of the parent material; in fact, some stages (e.g., stage 12) involve principally different chemical elements compared to preceding stages. In general:

11. Lack of stoichiometry between successive stages of the sequence arises through processes of eluviation and illuviation.

Thus a small residue of the parent material may constitute the bulk of a successive stage through eluviation of all else. This is analogous to the persistence of a few short threads as the "successive stage" after dissolution of a bag of sugar in which the threads had originally been a scarcely noticeable impurity. A great reduction in lithological volume accompanies such a transformation, a familiar example being the weathering away of several feet of limestone to produce a few inches of earthy material. This "simple residue" principle may be continued successively. In the analogy, the threads might consist of carbon pigment and cellulose, and the cellulose decompose leaving the trace of carbon pigment as the bulk of the next stage. Of the earthy residue of the limestone, for example, the antigorite (stage 4, iron-magnesium analogue of kaolinite) might decompose, leaving albite and illite present in the successive stage.

In addition to this "simple residue principle," the coarser minerals on the one hand, and the colloidal products of decomposition on the other, may not be entirely recovered in the residue, but in themselves give rise to (be the "successive parent materials" of) successive stages. In the analogy, the cellulose might give rise to a carbonaceous residue, almost as resistant as the carbon pigment, and might be considered as belonging to the same stage. In the earthy product of the limestone decomposition, some of the magnesium associated with the calcium carbonate might combine with a portion of the albite decomposition product and result in sericite–illite or montmorillonite. At the same time some of the original illite might weather into mica-intermediate and montmorillonite. The distribution curve of mineral composition of the colloid would thus advance through stages 5, 6, 7, 8, and 9. Under acid conditions, albite or illite may alter in part directly to kaolinite, and some of the montmorillonite be desilicated to kaolinite. The colloid thus advances through stages 7, 8, 9, and 10.

By the time stages 10 and 11 are becoming prominent, wherein no isomorphous substitution of iron is possible, the goethite–hematite (stage 12) content will have begun to be built up appreciably, as iron is released from weathering of the minerals of earlier stages. Hematite and goethite thus build up in bauxites and laterites. The basis of the occurrence of three to five minerals as a distribution curve for a colloid at the later stages (first generalization) is thus apparent.

In addition to this "successive parent material principle," additional chemical sources are available through illuviation from other horizons. This is the complement of the "simple residue" process going on in the source horizon. The occurrence of gibbsite as a primary weathering product from feldspar and mica was noted by Alexander, Hendricks, and Faust (1), and they advance the principle that this gibbsite may normally be resilicated to kaolinite in the zone near the unweathered parent rock surface. This is an important observation bearing on the mechanism of clay formation. Harrison (11) postulated resilication of gibbsite through rise of silica through the ground water from freshly decomposing silicates.[9] It appears probable that resilication can proceed on to the montmorillonite and mica stages under some circumstances, particularly in sediments. That mineral weathering in soils is dominantly an open system rather than a closed system within individual mineral grains is apparent from these reactions brought about by illuviation.

Desilication

Reversal of these silication equations, giving desilication (figure 6), in advancing stages of weathering is believed by the writers to be a dominant mechanism for the chemical transformations within the colloidal fraction between stages 7 to 11. In this range, minerals of each stage may be the parent material of those in successive stages. Simultaneously, the early-stage primary minerals of the silt and sand such as ferromagnesians and feldspars may, through binary weathering reactions, also form one or several of the minerals of stages 7 to 11, with accompaniment of eluviation of the excess iron and silica. In contrast, quartz undergoes gradual solution and eluviation in much the same way as gypsum and calcite, but more slowly (lower k_m factor).

Why desilication proceeds at an accelerated rate as acidity increases has been a much debated point, but the experimental fact is well established that kaolinite and gibbsite form under increased acidity associated with increased weathering intensity and time, at the expense of minerals of higher silica content. Chemically, the solubility of silica would be expected to decrease as the acidity increased. Various proposals have been advanced (cited in reference 14) that an isolated alkaline condition such as in the weathering shell of feldspar grains must intervene to remove the silica, but this proposed mechanism is inconsistent with the acidity of the soil systems actually existing, through which the silica must move. Moreover, ample basis exists for the assumption that desilication occurs

[9] See footnote 8.

in the acid regime of weathering. First, the increase of soil acidity to pH 4, though representing "extremely acid" soil conditions with reference to plant growth, represents only a *very slight* degree of acidity in the chemical system for insolubilizing silica. Marked insolubility of silica is brought about at negative pH values, which are 4 to 6 magnitudes more acid than pH 4. Second, from the viewpoint of relative stability, kaolinite, gibbsite, and boehmite, while structurally hydroxyl compounds, respond chemically more like insoluble weak acids or acid anhydrides, weaker than silicic acid. As such, they should tend to form in acid systems. In line with this property, the gibbsite and boehmite are solubilized more easily by treatment with alkali than with acid. The explanation of increasing depletion of silica ("laterization") associated with increasing acidity

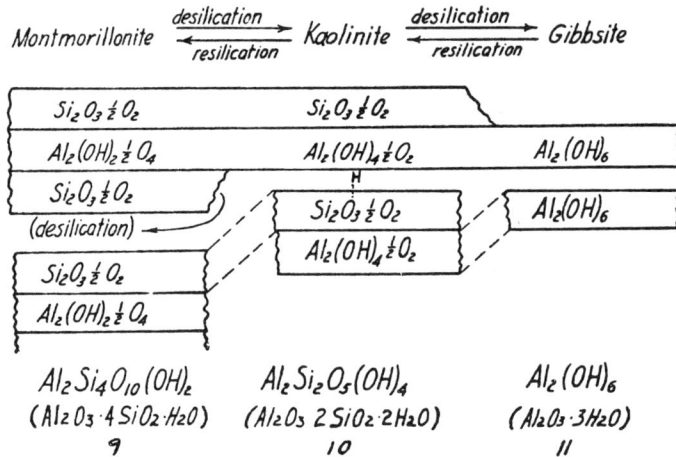

Fig. 6. Reversible silication reaction of the gibbsite layer of stages 9, 10, and 11. (Ionic substitution not shown in montmorillonite formula for the sake of simplicity.)

lies in the fact that both factors result *independently* from increased weathering (leaching, etc.).

The possibility of hydrogen bonding between crystal plates of minerals of stages 10 and 11 also tends toward more compact and more stable crystals, and this hydrogen bond is created in an acid regime by the desilication of one surface of montmorillonite to form kaolinite or of the second surface of kaolinite to form gibbsite. The fundamental basis for the continuously decreasing ratio of silica to sesquioxides and bases in colloids, with advance in weathering intensity, is apparent in the procession in the weathering sequence through ratios of infinity to zero, as shown in figure 2.

Podzolization

The dominant factor in the chemical transformations in podzolization is reduction, or deficiency of oxygen. Reducing conditions may be created by accumulations of organic matter, either because of low average temperatures, or because of accumulation of excess water and exclusion of air. Reduction mobilizes the iron released by weathering by converting it to ferrous compounds and frequently to the form of organic complexes. The mobilized iron usually is deposited in deeper horizons where the oxidation potential is higher, owing to increased pH value and calcium saturation. In podzolization, therefore, one end-product of weathering (iron oxide) is removed from the site of weathering, and to this extent the A_2 horizon represents a reversal of the normal weathering equation (oxidation considered positive). However, in other respects the mineral weathering sequence is followed, viz., disappearance of feldspars and low content of quartz in the colloid, and the occurrence of a distribution curve of the illite, mica-intermediate, montmorillonite, and kaolinite in the colloid of podzols, even in the A_2 horizon. Desilication of the lattices in the clay-size particles and succession through the sequence apparently takes place in the A_2 horizon. The ashy appearance is considered to represent prominence of the gray silt and sand grains, resulting from removal of coloring agents and migration of some of the colloidal clay particles out of the A_2 horizon. But it is not considered to represent precipitation of silica, or even an absence or retardation of the normal rate of dissolution and leaching of quartz and other forms of silica, particularly from the clay-size particles. Podzolic soils are frequently acid, but may be nearly neutral so far as the dominant chemical processes of podzolization are concerned. Podzolization has been reported at pH values of 6.5 or above, as might have been predicted.

Laterization

From the point of view of the weathering sequence, laterization represents the end-product of intensive desilication under conditions of intensive leaching, usually in the presence of slightly acidic solutions (e.g., pH 4), as discussed above. The process of laterization differs from that of podzolization in having good oxidation, which preserves the iron in the form of goethite and hematite (stage 12) and the aluminum in the form of gibbsite and boehmite (stage 11). Continuity and basic similarity of podzolic and lateritic weathering processes is intimated in the literature by recognition of previously formed "lateritic" soils as red and yellow "podzolic" soils.

The fundamental thesis of the concept of a mineral weathering sequence is an integration of the various binary weathering reactions (stoichiometric systems) such as ferromagnesium mineral → hematite, volcanic ash → montmorillonite, or

$$\text{feldspar} \begin{array}{c} \nearrow \text{sericite--illite} \\ \rightarrow \text{kaolinite, or} \\ \searrow \text{gibbsite} \end{array}$$

These processes may occur singly, simultaneously, or successively, according to circumstances. The integration occurs as a phenomenon of nature, but its definition in the weathering sequence presented could be achieved only through observation of a large body of data for colloids from diverse sources.

Thus, the integrated sequence function is considered to be continuous, and therefore the gradation from podzolization to laterization must be a continuous function, with all intermediate soil conditions to be expected (podzol-podzolic-lateritic-laterite sequence). One illustration of continuity will be presented centering around the extinction function of quartz in approaching the laterite (table 2). It will be noted that the quartz content of the soil decreases systematically with increasing weathering intensity, and that the percentage of quartz in the finer size fractions disappears more quickly than for the soil as whole, but continuously. In the Nipe clay soil, the quartz content is 10 or 15 per cent (as determined by x-ray diffraction analysis of the total soil ground to sufficient fineness by light crushing for a few minutes in an agate mortar). This establishes the fact that a considerable amount of silica in the upper 36 in. of soil has been available to be weathered into the finer sizes. However, weathering intensity

TABLE 2
Extinction of quartz in approaching the laterite

SOIL AND SOURCE (B HORIZON IN ALL CASES)	PERCENTAGE OF QUARTZ IN VARIOUS SIZE FRACTIONS			PERCENTAGE OF STAGES 11 AND 12
	Fine clay $<0.2\ \mu$	Coarse clay $0.2\text{--}2\ \mu$	Whole soil	Fine clay $<0.2\ \mu$
Miami silt loam (Wisconsin)	5–8	30–40	60–80	<2
Cecil clay (Alabama)	<5	5–10	20–40	10
Catalina clay (Puerto Rico)	<5	5–10		15
Fannin sandy loam (North Carolina)	<5	5–10	20–40	20
Nipe clay* (Puerto Rico)	0	0	10–15	85
Laterite (Haiti)	0	0	0	100

* Sandy A horizon assumed lost by erosion.

(leaching in so far as quartz is concerned) has been great enough to prevent any appreciable accumulation of quartz in the size range below $2\ \mu$ diameter. According to the particle-size function (generalization 8), quartz cannot or scarcely can reach clay size under a weathering "intensity $\times$ time" factor of sufficient magnitude to produce a laterite.

Later stages of the weathering sequence increase reciprocally with the quartz decline, particularly stages 11 and 12 (right-hand column of table 2). In the Fannin and Nipe soils, the gibbsite content of 10–20 per cent fitting between kaolinite and hematite in the distribution curve helps to establish the position of gibbsite in the sequence and substantiate the desilication reaction.

The necessity of eluviation and illuviation in building up the alumina and iron content at the close of the sequence has been stressed. Evaporation of illuviating ground waters (11) and oxidation of reduced iron contained undoubtedly has sometimes been a major contributing factor in the enrichment of laterites (5, 13, 14). Laterite in fact has been termed (5) a "fossil illuvial horizon of an ancient soil." To the extent to which illuviation has been a factor, the laterite is a counterpart (on a large scale) of the illuvial B horizon of the podzol. Moreover, a predominance of easily weathered minerals (high k_m, as in basic rocks) favors rapid progression through the weathering sequence, and from a practical standpoint may be a dominating factor (5, 9, 13, 14) in the development of some of the great laterite deposits.

The enrichment of the colloids in alumina and iron has been spoken of as a tropical weathering process (12, p. 36). The weathering sequence view emphasizes the depletion of silica and enrichment in alumina and iron as a trend in weathering generally. The accumulation of a quartz-rich A horizon on the surface of senescent laterites would appear to contradict the concept of hematite as the final weathering stage. Quartz is listed in weathering series for coarse particles as the most advanced (resistant) stage. However, this seeming paradox arises from the fact that quartz is perhaps the silicate mineral most resistant to *physical* weathering, being much more resistant than mica, for example. When it is in sufficiently fine grain size, chemical weathering takes precedence over physical weathering and the mica turns out to be more resistant. Sandy A horizons of senescent lateritic and laterite soils may develop through rather complete mechanical eluviation of the colloidal constituents.

Weathering sequence in relation to the "normal" soil

The concept of the "normal" soil advanced by Marbut implies that the soil profile is in equilibrium, being lost by erosion as fast as formed through weathering. The concept of weathering sequence operates to a considerable extent satisfactorily within this frame of reference. In agreement with Marbut's view, the sequence transcends the parent material as the key to trend of weathering of the colloid. In general:

12. The *course* of weathering (the sequence) is unaffected by the parent material, but the stage existing (the mineral content) at any time is influenced by the parent material to the extent that the "intensity × time" product has been insufficient to complete transformation of the source material.

However, the concept of "normal" soil fails to describe the young soils which are common (*normal*) in Southern Ontario and so, according to Marbut's view, these are azonal soils. This terminology fails to fit these soils which, from a broad view, are truly both *normal* and *zonal*, but this discrepancy may be passed over as being only a fault in terminology. A more fundamental question is whether virgin profiles of the so-called "normal" soils are truly at equilibrium, i.e., at a steady state. To answer this requires conjecture as to what they would be like in another 20,000 to 100,000 years or more. It is important to note that the weathering sequence view as presented is *equally applicable* whether the soils are at equilibrium ("normal"), or whether they are continually moving into more advanced stages. The very old soils of the Southern Appalachian mountains and Piedmont plateau, with their high kaolinite, gibbsite, and hematite contents (6), probably should be classified as lateritic or early-stage laterites rather than as red podzolic soils. Likewise, the old laterite soils on the peneplain of western Australia (4, 5, 15) are in still further advanced stages of weathering, as a result of a large "intensity × time" factor. Comparison of these soils with younger soils in the same regions suggests that a true equilibrium, i.e., a steady state, has *not* been reached by soils generally, and that *most soils* are even now slowly *advancing further through the weathering sequence.*

SUMMARY

The stability series or weathering sequence of minerals present in the colloids of soils and sediments is considered both from the standpoint of the minerals found and from the standpoint of the basis in crystal chemistry for the sequence. The sequence of thirteen stages is represented by the type minerals: gypsum, calcite, hornblende, biotite, albite, quartz, illite, mica-intermediate, montmorillonite, kaolinite, gibbsite, hematite, and anatase (corundum). The weathering stage of a colloid is considered to be a resultant of intensity factors (temperature, moisture transfer, acidity, and oxidation–reduction) and capacity factors (particle size and specific nature of the minerals), together with time. The following generalizations are made on the basis of observed data:

1. From three to five minerals of the weathering sequence are usually present in the colloid of any one soil horizon, one or two minerals being dominant and other adjacent minerals in the sequence decreasing in amounts with remoteness in the sequence.

2. The percentage of minerals of the early stages of the weathering sequence decreases, and of the successive members increases, with increasing intensity of weathering.

3. Intermediate stages may occasionally be absent, giving a bimodal curve, and secondary deposits such as calcite or gypsum may occur out of sequence.

4. The weathering equations are considered reversible, moving largely to the right in soils, and to the left in sedimentary deposits. However, alluvial and aeolian sediments which remain exposed to continued weathering continue the sequence as of their parent soils.

5. The mineralogical composition of the soil colloids varies according to three analogous sequences: *viz.*, according to geographic (climatic) variations, particle surface function, and proximity to surface of the soil.

6. The stability factor of the minerals is a resultant both of crystal structure and of the specific isomorphous elements. Lack of stoichiometry between successive stages arises through processes of eluviation and illuviation. The course or direction of the sequence is unaffected by parent material, although the stage at hand may be.

Podzolization and laterization are considered to differ principally in the degree of oxidation and in summation of "weathering intensity × time", but to be following through the sequence as otherwise fundamentally similar desilication processes in an acid regime. It is suggested that soils in general, instead of being at a steady state as embodied in the concept of the "normal" soil, may be even now slowly advancing in the weathering stage.

Helpful suggestions and criticisms of this manuscript offered by Professors P. D. Krynine and J. C. Griffiths of The Pennsylvania State College and by Professor R. J. Muckenhirn of the University of Wisconsin are acknowledged with gratitude.

REFERENCES

(1) ALEXANDER *et al.*: Soil Sci. Soc. Am. Proc. **6**, 52 (1942).
(2) BRAY: Soil Sci. **43**, 1 (1937).
(3) BYERS: Bull. Am. Soil Survey Assoc. **14**, 47 (1933).
(4) CAMPBELL: Mining Mag. **17**, 67, 120, 171, 220 (1917).
(5) CARROLL AND JONES: Soil Sci. **64**, 1 (1947).
(6) COLEMAN, MEHLICH, AND JACKSON: Manuscript, University of Wisconsin, 1947.
(7) FITTS: M. Sc. Thesis, University of Nebraska, 1937.
(8) FOX: Records Geol. Survey India **69**, 389 (1936).
(9) HANLON: J. Roy. Soc. N. S. Wales **78**, 94 (1945).
(10) HONESS AND JEFFRIES: Sedimentary Petrol. **10**, 12 (1940).
(11) HARRISON: Imp. Bur. Soil Sci. Harpenden, England (1933).
(12) JENNY: *Factors of Soil Formation*. The McGraw-Hill Book Company, Inc., New York (1941).
(13) MARTIN AND DOYNE: J. Agr. Sci. **17**, 530 (1927); **20**, 135 (1930).
(14) REICHE: Univ. New Mex. Pubs. Geol. No. 1, Albuquerque, 1945.
(15) SIMPSON: Geol. Mag. **9**, 399 (1912).
(16) WILLIS AND JACKSON: Manuscript, University of Wisconsin, 1947.

3

Copyright © 1955 by the Soil Science Society of America
Reprinted from *Soil Sci. Am. Proc.* **19**:334–339 (1955) by permission of the Soil Science Society of America

Chemical and Clay Mineral Properties of a Red-Yellow Podzolic Soil Derived from Muscovite Schist[1]

C. I. Rich and S. S. Obenshain[2]

ABSTRACT

A study was made of a Red-Yellow Podzolic soil with particular regard to the properties of dioctahedral vermiculite, one of its major clay minerals. The soil, Nason silt loam, which is derived from a muscovite schist residuum, was found to be nearly devoid of exchangeable calcium and low in other bases. Although the cation exchange capacity of the B_2 horizon was 25 me. per 100 gm. soil, this horizon contained only 0.08 me. Ca per 100 gm. Clay minerals present were kaolinite, dioctahedral vermiculite, and regularly and randomly interstratified illite-vermiculite. The 14.7A basal spacing of vermiculite from the C_1 horizon moved to 10.5A when the clay was K saturated, but the effectiveness of K saturation decreased from the C_1 horizon through the A horizon where there was only slight collapse. Boiling the clay from the B_2 horizon for 102 hours in $1N$ KCl caused a change of the 14.7A basal spacing to only 14.2A. However, treatment of the clay in $1N$ KCl plus $0.1N$ HCl or treatment with $1N$ NH_4F altered this spacing to near 10A. The difficultly collapsed mineral had a high internal surface, high base exchange capacity, and low divalent cation content. Easily collapsed dioctahedral vermiculite was made difficultly collapsed by repeated Al saturation and drying. Heat treatment at 800°C. collapsed the 14.7A spacing of the mineral in all horizons to 10.3A. Glycerol solvation caused no increase of the 14.7A spacing. These results together with D.T.A. data support the theory that non-exchangeable Al in the interlayer position in vermiculite restricts collapse of the mineral on K saturation.

Horizon	Depth	Description
A	0 to 9 in.	Yellowish-brown (10YR 5/4) very friable silt loam with a weak fine granular structure. The top 2 inches are stained darker with organic matter. It is slightly hard, almost white, and contains many small brown, white, and gray quartz gravel. It gradually grades into:
B_1	9 to 13 in.	Reddish-yellow (7.5 YR 6/8) friable light silty clay loam with a weak fine subangular blocky structure. It is moderate hard when dry, and contains some small quartz and schist fragments and mica flakes.
B_2	13 to 22 in.	Predominantly yellowish-red (5YR 6/6), mingled with yellowish-brown and red, friable clay or silty clay; moderately medium subangular blocky structure. It is hard when dry, contains few roots, schist fragments, quartz gravel, and mica flakes. Mica flakes increase with depth and the soil is slick when wet.
B_3[3]	22 to 36 in.	Mingled red (2.5YR 4/6) and reddish-yellow (7.5YR 6/8) friable clay; partially disintegrated schist increases with depth.
C_1[3]	36 to 42 in.	Same color as above with more rock showing. Some of the soil material shows rock forms and is still consolidated.

Nason silt loam is a Red-Yellow Podzolic soil well known locally for its low fertility and for the large amount of lime required to change its pH significantly. Tatum silt loam is developed from the same parent material as Nason, and the two soils have the same catena relationship as the Cecil and Appling soils, respectively. Nason and Tatum soils are located in the Piedmont Plateau and are developed from the residuum of Wissahickon schist (5). A laboratory study is underway to understand better the genesis and cause of the low fertility and high lime requirement of these soils. The high content of parent material muscovite and its alteration to clay minerals was of particular interest from the standpoint of rock weathering and soil genesis.

Profile samples of both Nason and Tatum were collected in each of five counties of Virginia in cooperation with the correlation staff of the Soil Conservation Service. This is a report of the results of this study for one profile of Nason which is generally representative of the other four profiles. The soil was sampled in Fluvanna County and was described as follows:

---

[1]Contribution from the Virginia Agr. Exp. Sta., Blacksburg, Va. This project was supported in part by grants from the Old Dominion Foundation and The Smith-Douglass Fertilizer Co. The authors also wish to acknowledge the assistance of H. C. Porter in collecting the soil samples. Presented before Division V, Soil Science Society of America, St. Paul, Minn., Nov. 9, 1954. Rec. for publication Oct. 14, 1954.

[2]Associate Professor and Professor of Agronomy, respectively.

[3]Horizons B_3 and C_1 were classified C_1 and C_2, respectively, in the field but reclassified later on the basis of clay distribution (table 1).

[4]Jackson, M. L. Soil Analysis—Chemical and Physicochemical methods. Mimeo. Manuscript, Madison, Wis. (1949).

Methods

Soil fractions were segregated according to the method of Jackson[4] without prior removal of free iron oxides. Samples of the clay separates were Mg-saturated by the usual centrifuge method. Other clay samples were K-saturated by washing twice with $1N$ KCl and then digesting (100°C.) a clay-$1N$ K acetate suspension (1:200) for 6 hours followed by ethanol washing.

X-ray (Cu Kα or Co Kα) diffraction patterns were obtained with a Geiger counter type (General Electric) x-ray spectrogoniometer and in some instances with a powder camera. Oriented specimens of clay separates were obtained by drying slowly, at room temperature, suspensions of 50 mg. clay in 0.5 ml. of water on glass slides. Interpretations of the x-ray patterns were based in general on the monograph edited by Brindley (3), the paper by MacEwan (10), and in reference to the dioctahedral vermiculite the report by Brown (4).

D.T.A. equipment used included a stainless steel sample block which was heated in a vertically mounted furnace at the rate of 12.5°C./min. Furnace temperature control and drawing of thermograms were automatic.

Chemical analysis of the whole soil samples followed the methods of Peech *et al.* (12) modified to separate the R_2O_3 in the base exchange procedure. Al was determined polarographically. Flame photometric methods were used for Na, K, Ca, and Mg. Total chemical analysis of the clay fractions was made according to the methods of Corey and Jackson (6) modified to include flame photometric analysis of Ca and Mg after removal of the R_2O_3. Also Al, Ti, and Fe were determined polarographically. Free iron oxides were removed according to the method of Aguilera and Jackson (1). However, Fe was determined polarographically without removal of citrate. Surface determinations were made by the method of Vanden Heuvel and Jackson (15). Potassium fixation was determined by adding a solution of KCl to Mg saturated clay at the rate of 400 me. per 100 gm. clay. The mixture was then dried four times at 110°C. with intervening wettings. The clay was then washed five times with ammonium acetate ($1N$, pH 7.0) and fixation calculated from the difference, in K removed, between the sample with clay and a blank treated similarly.

Estimates of the percentage of minerals in the clay fractions were based on the following associations: free iron oxides-chemical determination; quartz—4.26A peak height; kaolinite—D.T.A.

(trough area for the 565°C. endothermic reaction); vermiculite—internal surface measurements; illite was estimated by difference.

Results and Discussion

MECHANICAL ANALYSIS

The results of the mechanical analysis (table 1) indicate that the soil is well developed. The percentage of clay increased from 14.6% in the A horizon to 61.2% in the B_3. Strong weathering of the parent material also is indicated by high clay content (46.6%) in the C_1 horizon.

CHEMICAL PROPERTIES OF THE WHOLE SOIL

The Nason soil contained very small amounts of exchangeable bases, particularly Ca, in comparison with the high exchangeable H content (table 2). Although the B_3 horizon had a cation exchange capacity of about 25 me., the content of exchangeable Ca was only 0.08 me. The maximum content of exchangeable Mg tended to occur at a lower depth in the profile than Ca, possibly because of the greater mobility of Mg. The distribution of K or Na relative to other cations was not consistent in the five profiles.

Although the soil is indicated to be nearly H saturated, exchangeable Al may account for a large part of the measured "exchangeable H". Paver and Marshall (11) found that clays on treatment with neutral salts liberate Al. With Nason silt loam the amount of Al removed by ammonium acetate was greatest in the C_1 horizon, and the amount of Al removed by non-buffered NaCl increased with depth. The ratio—Al released by NaCl : percentage clay—also increased with depth except for a slight reversal between the A and B_1 horizons. Assuming that the amount of Al removed from the sand and silt fractions was negligible, the results indicate that Al was removed more easily by NaCl from the clay from the B_2, B_3, and C_1 than that from the A and B_1 horizons. The anomalous results for the A horizon may be due to complexing of Al by organic matter. Other data indicate that Al has a major role in exchange reactions and in the alteration of clay minerals of this soil.

CLAY MINERAL COMPOSITION

One of the principal clay minerals in Nason silt loam appears to be an alteration product of muscovite, one of the major minerals in the parent rock. This clay mineral has a 14.7A basal spacing and was identified as dioctahedral vermiculite based on the properties of a similar mineral described by Brown (4) together with other evidence, part of which is presented here. A portion of the weathering sequence of muscovite is indicated by the x-ray diffraction patterns in figure 1. The muscovite in the parent rock is largely unaltered as indicated by the sharp 10A peak and the high (6.08% K_2O) K content of the rock. However, the appearance of the rock and the presence of kaolinite (7.2A peak) indicate considerable weathering. Fresh samples of the rock under this profile were not obtained. The rock is weathered strongly to a depth of at least 20 feet and is easily crushed by hand. The principal minerals in the Wissahickon schist described by Cloos and Hietanen (5) are muscovite, quartz, and albite.

Kaolinite found in the rock from the C_1 horizon may be derived from albite because only a small amount of feldspar was evident from the x-ray pattern and kaolinite was not reported by Cloos and Hietanen (5). The rock fragment shows alteration of the muscovite since two peaks at 10.2 and 11.3A appear. These are interpreted as representing two portions of the

Table 1.—Mechanical analysis of Nason silt loam, Fluvanna County, Va.

Horizon	Depth, inches	Percentage of size classes (diameters in microns)								
		Sand 2000–50	Silt				Clay			
			50–20	20–5	5–2	Total	2–0.2	0.2–0.08	<0.08	Total
A	0–9	24.9	21.0	30.8	6.8	58.6	8.6	4.4	1.6	14.6
B_1	9–13	14.7	16.1	26.5	9.5	52.1	17.5	11.9	2.5	31.9
B_2	13–22	8.8	1.8	24.1	7.0	32.9	30.6	21.5	5.4	57.5
B_3	22–36	10.2	7.2	16.9	4.2	28.3	28.3	27.0	5.9	61.2
C_1	36–42	15.3	14.4	19.2	4.2	37.8	21.3	20.9	4.4	46.6

Table 2.—Chemical properties of Nason silt loam.

Horizon	pH	Organic matter, %	Exchangeable cations, (me./100 gm. soil)							Al removed by N salt solutions and pH of soil-salt solution mixtures†			
			H^+	K^+	Na^+	Ca^{++}	Mg^{++}	Total	C.E.C.*	NH_4 acetate		NaCl	
										pH	Al^{+++} me./100 gm.	pH	Al^{+++} me./100 gm.
A	4.68	1.84	7.53	0.02	0.08	0.09	0.04	7.76	5.51	6.12	0.18	3.70	1.4
B_1	4.69	1.30	12.19	0.05	0.09	0.15	0.11	12.59	11.50	5.87	0.09	3.58	1.6
B_2	5.12	0.82	19.29	0.15	0.07	0.13	0.84	20.48	20.70	5.73	0.10	3.58	10.3
B_3	5.14	0.26	24.19	0.15	0.08	0.08	0.64	25.14	23.20	5.70	0.01	3.62	12.7
C_1	5.08	0.25	22.23	0.07	0.10	0.06	0.31	22.77	20.80	5.71	0.73	3.58	12.9

*Cation exchange capacity—determined by analysis of adsorbed NH_4. See literature citation (12).
†100 ml. of 1N NaCl (pH 7.0) was leached through 10 gm. soil (first 20 ml. stood overnight with soil); NH_4 acetate leaching was the same used to extract exchangeable bases; pH obtained after 1:2 soil-solution mixtures had stood overnight.

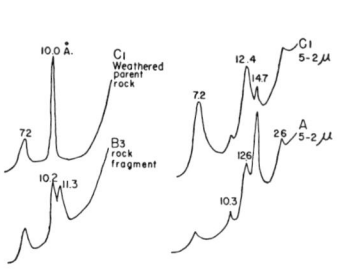 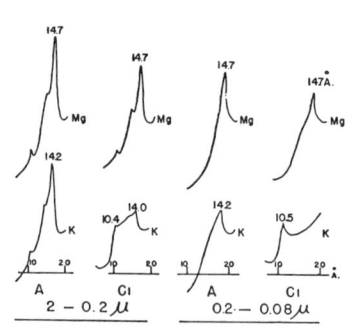 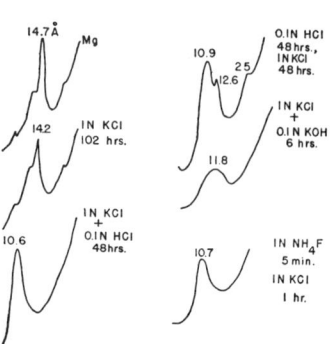

FIG. 1.—Smoothed tracings of x-ray diffraction patterns of selected oriented specimens indicating the weathering sequence of muscovite in Nason silt loam. The 5–2 μ fractions were Mg saturated.

FIG. 2.—Differential effect of K saturation on the basal spacing of the 14.7A mineral obtained from the A and C₁ horizons of Nason silt loam as indicated by tracings of x-ray diffraction patterns.

FIG. 3.—Smoothed x-ray diffraction pattern tracings for the 2–0.2 μ fraction of the B₁ horizon showing the influence of chemical treatments on the basal spacing of the 14.7A mineral. Clay-solution ratio was 1:200 and temperature was 100°C. during treatments.

altered muscovite each having different proportions of expanded units (14.7A) randomly placed between 10A units. On heating this specimen at 300°C. for 2 hours, a single strong peak at 10A replaced the 2 smaller peaks shown. Further weathering is indicated by a shifting of the peak for the predominant random mixture toward 14A and also the appearance and strengthening of segments in which all unit layers are expanded, that is, the 14.7A mineral. The random mixture is indicated by non-integral higher order reflections. The 25–26A reflection is indicative of a regular sequence with expansion of only alternate mica layers which may be associated with the two-layer muscovite structure (7). The second order reflection (12.6A) is evident in figure 3 but usually is superimposed on the reflection from random mixtures of 10 and 14.7A units.

The 14.7A mineral was studied in more detail because it was found that K saturation had a variable effect on the basal spacing depending on the position of the sample in the profile. This suggested a vermiculite to chlorite weathering sequence or a blocking action by some non-exchangeable cation. The variable effect of K saturation and other properties are shown by the following results:

Influence of K saturation.—Barshad (2) and others found that K saturation of vermiculite derived from biotite caused the collapse of the 14.3A basal spacing to about 10A. The 14.7A mineral in the C₁ horizon of this soil reacted similarly (figure 2). However, the 14.7A mineral from samples of upper horizons showed a progressive resistance to collapse on K saturation.

In 8 of the 10 profiles of Nason and Tatum soils, the 14.7A mineral from the C horizons was more easily collapsed than that from the B horizons. In the other two soils this mineral in all horizons did not collapse on K treatment. These two soils differed also in containing considerable gibbsite indicating a higher degree of weathering (8).

In the soil described here, the 14.7A mineral from the A horizon was affected little by K saturation but in A₁ horizons of other soils where the percentage organic matter was high (4 to 6%) the mineral was readily collapsed by K saturation. This indicates that organic matter not only did not prevent vermiculite collapse, but also may have had a complexing effect on Al or Fe and thus may have reduced the amount of these cations in interlayer positions.

Stephen (14) has shown that chlorite weathers to vermiculite in some soils. The formation of normal chlorite from vermiculite in the Nason soil did not seem likely because this would imply a reversal of the probable weathering sequence in a soil where divalent bases are particularly limited. Such bases would be needed to fill out the brucite layer in chlorite.

Glycerol solvation.—Although the clay was able to adsorb considerable glycerol (table 4) the 14.7A mineral did not expand. Thus, little if any montmorillonite was present.

Powder camera x-ray patterns.—The dioctahedral nature of the 14.7A mineral is shown by powder photographs of clay samples heated at 500°C. to destroy kaolinite. Strong (060) reflections (10) were present for clay (0.2–0.08μ) from the A, B₂, and C₁ horizons at 1.492, 1.496, and 1.498A, respectively. No other lines were present in this region. Powder photographs also indicated that in the coarser separates the 14.7A mineral was dioctahedral although quartz in some separates interfered with this method (10).

Table 3.—Influence of heat treatment on the basal spacing of the 14.7A mineral (2–0.2μ) from Nason silt loam.

Temp. °C (>4 hrs.)	Source of clay and saturation			
	A horizon		C₁ horizon	
	K	Mg	K	Mg
	Ångstrom units			
25°	14.2	14.7	*	14.7
300	12.1	13.4	10.3	11.9
600	10.9	10.9	10.3	10.3
700	10.5	10.5	10.3	10.3
800	10.3	10.2	10.3	10.2
950	10.3†	None	10.3†	None

*See figure 2.
†Weak.

Table 4.—Cation exchange capacity, K "fixation", and internal surface of clay fractions from Nason silt loam.

Horizon	C.E.C.,* me./100 gm.		K fixation, me./100 gm.	Internal surface, M.²/gm.	
	2–0.2μ	0.2–0.08μ	2–0.2μ	2–0.2μ	0.2–0.08μ
A	29	33	9	306	†
B_1	24	28	20	349	303
B_2	25	30	39	194	324
B_3	25	28	41	208	371
C_1	37	32	—	232	320

*Cation exchange capacity determined by K-NH₄ exchange in clay (wet condition).
†Insufficient sample.

Chemical treatment of clay to remove interlayer Al (and Fe).—In order to gain information on the possible effect of Fe, as well as Al, in preventing collapse of the 14.7A mineral on K saturation, additional experiments were conducted. Removal of free iron did not alter the variable effect of K saturation, and in this regard it was noted that ease of collapse was not related to the free iron content. Figure 3 shows the results of attempts to collapse the mineral from the B_1 horizon. Boiling the clay in 1N KCl plus 0.1N HCl for 48 hours or boiling in 1N NH₄F (pH 7.0) for 5 minutes apparently were effective in removing interlayer material without undue destruction of the vermiculite. Prolonged boiling in NH₄F was destructive. Boiling in the KOH-KCl solution evidently removed interlayer material but also caused considerable destruction of the mineral as evidenced by the diffuse x-ray pattern and because about one-fifth of the total Al was brought into solution. Tetrahedral Si may be removed by the alkali treatment. However, on treatment of the residue with Al or Mg, a 14.5A peak was again obtained in general agreement with the results of Brown (4).

Treatment of the clay with acid prior to K saturation was not as effective as simultaneous KCl-HCl treatment. The 25A peak and its second order reflection at 12.6A were present after the divided HCl-KCl treatment. The results of the acid and NH₄F treatments indicate the presence of either Fe or Al in the interlayer position although these treatments would favor removal of Al because of the greater complexing ability of NH₄F for Al and the greater solubility of Al in weak acid as compared to Fe (ferric). Also the clay remained red after NH₄F treatment indicating that much of the iron oxide remained. The results of the KOH-KCl treatment indicate Al removal rather than Fe because of the insolubility of Fe in alkali.

Heat treatment.—Heat treatment caused the collapse of the 14.7A mineral in all horizons to 10.3A (table 3); however, lower temperatures were needed for K saturated clay, and the variable resistance to collapse noted at room temperature was evident at higher temperatures. The simple x-ray patterns for heated clays indicated that the 14.7A value was due to only one mineral.

Aluminum fixation.—Aluminum fixation in the interlayer positions of the easily collapsed 14.7A mineral may prevent this collapse. However, it was found that washing the clay with 1N AlCl₃ (pH 2.8), without drying, in a manner parallel to Mg saturation, was not effective. Drying the clay at room temperature after Al saturation was effective in inhibiting collapse on K saturation. With repeated saturation and drying cycles, the effect became more pronounced. When the clay suspension was adjusted to pH 7.0 (after Al saturation and removal of excess Al), one cycle of Al saturation and drying was more effective than five cycles without neutralization. These results were obtained with clay from the C_1 horizon of the Nason soil reported here as well as NH₄F treated clay from the B_1 horizon. Figure 4 gives the results for the clay from the A_1 horizon of a Tatum silt loam (Orange County) which contained a similar easily collapsed 14.7A mineral in the A and C_1 horizons, but the mineral in the B horizon was not affected by K saturation. In similar experiments with Fe no blocking action was found.

These results suggest that basic aluminum ions may be fixed more readily than Al^{+++} and that drying may bring about a more stable association of the hydrated Al and the tetrahedral layers. Paver and Marshall (11) found that Al was much more effective than Fe in reducing the cation exchange capacity of clay. They suggested that the superiority of Al in this respect may be due to its capacity for forming Si-O-Al linkages.

Cation exchange capacity and K fixation.—The cation exchange capacity of the clay fraction (table 4) differed little according to the soil horizon. However, if the composition of the clay fraction is considered (table 6) and kaolinite assumed to have a C.E.C. of 10 me. per 100 gm., differences are evident for the 14.7A mineral. In the C_1 horizon the C.E.C. is calculated to be 65 me. for the 14.7A mineral in the 0.2 to 0.08 μ fraction, whereas in the A_1 horizon the value is 50 me. The low exchange capacity for this mineral compared to 100 to 150 me. for vermiculite derived from biotite (2), and also the decrease in exchange capacity from the C to A horizons may be due, in part, to non-exchangeable Al. The low values probably are not due to K fixation in the K-NH₄ exchange since the clay was not dried, and Na-NH₄ and Ca-Na exchange gave similar values for the C_1 horizon. However, when the clay

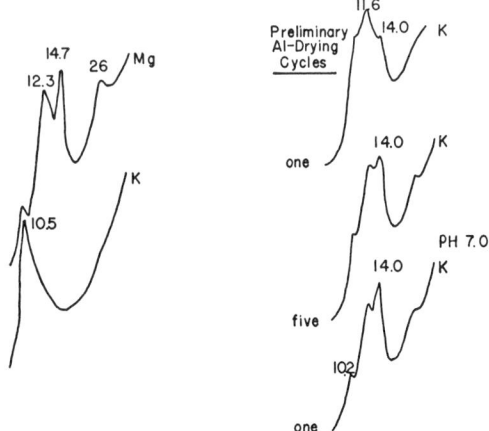

FIG. 4.—Addition of non-exchangeable interlayer aluminum by Al saturation (plus drying at room temperature or neutralization and drying) of the 2–0.2 μ fraction from the A_1 horizon of Tatum silt loam as indicated by smoothed x-ray diffraction tracings.

was dried prior to K extraction, 5 to 6 me. K were fixed in the 0.2 to 0.08 μ fraction from all horizons. In the 2 to 0.2 μ fraction this procedure fixed 7.3 me. K in the sample from the B_3 horizon and only 2.1 me. K in the sample from the A horizon. In the presence of excess K and with heating, this fraction was able to fix about five times as much K. Stanford (13) showed that treatment of bentonite with NaF increased the K fixation and suggested that Al in the interlayer position partially blocked fixation. The same mechanism may be operative in these soils.

Internal surface.—X-ray diffraction patterns showed that internal glycerol sorption formed a monomolecular layer, and the internal surface was calculated (15) on this basis. Rate of glycerol sorption was much slower for the 2 to 0.2 μ fraction from the A horizon than that from the C_1 horizon. The internal surface covered by glycerol with 3 ternary solution washes was 89 and 215 M.²/gm. for the A and C_1 horizons, respectively, whereas after sorption over a 16-hour period from a fourth ternary solution the values were 306 and 232 M.²/gm., respectively (table 4). The high internal surface indicates that the 14.7A mineral in all horizons is vermiculite and not chlorite, although the blocking action of some interlayer cation is again indicated by the slow sorption of glycerol.

Total chemical analysis.—Table 5 gives the total chemical composition of the 0.2 to 0.08 μ fraction. The high Al and Fe content and the low amounts of other bases is evident. A major portion of the Fe exists as free iron oxides. Although the percentage of kaolinite was reduced from about 40 to 20% between the C_1 and A horizons the percentage of Al did not decrease significantly. This is further evidence supporting the accumulation of Al by vermiculite in the course of soil development. Aluminum occupying exchange positions may amount to a few percent, the amount depending on whether the ions are Al^{+++}, $Al(OH)^{++}$ or $Al(OH)_2^+$. Further Al substitution in the tetrahedral layer would permit more Al in exchange positions in order to balance the charge. The decrease, with depth, of water removed between 110° and 300°C. and the increase of water removed between 300° and 950°C. supports the other data showing the changes with depth of the proportion of kaolinite and vermiculite.

Differential thermal analysis.—Differential thermal analysis of clay from the C_1 horizon indicated typical kaolinite reactions plus low temperature endothermic reactions similar to those of vermiculite (2). When the clay was Mg saturated, troughs at 150° and 280°C. were present (figure 5), whereas when K saturated, these troughs disappeared. Both vermiculite and kaolinite trough areas diminished as the surface horizon was approached but a broad endothermic trough at

Table 6.—Estimated mineral composition of clay fractions of Nason silt loam.

Horizon	Percentage composition*				
	I	V	K	Q	G
2–0.2μ					
A	37	39	8	10	6
B_1	25	44	17	8	6
B_2	38	25	25	4	8
B_3	36	26	26	4	8
C_1	31	29	28	4	8
0.2–0.08μ					
A	—	(62)†	19	2	14
B_1	—	41	24	0	15
B_2	—	44	34	0	16
B_3	—	50	36	0	16
C_1	—	43	39	0	12
Estimated error (±)	15	10	5	2	2

*I, illite (10A layers in mixtures and in uniform particles); V, dioctahedral vermiculite [14.7A (Mg) layers, in mixtures and in uniform particles]; K, kaolinite; Q, quartz; G, "free" iron oxides.
†Determined by 14.7A peak height relationship.

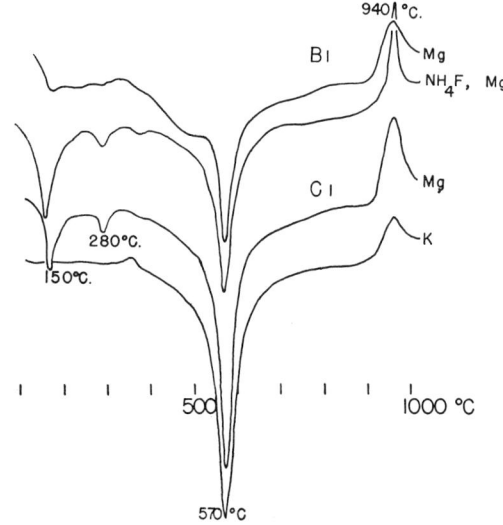

FIG. 5.—Differential thermal analysis tracings indicating the influence of NH_4F treatment on the 2–0.2 μ fraction from the B_1 horizon and effect of K saturation of the 2–0.2 μ fraction from the C_1 horizon of Nason silt loam. Free iron was removed before analysis.

Table 5.—Chemical analysis and heating weight loss of 0.2–0.08 micron fractions of Nason silt loam.

Horizon	Percentage on 110°C. basis												
	SiO_2	Al_2O_3	Fe_2O_3	TiO_2	K_2O	Na_2O	MgO	CaO	110°–300°C H_2O	300°–950°C H_2O	Total	Fixed Fe_2O_3	Free Fe_2O_3
A	35.2	28.5	15.7	0.53	0.92	0.14	0.87	0.15	8.12	9.26	99.4	2.3	13.4
B_1	36.7	29.0	16.9	0.57	0.86	0.14	0.67	0.12	6.54	9.48	101.0	2.5	14.4
B_2	35.0	29.7	16.7	0.55	0.85	0.09	0.73	0.12	6.24	9.67	99.6	2.0	14.7
B_3	36.5	29.9	16.2	0.72	0.90	0.08	0.66	0.06	5.68	9.70	100.5	1.6	14.6
C_1	39.3	30.0	13.7	0.60	0.92	0.06	0.68	0.11	4.71	10.64	100.7	2.3	11.4

490°C became more evident. This reaction may be loss of OH groups associated with interlayer Al. When the clay from the B_1 horizon was treated with NH_4F before Mg saturation, the vermiculite low temperature endothermic troughs were greatly enhanced; this is in agreement with the internal surface and x-ray evidence. The broad endothermic trough at 490° disappeared but a slight endothermic trough at 350°C. appeared.

The data indicate that the 14.7A mineral in all horizons is a dioctahedral vermiculite having a variable content of non-exchangeable Al in the interlayer positions. The identification of the type which does not collapse on K saturation as vermiculite rather than chlorite is based on the following evidence: *(1)* low Mg and fixed Fe and high Al content, *(2)* collapse to 10.3A on heating to 800°C, *(3)* weak second and third order reflections, *(4)* high internal surface and exchange capacity, *(5)* resistance of the basic unit to acid treatment, *(6)* chlorite is normally trioctahedral whereas this mineral is dioctahedral, *(7)* the D.T.A. low temperature endothermic reactions indicate vermiculite, and *(8)* the mineral can be altered by acid or NH_4F treatment to a mineral which collapses on K saturation.

A summary of the clay mineralogical analysis is given in table 6. Incomplete glycerol sorption by vermiculite in the A horizon 2 to 0.2 μ fraction is likely since the x-ray patterns indicated a higher content of vermiculite in the A horizon than the B_1. Incomplete glycerol sorption is also evident in the 0.2 to 0.08 μ fraction of the B_1 horizon. The mineral constituents of the fine clay fraction (<0.08 μ) appeared to be quite similar to those of the medium clay in all horizons.

The increase with depth of the proportion of kaolinite as well as the decrease of vermiculite was evident in all 10 Nason and Tatum profiles. The proportion of vermiculite also decreased with depth in the profile described by Brown (4). Changes with depth in the proportion of clay minerals in the Nason and Tatum soils suggest that dioctahedral vermiculite was more resistant to weathering than kaolinite. Alternate explanations or contributing factors may be: *(1)* lack of source material (feldspars) for kaolinite in the upper horizons, and *(2)* conditions for the formation of kaolinite are more favorable in the less acid B_2, B_3, and C_1 horizons.

In the near-absence of exchangeable bases, interlayer non-exchangeable Al may have stabilized the vermiculite in this soil and decreased its rate of weathering. This low rate of weathering, together with the abundance of a source mineral, muscovite, apparently produced a clay in which dioctahedral vermiculite is a major constituent. Muscovite is commonly found in soils and because many soils of the Red-Yellow Podzolic group are well developed and contain small amounts of exchangeable bases, dioctahedral vermiculite may be present in other soils of this group. Even small amounts of this mineral are agriculturally significant because of its high cation exchange capacity and its ability to fix potassium.

Literature Cited

1. Aguilera, N. H., and Jackson, M. L. Iron oxide removal from soils and clays. Soil Sci. Soc. Amer. Proc. 17:359 (1953).
2. Barshad, I. Vermiculite and its relation to biotite as revealed by base exchange reactions, x-ray analysis, differential thermal curves, and water content. Amer. Min. 33:655 (1948).
3. Brindley, G. W. (ed.). X-ray identification and crystal structure of clay minerals. The Mineralogical Society, London (1951).
4. Brown, G. The dioctahedral analogue of vermiculite. Clay Min. Bul. 2:64 (1953).
5. Cloos, E., and Hietanen, A. Geology of the "Martic Overthrust" and the Glenarm series in Pennsylvania and Maryland. Geological Society of America Special Papers Number 35 (1941).
6. Corey, R. B., and Jackson, M. L. Silicate analysis by a rapid semimicrochemical system. Anal. Chem. 25:624 (1953).
7. Hendricks, S. B., and Jefferson, M. E. Polymorphism of the micas. Amer. Min. 24:729 (1939).
8. Jackson, M. L., *et al.* Weathering sequence of clay size minerals in soils and sediments. I. Fundamental generalizations. Jour. Phys. Colloid. Chem. 52:1237 (1948).
9. ———, *et al.* Weathering sequence of clay size minerals in soils and sediments. II. Chemical weathering of layer silicates. Soil Sci. Soc. Amer. Proc. 16:3 (1952).
10. MacEwan, D. M. C. Some notes on the recording and interpretation of x-ray diagrams of soil clays. Jour. Soil Sci. 1:90 (1949).
11. Paver, H., and Marshall, C. E. The role of aluminum in the reactions of the clays. Jour. Soc. Chem. Ind. 53:750 (1934).
12. Peech, M., *et al.* Methods of soil analysis for soil-fertility investigations. U.S.D.A. Circular 757 (1947).
13. Stanford, G. Fixation of potassium under moist conditions and on drying in relation to clay mineral. Soil Sci. Soc. Amer. Proc. (1947) 12:167 (1948).
14. Stephen, I. A study of rock weathering with reference to the soils of the Malvern Hills II. Weathering of appinite and "ivy-scar rock". Jour. Soil Sci. 3:219 (1952).
15. Vanden Heuvel, R. C., and Jackson, M. L. Surface determination of mineral colloids by glycerol sorption and its application to interstratified layer silicates. Presented before Div. I, Soil Science Society of America, Dallas, Tex., Nov. 19, 1953.

Reprinted from *Soil Sci. Plant Nutr. (Tokyo)* **8**:22-29 (1962)

IMOGOLITE IN SOME ANDO SOILS*

Naganori YOSHINAGA and Shigenori AOMINE**

Faculty of Agriculture, Kyushu University, Fukuoka,
RECEIVED SEPTEMBER 15, 1961

It was reported in a preceding paper[7] that the Ando soils from Uemura, Choyo, and Kawanishi contained an unknown mineral colloid which was distinctly different in some respects from coexisting allophane. In the Uemura soil, this clay fraction made up more than 20 per cent of the total clay and more than 6 per cent of the soil[7].

Although its chemical and mineralogical properties are not well known at present, it is considered that this mineral has a more ordered structure than allophane. Its occurrence is also considered to be fairly common in most Ando soils and weathered pumices[5]. Therefore, this mineral was tentatively designated as imogolite by the present authors[7].

The objective of this paper is to describe the properties of imogolite as examined by the techniques of electron microscopy, x-ray diffraction, thermal analysis, infrared absorption, and chemical analysis.

Materials

The samples used in this study were the -0.2μ or -0.08μ clay fractions separated in an acid medium from the Uemura, Choyo, and Kawanishi soils[7]. Since allophane has been removed almost exhaustively from the soils prior to the separation of these clay fractions[7], it may be taken that these fractions are free from allophane. Obtained imogolite clay was quite different in appearance from allophane.

Flocculating with NaCl, the imogolite clay formed a voluminous floccule throughout which innumerable small air bubbles were adsorbed; it looked like a drifting cloud (Figure 1). On the contrary, the allophane clay separated from the same soil formed small flakes in NaCl solution and settled readily to the bottom of a container (Figure 1). When collected in a centrifugal tube, imogolite gave a jelly-like appearance, while allophane gave an appearance of white paste.

On drying with a motor fan after washing with water, alcohol, and acetone successively, about 500 mg of imogolite clay required more than 5 hours until apparent air-dryness was attained, while about 15 minutes were sufficient for allophane clay. The air-dried clay cake of imogolite was somewhat greyish, and showed weak elasticity, whereas that of allophane was white, and loose and fragile.

On pulverizing the air-dry clays in an agate mortar, allophane adhered to the mortar and pestle, but imogolite did not.

The imogolite clay, once air-dried, did not swell in water and also in certain neutral or alkaline salt solutions (NaCl, $CaCl_2$, $CaAc_2$, Na_2CO_3). This was the case with the allophane clay. But, when immersed in a salt solution of about pH 3.5 to 4.0 (NaAc, $CaCl_2$), the former easily recovered the property of swelling, while the latter remained just as it was.

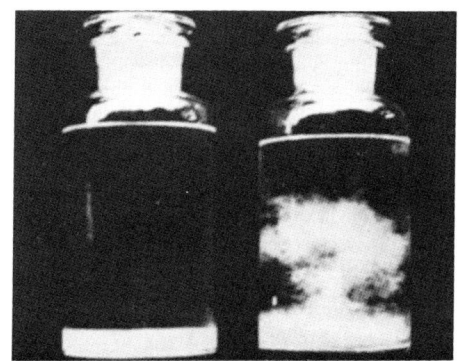

Allophane   Imogolite

Fig. 1. *Imogolite and allophane clays flocculated with normal NaCl solution (-2μ, Uemura).*

Electron Micrograph

The specimens for electron microscope observation were prepared from -0.2μ clay which had been preserved in a normal NaCl solution, according to the procedure described by JACKSON

---

* The results were used by the senior author in partial requirement for the degree of D. Agr., The Kyushu University, 1961.

** Senior author's present address is Faculty of Agriculture, Ehime University, Matsuyama, Japan.

(3, 420pp.). Some specimens were shadowed with chromium at an angle of about 30 degrees.

Electron micrographs were shown in Figure 2. The clays were all composed mainly of threadlike particles of relatively uniform size. Their diameter of cross section is about 100 to 200 Å. The similar shape of particles was found by AOMINE and JACKSON[1] in the Sakae and Choyo clays. The relative uniformity in shape seems to suggest some regularity of atomic arrangement with in the imogolite particles. It is easily supposed that such a shape of particles is responsible for the voluminous floccule of the clay in NaCl solution.

Occasionally the particles of somewhat fluffy appearance were observed in the imogolite clays (not shown). The very similar particles were found by AOMINE and YOSHINAGA[2] in the clay fractions of the Uemura and Kawanishi soils. Although it is not known whether these particles are imogolite or not, the shape is clearly different from that of allophane[7]. More than 12 fields were observed with each specimen; nevertheless, such a shape of particles was not found in the allophane fraction.

X-ray Diffraction Analysis

X-ray diffraction analyses were carried out on parallel and random orientation specimens using a Geiger counter x-ray diffractometer "Geigerflex". The flat specimens of Mg- and K-saturated and Mg-saturated and ethylene glycol solvated clays were prepared according to the procedure of JACKSON (3, 184 pp.). The K-saturated specimen was heated at various temperatures up to 500°C for two hours, and x-rayed immediately and after being allowed to stand in the air for 15 minutes, one hour, and two months. The random orientation specimens of the clays which were air-dried and heated at 1000°C and at about 1200°C for two hours were prepared according to the procedure of MCGREERY[6].

The diffraction patterns obtained with the random orientation specimens of air-dried, Na-saturated -0.2μ clays were shown in Figure 3. The patterns resemble each other closely; all samples exhibited several broad peaks and diffuse bands. Although the peaks at 3.26 Å and 2.26 Å and the band ranging from 4 to 4.5 Å have been observed similarly, but in a less intensity, in the samples of allophane separated from the same soils[7], the general feature of patterns is quite different from that of allophane. It may be considered that imogolite has a more ordered structure than allophane. However, it is doubtful that the band ranging from 4 to 4.5 Å is an evidence of layer silicate minerals, because the samples of allophane also have exhibited the similar diffraction band[7].

The diffraction patterns obtained with the flat

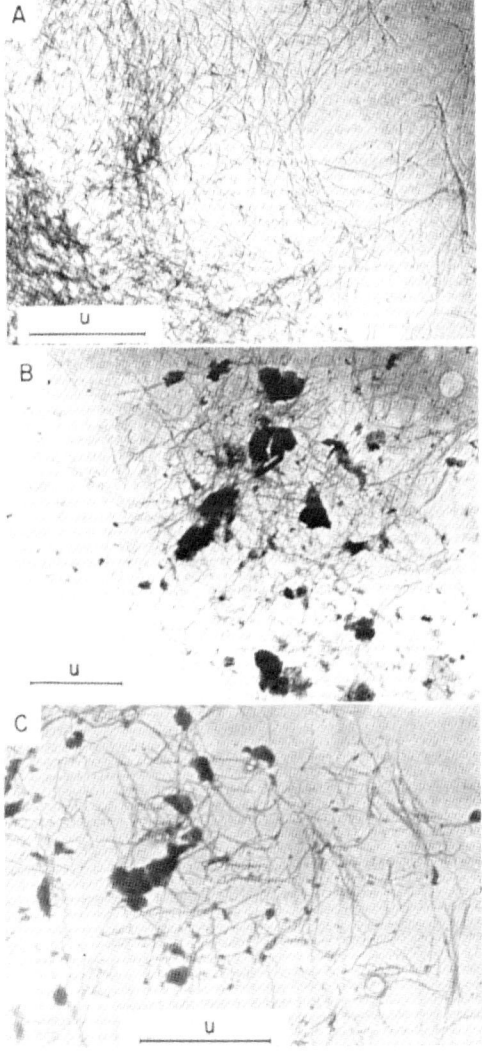

Fig. 2. *Electron micrographs of -0.2μ fractions. A—Uemura; B—Choyo; C—Kawanishi.*

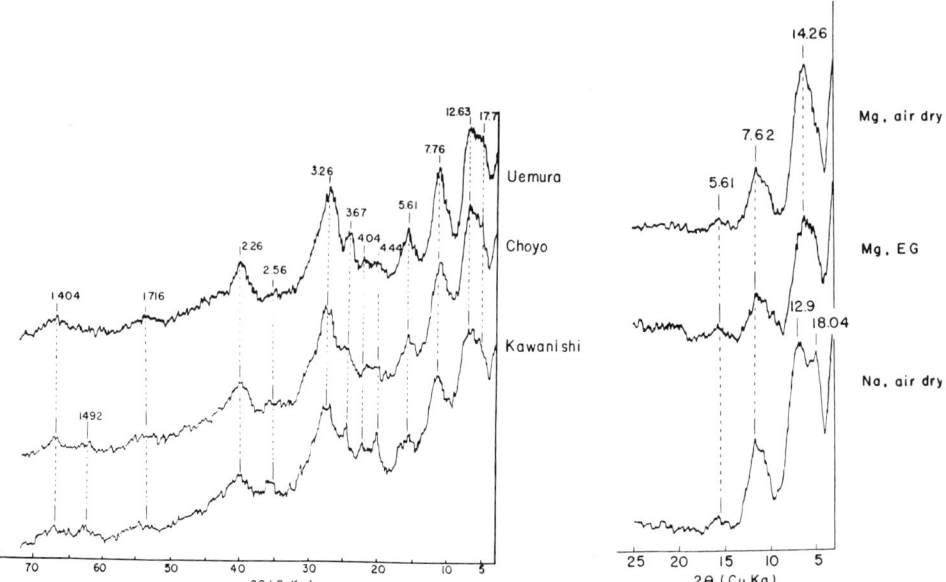

Fig. 3. X-ray diffraction patterns of air-dried, Na-saturated −0.2 μ clays in random orientation.

Fig. 4. X-ray diffraction patterns of the Uemura −0.08 μ clay.

specimens were shown in Figures 4, 5, and 6. The magnesium-saturated, Uemura −0.08 μ clay (Figure 4) gave broad, relatively intense two diffraction peaks at 14.26 Å and 7.62 Å and a very weak diffraction at about 5.6 Å. Ethylene glycol (EG) solvation caused no noticeable change in the pattern. Sometimes the 14 Å broad peak gave two diffraction maxima at about 13 Å and 18 Å as was seen in the pattern of Na-saturated clay (Figure 4), indicating that the 14 Å broad peak is composed of two diffraction components. It seems impossible, however, to attribute the 18 Å maximum to "lattice expansion" caused by the hydration of Na-clay, because the heated specimen produces a very sharp and strong peak at about 17 to 19 Å (Figure 5).

Potassium saturation of the specimen caused a slight decrease of the 14.26 Å spacing (Figure 5). It is doubtful, however, that this shift of spacing is a "lattice contraction" as seen in the usual expandable layer lattice minerals. Subsequent heat treatments caused an interesting behavior of this peak. Upon heating to 100°C, the peak suffered a noticeable change and produced a strong and sharp 18.8 Å peak with an inflection at about 13 Å. This would be another indication

that the broad 14 Å peak of Mg-clay is composed of two diffraction components. The 18.8 Å peak increased its intensity progressively with temperature up to 300°C with concurrent decrease of the spacing to 17.63 Å, but was greatly weakened at 350°C and almost disappeared at 400°C leaving a slight inflection at about 17 Å. On the other hand, the broad peaks at 7.8 Å and 5.5 Å hardly underwent any change up to 300°C, but disappeared abruptly at 350°C. These evidences indicate that the structure of imgolite is decomposed almost entirely by heating at 350°C to 400°C.

The clay heated from 100°C to 250°C regained the intensity of reflection at about 13 Å when exposed in the air. This recovery would be attributed to the rehydration of the heated clay, though the mechanism of water retention by imogolite is not known. When heated to 300°C, the recovery became very little. This may suggest that, if the water of original structure is removed completely or nearly so, the ability of rehydration would also be lost almost entirely. A slight recovery of intensity was also noticed for the 17 Å inflection of the specimen heated at 400°C and 450°C, indicating a partial revival

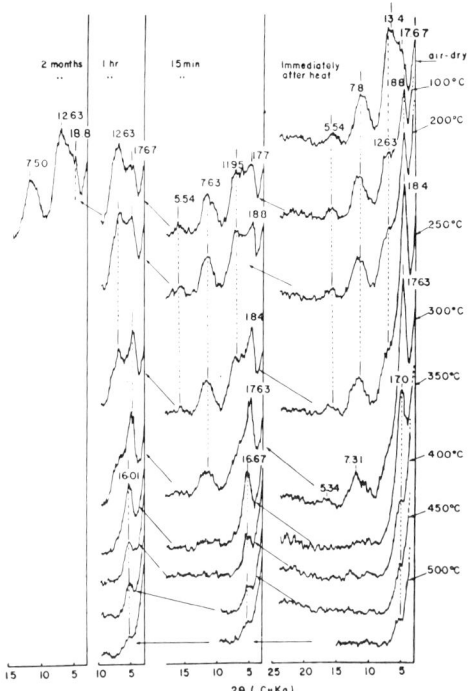

Fig. 5. *X-ray diffraction patterns of the Uemura, K-saturated* -0.08μ *clay.*

of imogolite structure.

The diffraction patterns of the clays from Choyo and Kawanishi (Figure 6) were essentially the same as those of the Uemura clay. It may be considered that these clays also consist principally of imogolite. The broad diffraction peak at about 12 Å in the heated Kawanishi clay may probably be attributed to the contamination of chlorite-vermiculite intergradient[7]. Fairly distinct 4.44 Å peak exhibited by the random specimen (Figure 3) may establish the presence of this mineral.

Although no definite conclusion can be reached as yet concerning the structure, the diffraction patterns given by imogolite seem to be made up of two types of diffractions, the broad peak at about 14 Å which consists of 18 Å and 13 Å diffraction components, and the others. A little decrease of 18.8 Å spacing of K-saturated clay caused by heat treatment is somewhat similar to the lattice contraction of layer silicate minerals. In contrast, the 7.8 Å and 5.5 Å peaks are hardly affected by heat treatment; they would be compared to hk0 reflections. Thus, imogolite seems to bear some structural resemblance to the layer silicate mineral. However, the data presented above, together with the absence of prism zone reflections in the patterns of randomly oriented specimen heated at 300°C (not shown in figure), suggest very little possibility of interstratification of common silicate layers. Data of thermal (Figures 8 and 9), infrared absorption (Figure 10), and chemical (Tables 1 and 2) analyses also rule out this possibility. Also the possibility of already-known chain-structure minerals, such as attapulgite, sepiolite, and palygorskite, is very little if any.

KANNO *et al.*[5] have recently investigated gel-like substances separated from the Kitagami and Akutsu pumice beds. Data of x-ray diffraction, thermal, infrared absorption, and chemical analyses presented by them are nearly the same as those obtained with imogolite clays. It is almost doubtless that imogolite and their gel-like substances are essentially of the same kind. These authors attributed the broad diffractions of the gel-like substances to poorly crystallized montmorillonite. Imogolite, however, is apparently different from montmorillonite.

The diffraction patterns of fired clays were shown in Figure 7. The clays heated at 1000°C gave the patterns which were similar to those of allophane separated from the same soils[7]. Their faint diffractions at 2.38 Å and 1.973 Å

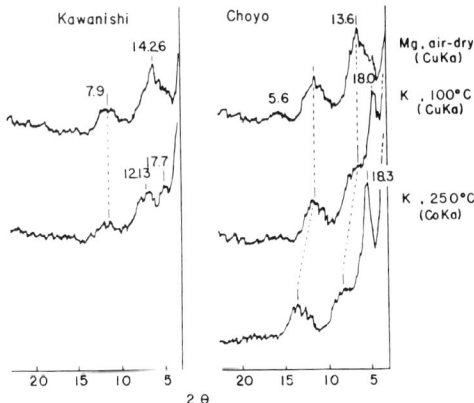

Fig 6. *X-ray diffraction patterns of* -0.2μ *fractions of the Choyo and Kawanishi soils.*

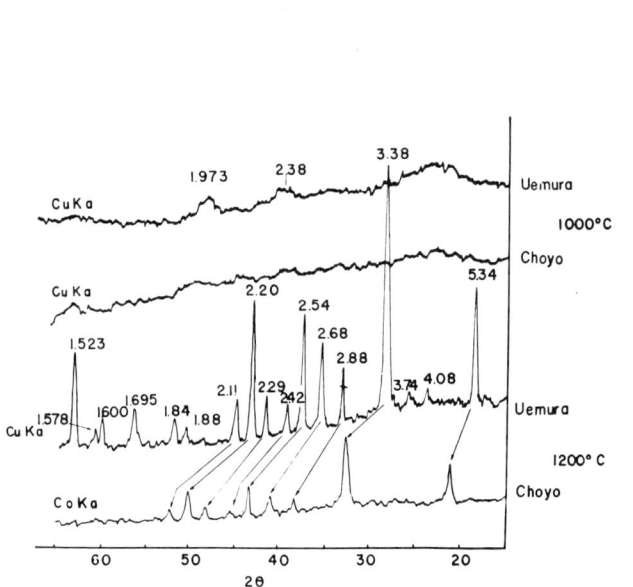
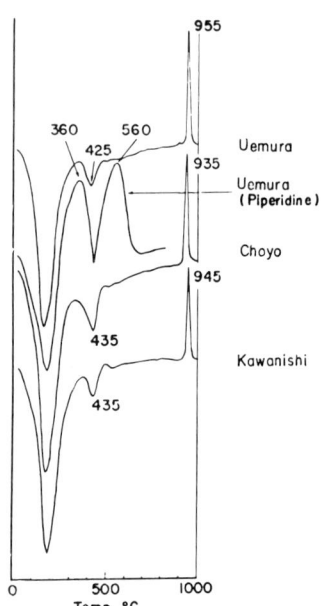

Fig. 7. X-ray diffraction patterns of clays heated at 1000°C and 1200°C. The product at 1200°C is mullite.

Fig. 8. Differential thermal curves of -0.2μ Na-clays.

would be attributed to the structural remnants of gamma-alumina. Upon heating to 1200°C, however, the clays yielded strong diffraction peaks for mullite. This forms a striking contrast to allophane which produces only alpha-alumina at the same temperature[7]. The difference in high temeprature phases may be due to differences in chemical composition and in ionic arrangement between these two materials.

Thermal Analysis

Differential thermal and dehydration analyses were carried out on -0.2μ size fractions by the same method and with the same equipments as used for allophane[7].

Differential thermal curves obtained were shown in Figure 8. A strong endothermic peak is observed in all samples at about 170°C to 190°C. This peak is apparently due to a large amount of hygroscopic water (Table 1). The exothermic peak observed between 935°C and 955°C is probably due to the formation of gamma-alumina. These two peaks were observed similarly in the samples of allophane separated from the same soils[7]. But the exothermic reaction increased its intensity remarkably as compared with allophane, with simultaneous shift of the peak toward high temperature by about 20 to 40°C. This may probably be connected with the difference in the "inner structure" between imogolite and allophane.

Besides the above two peaks, all samples exhibited an endothermic reaction at about 425°C to 435°C. This endotherm has never been observed in the samples of allophane[7]. It is undoubtedly a characteristic of imogolite. The peak temperature coincides fairly well with the temperature where the diffraction lines of imogolite disappear, taking into account the possible delay of peak temperature in differential thermal curves. It seems probable, therefore, that this endothermic peak is indicative of the destruction or dehydroxylation of imogolite.

AOMINE and YOSHINAGA[2] noticed the similar endotherm in nondeferrated clay fractions from the same three soils and their neighboring horizons, and KANNO et al.[5], in the gel-like substances separated from pumice beds. These investigators supposed that the endotherm was attributed to the presence of impurities such as hydrated oxides of iron or aluminum or aluminum

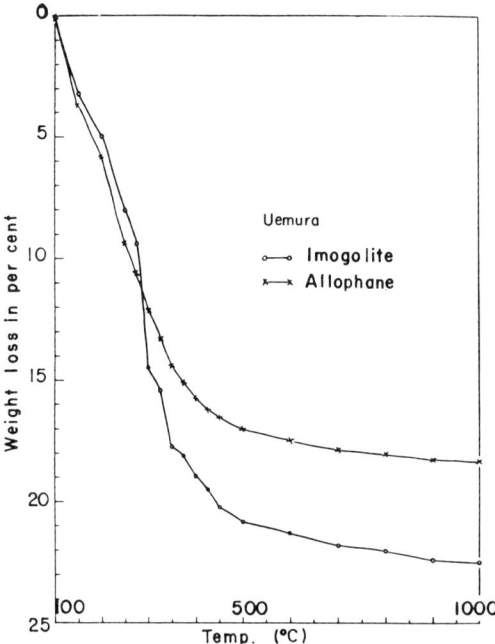

Fig. 9. *Dehydration curves of* $-0.2\,\mu$ *Na-clays of imogolite and allophane. The curve of allophane was reproduced from the preceding paper[7].*

compounds. However, the imogolite is a purified sample[7], and the peak intensity is nearly equal for all samples. Therefore, it would be reasonable to consider that the peak is a characteristic of imogolite, but not being due to the impurities.

The integral dehydration curve of imogolite, together with that of allophane, was shown in Figure 9. The curve of imogolite shows an inflection indicating a rapid dehydration between 275°C and 350°C. This temperature range also coincides fairly well with that of the disappearance of diffraction peaks in x-ray analysis (Figure 5), taking into account a long time of heating in the dehydration analysis.

The results of thermal analyses indicate that imogolite hold structural water which is released in the neighborhood of 300°C.

Infrared Absorption

Infrared absorption measurement was run with the same method as used for allophane[7].

The curve of imogolite showed four pronounced (2.7μ, 2.9μ, 6.2μ, 10.2μ) and one faint (7μ) absorption maxima (Figure 10). The positions of these maxima are entirely the same as those of allophane. KANNO *et al.*[5] obtained

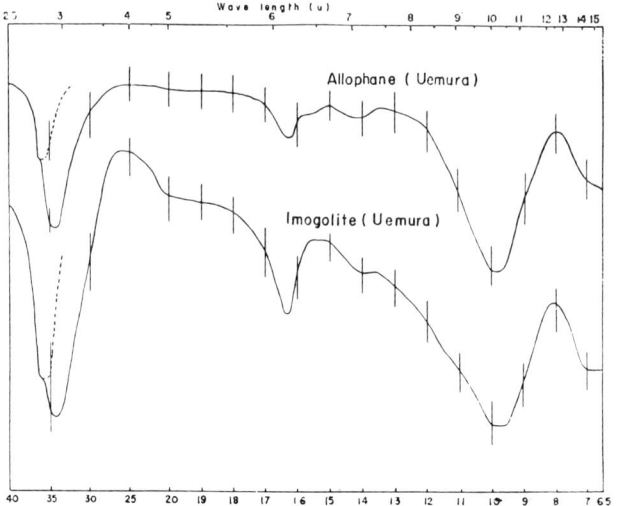

Fig. 10. *Infrared absorption curves of* $-0.2\,\mu$ *Na-clays of imogolite and allophane. The curve of allophane was reproduced from the preceding paper[7].*

Table 1
Chemical Composition of the Uemura, $-0.2\,\mu$ Na-clay

	%*
SiO_2	29.71
Al_2O_3	47.60
Fe_2O_3	0.57
CaO	0.04
MgO	0.02
Na_2O	0.74
K_2O	0.13
TiO_2	1.04
P_2O_5	0.26
I. L. (H_2O+)	22.05
Total	102.16
$H_2O(-)$**	19.69
SiO_2/Al_2O_3	1.06
H_2O/Al_2O_3***	4.96

* Oven-dry basis.
** Equibrated at 50% relative humidity.
*** Including $H_2O(-)$.

Table 2. Cation-exchange Capacity and CEC-delta value of the Uemura, -0.2μ Fractions

	CEC (pH 7)	CEC pH 3.5	CEC pH 10.5	CEC-delta value
	me/100g*	me/100g*	me/100g*	me/100g*
Imogolite	48	43	113	70
Allophane	151	49	178	129

* Oven-dry basis.

the similar pattern for their gel-like substances, and attributed the band ranging from 7.7 to 12.5μ (maximum; about 10.5μ) to the presence of allophane. However, it seems that the band comes from both imogolite and allophane in their pattern, since these investigators have not separated allophane from imogolite. Other absorption bands are apparently due to bonded (2.9μ) and unbonded (2.7μ) hydroxyls and absorbed water (6.2μ). It seems that imogolite is indistinguishable from allophane by infrared absorption method, and this suggests that imogolite is a low crystalline mineral in an intermediate stage from amorphous material (allophane) to crystalline clay mineral.

Chemical Composition, Cation-exchange Capacity, and CEC-delta Value

Cation-exchange capacity (CEC) was determined by the Ca saturation-EDTA titration method, and CEC-delta value, by the procedure proposed by AOMINE and JACKSON[1].

The chemical composition of the Uemura, -0.2μ Na-clay was shown in Table 1. The result shows that the clay consists principally of silica, alumina, and water. The molecular ratio of silica to alumina is 1.06, and that of water to alumina is 4.96; thus the ratio of silica : alumina : water is approximately 1 : 1 : 5. The content of titanium dioxide is about one per cent of the clay, but it is not known that this element takes part in the structure of imogolite. Iron is probably in a form of free oxide which has been difficultly soluble against the deferration treatment. The chemical composition is very similar to that presented by KANNO et al[5] for the gel-like substances of some pumices.

The CEC and CEC-delta values of imogolite were shown in Table 2. The data of allophane separated from the same Uemura soil[7] was added to the table for comparison. The CEC of imogolite clay was 48 me per 100g oven-dry clay. It is doubtful, however, that this is the absolute value for imogolite, because the CEC of allophane is greatly affected by past acid or alkali treatment[1], and this may be true to some extent with a low crystalline material such as imogolite (see CEC-delta value). The CEC-delta value was 70 me per 100g oven-dry clay. This value is about half that of allophane. The difference seems to have some significance in comparing these two minerals for "crystallinity". The low value for imogolite means a lower content of active OH radicals, namely some higher regularity of atomic arrangement within particles. On the other hand, however, this value is remarkably higher than that of common layer silicate minerals[1,4]. This indicates that the "crystallinity" of imogolite is low as compared with these minerals.

Summary

Three Ando soils contained two kinds of mineral colloids, allophane and an unknown mineral of low crystallinity. The latter mineral was designated tentatively as imogolite by the present authors. Fractionated imogolite clays exhibited an extraordinary volume in some aqueous salt solutions; their appearance was almost jelly-like.

In electron micrographs, imogolite appeared as thread-like particles of relatively uniform size.

X-ray diffraction patterns revealed that imogolite has somewhat definite "crystal structure" which bore some resemblance to the layer silicate mineral. This fact was substantiated by differential thermal and dehydration analyses. On the other hand, however, infrared absorption curve, chemical composition, and CEC-delta value bore a close resemblance to allophane.

Although the structure is not exactly known at present, it would be concluded that imogolite is an intermediate weathering product of a metastable state between amorphous material (allophane) and crystalline clay mineral, and that it is formed under a well-drained condition of the Ando soils. It is also suggested that this mineral occurs fairly widely in nature.

Acknowledgement

The authors wish to express their sincere thanks to Mr. TANIGUCHI of Faculty of Agriculture, Kyushu University for his help in the infrared absorption measurement, and to Mr. MIYAUCHI of Faculty of Medicine, Kyushu University and Mr. NANRI of Faculty of Agriculture, Kyushu University for electron micrographs.

References

1) AOMINE, S. and M. L. JACKSON, *Soil Sci. Soc. Am. Proc.*, **23**. 210 (1959).
2) AOMINE, S. and N. YOSHINAGA, *Soil Sci.*, **79**, 349 (1955).
3) JACKSON, M. L., *Soil Chemical Analysis-Advanced Course*, Pub. by the Author, Dept. of Soils, Univ. of Wis., Madison, Wis. (1956).
4) JENNE, E. A., Mineralogical, Chemical, and Fertility Relationships of Five Oregon Coastal Soils. A Thesis Submitted to Oregon State College (1961).
5) KANNO, I., Y. KUWANO, and Y. HONJO, *Advances in Clay Science*, Pub. by Clay Research Group of Japan, **2**, 355 (1960).
6) KLUG, H. P. and L. E. ALEXANDER, X-ray Diffraction Procedures for Polycrystalline and Amorphous Materials, John Wiley & Sons, Inc., New York, 300 (1954).
7) YOSHINAGA, N. and S. AOMINE, *Soil Sci. and Plant Nutrition*, **8**, No. 2, 6 (1962).

AMORPHOUS COATINGS ON MINERAL SURFACES[1]

R. C. Jones and G. Uehara[2]

ABSTRACT

Amorphous coatings on mineral surfaces too thin to be resolved by most electron microscopes were detected bridging particles suspended over holes in the supporting substrates. The coatings are gel-like in appearance, may flow when wet, shrink and become porous when heated, cement primary particles into large masses, and exhibit a coat-of-paint effect with curved solid-air interfaces. Amorphous gel coatings were found in alumino-silicate systems, in high aluminum soils, and on quartz surfaces.

Additional Index Words: gel-hulls, alumino-silicate system, aluminum system, silica system, electron microscopy, holey substrates.

WHILE THE ROLE played by amorphous materials in influencing soil behavior remains speculative, enough has been learned in the past few years to warrant re-examination of the topic. This paper presents new information which shows how amorphous materials are distributed in the clay fraction of soils. The evidence is illustrated through a series of high resolution electron micrographs.

---

[1] Published with the approval of the Director, Hawaii Agr. Exp. Sta. as Journal Series 1556 and with the approval of the Director, Hawaii Institute of Geophysics as contribution 502. Honolulu, Hawaii 96822. This work was supported in part by 211(d) Grant AID/csd-2833. Received Dec. 1, 1972. Approved May 31, 1973.

[2] Associate Soil Scientist, Dep. of Agronomy and Soil Science and the Hawaii Institute of Geophysics and Soil Scientist, Dep. of Agronomy and Soil Science, respectively. The authors wish to thank Dr. Richard Allen and the Pacific Biomedical Research Center for the use of their electron microscope.

METHOD

The electron microscope used for this study was a Hitachi HU-11A, operated at 100 kV with an emission current of less than 20 μA. Two anticontamination devices filled with liquid nitrogen were used at all times. Specimens were never exposed to the electron beam for more than 1 min. As soon as an area of interest was recorded on film, a new grid window was selected for the next exposure, thus preventing accumulation of contaminants of more than 10 Å on any one window of the grid.

A small quantity of soil was placed in a beaker containing distilled water and subjected to mild sonication with a Biosonik probe for 30 sec at a power of 300 W (20 kHz). No chemical dispersant was used to avoid formation of chemical artifacts. This suspension was washed through a 325-mesh screen. The concentration of the suspension was adjusted by visual inspection to provide adequate quantities of solids for sufficient coverage of the substrate. The suspension was sprayed onto the grid with a nebulizer. No attempt was made to separate clay from silt.

The carbon substrates were prepared in the following manner: A 7:1 mixture by volume of 2% formvar in ethylene dichloride solution to water was prepared. The mixture was emulsified with a probe-type sonicator. Immediately after sonication, one drop of the emulsion was placed on a clean water surface to form a film of Formvar containing numerous entrapped water droplets. A 400-mesh grid was inserted below the water surface and gently raised through the film. The grid, now coated with this film, was soaked twice for 30 sec in acetone to dissolve the thin Formvar film surrounding each water droplet. This operation produced a holey film. A carbon substrate was deposited on this film and the underlying Formvar film was dissolved with ethylene dichloride. The end result was a holey carbon substrate. (The method of preparing the holey carbon substrate was perfected by Georgianna Honea, a Junior Science Apprentice participating in a 10-week Student Scientist Training Program at the Univ. of Hawaii supported by the National Science Foundation.)

This type of substrate enabled us to observe amorphous materials stretched over the holes.

RESULTS

Figures 1A, B, and C illustrate the advantage of using holey substrates. Figure 1A shows a conventional electron micrograph of clay particles (kaolin) from an Aridisol. The same sample when viewed over a hole (Fig. 1B) shows a gel-like material bridging the particles. The light, fine-grained areas in Fig. 1B are free of substrate (the hole). Figure 1C shows clay-sized particles suspended over a hole and supported by a thin gel-like film which stretches from the particles to the substrate.

Some soils developed from volcanic ash (Andepts) contain a very high percentage of noncrystalline materials. Figures 2A and 2B are electron micrographs of a Hydrandept (Akaka series) and a Eutrandept (Waimea series), respectively. Figure 2A shows gel-like material covering or partially covering holes in the substrate. A high magnification micrograph of a gel-like material from the Eutrandept, stretched over a hole is shown in Fig. 2B.

Figure 3 is an electron micrograph of the clay fraction of a Gibbsihumox (Halii series). The individual crystals of gibbsite and geothite range in size between 50 and 200 Å in diameter. In most instances, these particles occur in clusters which are encased in gel-like masses. The light outlines which are Fresnel fringes (arrow A) surrounding many of the particles were caused by dehydration of the gel. This effect is typical of highly hydrated gel coatings.

Where there has been no separation of the gel from the particles, the light outlines are absent, for example at arrow B, Fig. 3. In some cases, the gels become porous and resemble sponge-like masses as a result of exposure to the vacuum and beam heat of the electron microscope. The Halii soil is highly aggregated and possesses excellent physical properties in the field.

To illustrate that gel-like coatings are not solely confined to soil materials, a single quartz crystal (from Hot Springs, Ark., Wards Natural Science Establishment) was finely ground in a tungsten carbide ball mill. The coarse fraction was removed by settling in distilled water.

Freshly crushed quartz particles revealed little in the way of gel coatings except for their tendency to aggregate. However, standing in water for long periods produced the gel coatings shown in Fig. 4A and B. Figure 4A shows a micrograph of clay size quartz particles which stood in water for 3 months. Note that the particles appear agglutinated and some enveloped in a gel. Some gel-like material can be seen smeared onto the substrate. After standing in water for 6 months, the gel coated quartz particles have coalesced to form a coat-of-paint effect (Fig. 4B).

DISCUSSION

Approximately 50 soil samples and geologic materials have been examined by high resolution electron microscopy. About half of them contained some gel-like material. In some Andepts, the major portion of the soil consisted of noncrystalline material. In other soils, which consisted predominantly of crystalline minerals, the gel-like materials occurred as coatings on the crystalline particles and not as discrete amorphous bodies.

As early as 1952, the existance of a surface coating on freshly ground quartz was surmised by a number of workers who studied the silicosis problem in the British coal mining industry. The first of a series of papers dealt with a "high-solubility layer" on siliceous dusts. Clelland, Cumming, and Ritchie (1952) reported chemical evidence for a "high-solubility layer" on the surfaces of dust particles from "rock crystal," silica sand, fused amorphous silica, olivine, and orthoclase feldspar. Clelland and Ritchie (1952) and Dempster and Ritchie (1952) asserted that the "high-solubility layer" was not a hydrated silica but rather a vitreous layer of the Beilby type (to be described later) that developed as a result of mechanical crushing and grinding. Estimates of the mean thickness of the "high-solubility layer" vary from 0.03 to 0.15μ (Gibb, Ritchie, and Sharp, 1953; Dempster and Ritchie, 1952; Nagelschmidt, Gordon, and Griffin, 1952). Gordon and Harris (1955) concluded that the amorphous layer on quartz was not of a definite thickness but consisted of a gradual transition from a non-crystalline exterior to a crystalline core. This conclusion was based on X-ray diffraction analysis which showed that particles smaller than 0.5μ seemed to display an X-ray amorphous character with virtually no crystalline core. Recently, Rieck and Koopmans (1964) used peak intensities of Debye-Scherrer lines of finely-ground quartz to conclude that the thickness of a

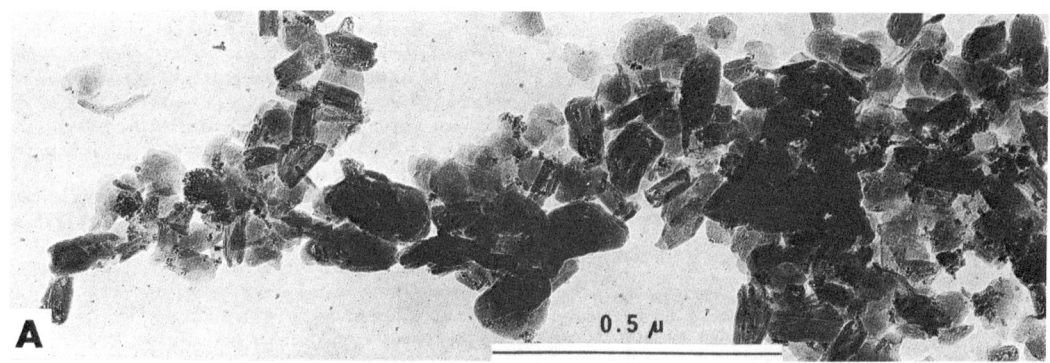
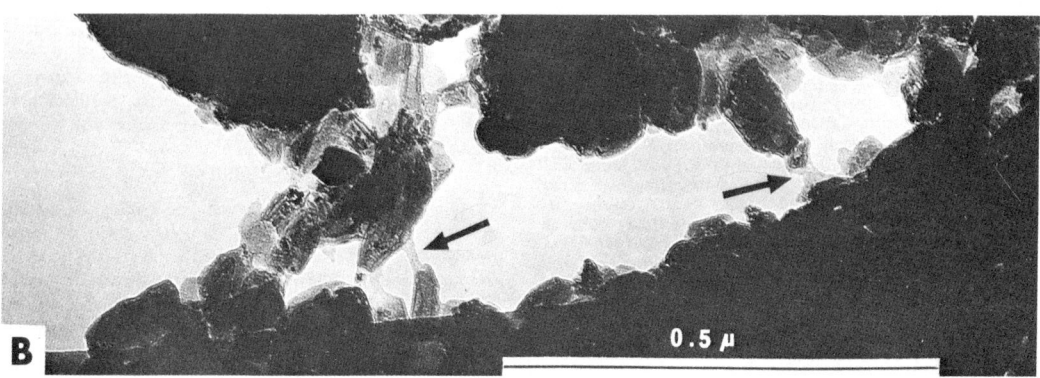
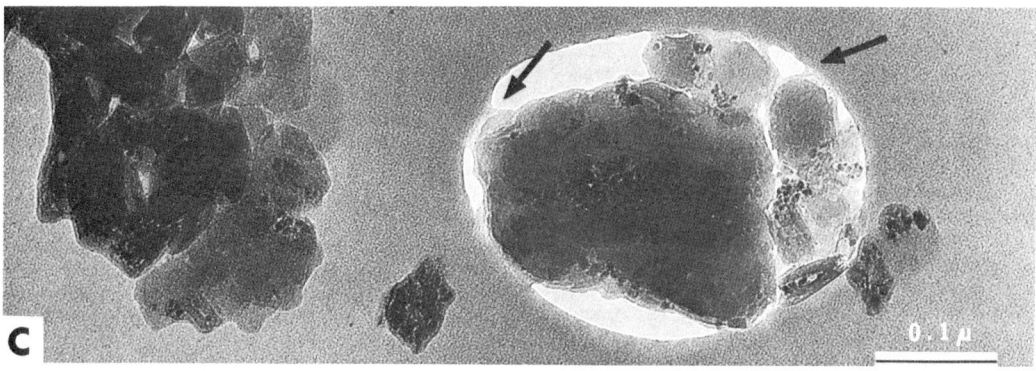

Fig. 1—Transmission electron micrographs of Kaolin particles from a Hawaiian Aridisol: A. Particles mounted on conventional carbon substrate. Prepared in this manner, this specimen fails to show gel-hulls; B. Same material as in Fig. 1A, mounted on holey substrate. Light, fine-grained area represents the hole. All materials in upper portion of this figure are suspended over the hole. Gel coattings (arrows) hold particles in place; and C. Even at higher magnification (210,000X), the gel coatings, clearly visible along perimeter of hole (arrows), is not discernible on substrate surrounding the hole.

"disturbed layer" was at least 0.4μ. The existence of such a layer on quartz is confirmed by Fig. 4A and 4B.

X-ray diffraction techniques were extended to the study of kaolinite by Engelhardt (1955) who estimated the thickness of the amorphous coatings to be 1 to 10 Å. Van der Marel (1966) stated that amorphous coatings on clay minerals may be formed by hydrolysis as a result of weathering and not necessarily as a result of grinding as had been

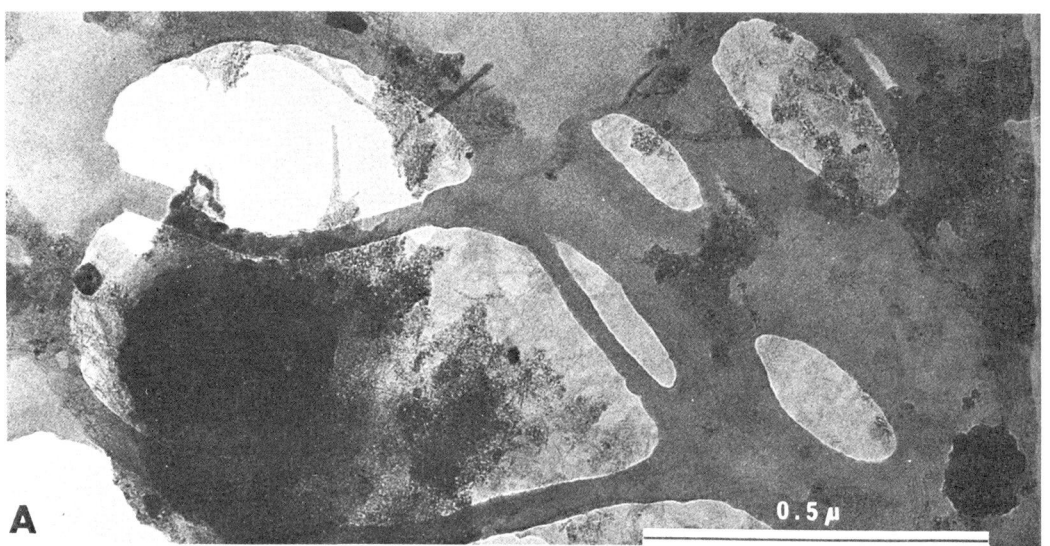

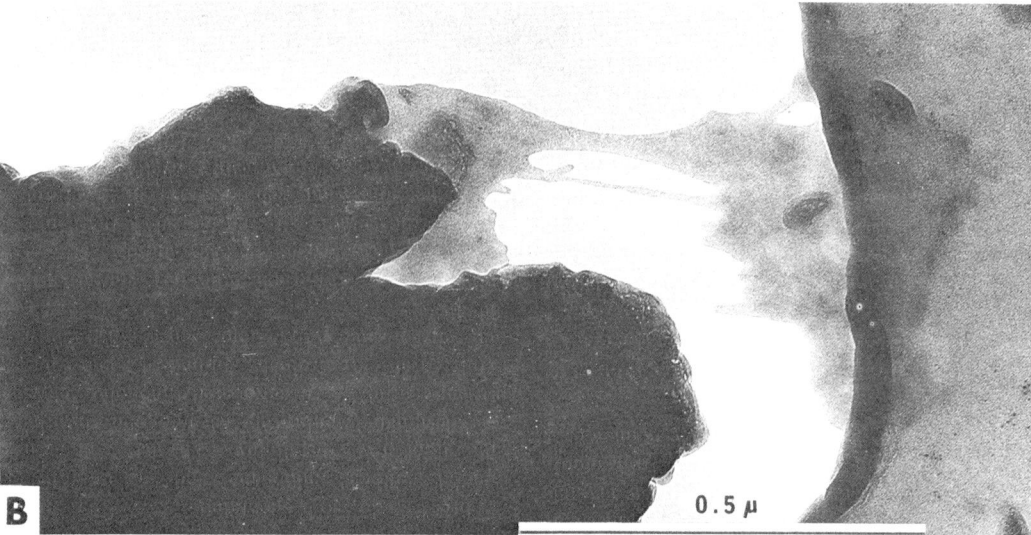

Fig. 2—Gelatinous material from Andepts stretched over holes in the substrate: A. Thin, gelatinous material from a Hydrandept covering holes in carbon substrate. The appearance of the material in the upper left hole suggests that in going from a wet, gel-like state to a dry elastic state, the membrane ruptured; and B. Noncrystalline material from a Eutrandept stretched over a hole. Soil materials from Hydrandepts (Fig. 2A) dry irreversibly, whereas material from Eutrandepts do not.

previously asserted in the literature. However, he referred to all surface coatings as Beilby layers regardless of their origin or composition. We prefer to call the naturally occurring surface coatings gel-hulls and to use the term Beilby layer for features associated with polished surfaces.

Beilby (1921, p. 129) studied polished metal surfaces by means of a microscope and noticed that these surfaces had a glossy film that seemed to flow into minute scratches and obscure all signs of metal's structure. French (1933) repeated Beilby's work and confirmed that there was indeed the appearance of a "super-cooled liquid" layer on the surface of highly polished metal which he called the Beilby layer. In "A Dictionary of Mining, Mineral, and Related Terms" (Thrush, 1968), the Beilby layer is defined

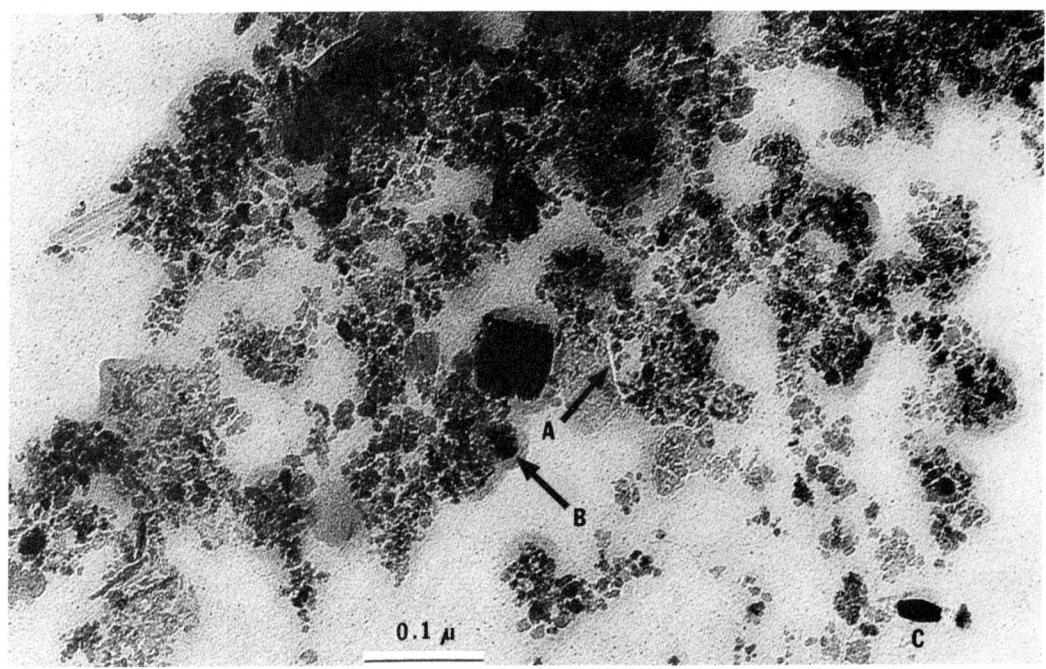

Fig. 3—The fine fraction of a Gibbsihumox mounted on a standard carbon substrate. Most of the black appearing particles contain a high percentage of iron. The lighter particles are mostly gibbsite and the barrel-shaped particle at C (lower right) is anatase. Since the particles absorb more heat from the electron beam than the gel or the substrate, gel dehydration and shrinkage occurs most frequently at the particle-gel-hull interface. Shrinking and cracking are typical when highly hydrated gels are dehydrated.

as, a "Flow layer resulting from incipient fusion during polishing mineral surfaces, and therefore, not characteristic of true lattice structure" and "The mirror-like surface layer, on all well-polished stones. . ." There is no mention of Beilby layers resulting from a chemical reaction.

There are several reasons why particle surface coatings have not previously been observed and measured on electron micrographs. The first and most likely reason was the limited capabilities of earlier electron microscopes. The absence of clear-cut evidence in support of surface coatings can be attributed to the low resolving power of electron microscopes available commercially as late as 10 years ago. Owing to the relatively high costs, most research organizations have been reluctant to replace these earlier prototypes with improved models which now permit routine resolution of 10 Å, and in more sophisticated instruments, 3Å or better. For example, Fieldes and Williamson (1955) suggested that positively charged noncrystalline aluminosilicate might exist as surface coatings. They, however, postulated that these coatings would exist as ultrafine particles beyond the resolving power of their electron microscope. An electron micrograph of finely particulate allophane occurring as a coating on an illite crystal has been published by Mitchell, Farmer, and McHardy (1964). Here again, because of limited resolution, the actual nature of the coating was not discernible.

Another and less obvious reason for difficulties with direct observation of surface coatings is the problem of specimen contamination and damage due to the electron beam. Particle growth in the electron microscope was observed by Cosslett (1947) to occur on zinc-oxide particles. Watson (1947) observed that carbon black particles grew in size with exposure to the electron beam and attributed the phenomenon to the accumulation of carbonaceous material from organic vapors in the microscope column. However, Cosslett (1947) disagreed with Watson (1947) and explained the particle growth as an "ejection" of adsorbed material and perhaps the metal itself (in the case of metallic oxides) and from the carbon black (in the case of Watson's observations). Burton, Sennett, and Ellis (1947) bombarded various salt samples with an intense electron beam and noted that the specimens became more electron transparent while "fine opaque particles" were being deposited around the specimen. They suggested that the interaction of the electron beam with the specimen induced ion migration. Evidence presented by Ennos (1953 and 1954) supported the carbon accumulation theory proposed by Watson (1947). Drawing on Ennos' findings, Noake, Hiroto, and Mizushima (1956) noted that intense electron irradiation for periods of 40 to 75 min on MgO smoke crystals produced an "interlayer" between the particle and the contaminant. They found that by dissolving away the specimen the interlayers were vacant. Although Noake et al. (1956) had no ready explanation for the gaps, they advanced the idea that there may be an electron beam reaction with the crystal as well as a buildup of contamination. Chute and Armitage (1968) also found this to be true for clay minerals. They showed that besides carbon buildup which grows linearly with time, there is an additional deposit which grows with the square root of time and apparently at the expense of the clay mineral. Since this type of rim growth was found to be particularly true in the case of kaolinite, Chute and Armitage postulated that hydroxyls diffused from the edges of the crystal and reacted with hydrocarbon vapors.

The materials shown in the electron micrographs presented in this paper reflect a minimum of electron beam damage and contamination. We can be sure of this on two counts. First, the emission current was held low and fine apertures were used in the condenser system so that the specimen current was never higher than absolutely necessary for focusing. A 100 kV beam potential was used for good penetration and minimum specimen heating. A large number of exposures were rapidly made and our interpretations of results is based on examination of micrographs rather than on observation of specimen in the microscope.

Secondly, the most convincing evidence for the authenticity of gel-hulls can be found in the micrographs themselves. In Fig. 2A and B, the noncrystalline material is stretched across holes in the substrate. This material must have been present on the grid prior to exposure to the electron beam. In Fig. 1B, particles of halloysite are suspended over holes in the substrate by gel-coatings. Here again the gel-coatings must have been present before the specimen was exposed to the electron beam. Soil particles possessing very thin gel-hulls fall through the holes in the substrate thus leaving the holes clear of particles.

Finally in Fig. 5, soil iron oxide particles cast shadows which originate some distance from the particle indicating the presence of a low density material. Close inspections of the interface between particle and shadow reveals the electron transparent material. Since the sample was shadowgraphed before placement in the electron microscope, we can conclude that this coating cannot be an electron beam contaminant (For the technique of shadowgraphing see Chapter 47, Electron Microscopic Techniques by J. A. Kittrick in C. A. Black (ed.) Methods of Soil Analysis Agronomy 9, Part 1, 1965.)

The occurrence of amorphous gel-like material as coatings on crystalline particles is particularly important. One can speculate that the chemical and physical behavior of porous bodies will depend heavily on amorphous materials if the amorphous fraction occurred as coatings on crystalline particles. The capacity of amorphous coating to act as viscous bodies when moist and elastic bodies when dry can have a pronounced effect on the porous body as a whole. Some amorphous materials, as in the case of the gel-like substance in Fig. 2A, irreversibly lose their viscous property upon drying. They are the high oxide systems. On the other hand, the high silicon system (Fig. 4A and 4B) appears to be able to alter reversible between the viscous and elastic states. These features may have strong implication in irreversible crusting of laterites, reversible crusting in irrigated soils of the arid regions and soil aggregation in general.

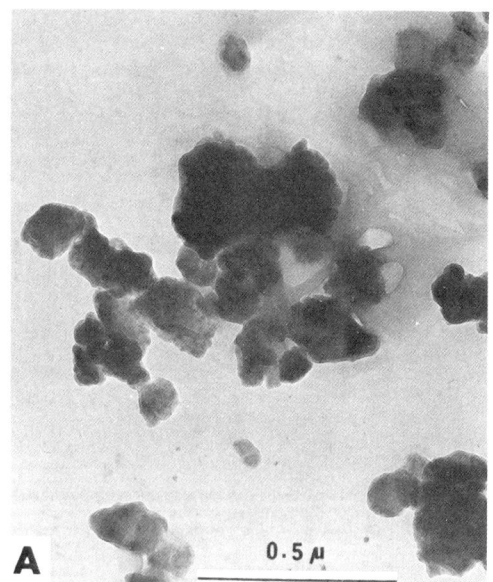

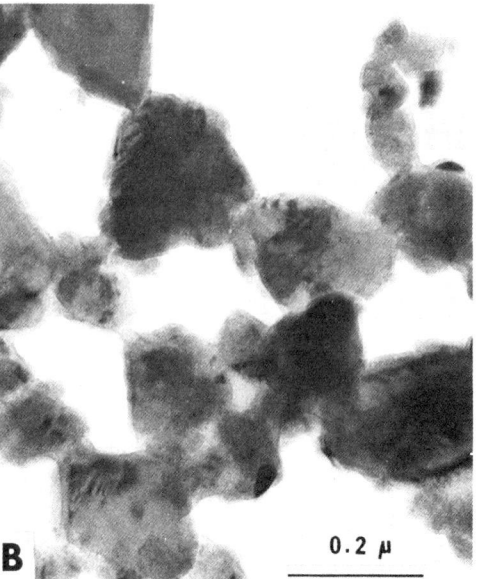

Fig. 4—Silica gel on the surfaces of finely-ground quartz. Freshly-ground quartz appeared relatively free of gel coating. After standing in water for 3 months in the dark, the quartz particles became encased in a gel which can be seen smeared over the particles and substrate (A). After standing for 6 months gel hulls coalesce to form a "coat of paint" effect (B).

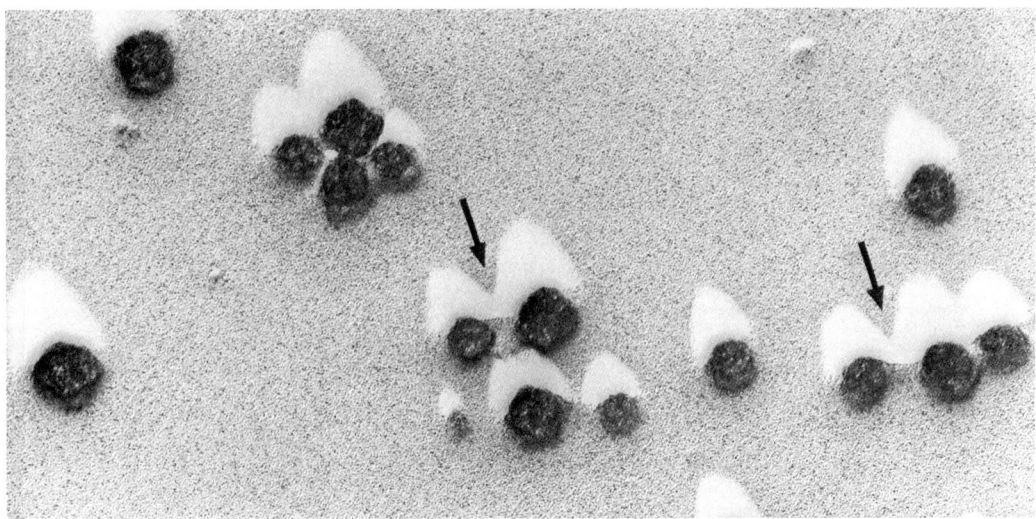

Fig. 5—Soil iron oxide particles shadowgraphed with platinum. The shadows cast from the particles originate approximately 130 Å from the surfaces of the particles, indicating the presence of an electron transparent coating. Particles slightly separated from one another cast shadows which over-lap because their gel coatings are in contact (arrows).

LITERATURE CITED

1. Beilby, G. 1921. Aggregation and flow of solids. Macmillan, London.
2. Burton, E. F., R. S. Sennett, and S. G. Ellis. 1947. Specimen changes due to electron bombardment in the electron microscope. Nature 160:565–567.
3. Chute, J. H., and T. M. Armitage. 1968. Alteration of clay minerals by electron irradiation. Clay Miner. 7:455–457.
4. Clelland, D. W., W. M. Cumming, and P. D. Ritchie. 1952. Physicochemical studies on dusts. I. A high-solubility layer on silicious dust surfaces. J. Appl. Chem. 2:31–41.
5. Clelland, D. W., and P. D. Ritchie. 1952. Physico-chemical studies on dusts. II. Nature and regeneration of the high-solubility layer on silicious dusts. J. Appl. Chem. 2:42–48.
6. Cosslett, V. E. 1947. Particle "growth" in the electron microscope. J. Appl. Phys. 18:844–845.
7. Dempster, P. B., and P. D. Ritchie. 1952. Surface of finely-ground silicon. Nature. 169:538–539.
8. Engelhardt, W. N. 1955. Über die Moglichkeit der quantitativen Phasenanalyse von Tonen mit Röntgeustrahlen. Z. Kristallogr. 105:430–459.
9. Ennos, A. E. 1953. The origin of specimen contamination in the electron microscope. Brit. J. Appl. Phys. 4:101–106.
10. Ennos, A. E. 1954. The sources of electron induced contamination in kinetic vacuum systems. Brit. J. Appl. Phys. 5:27.
11. Fieldes, M., and K. I. Williamson. 1955. Clay mineralogy of New Zealand soils. I. Electron micrography. N.Z. J. Sci. Tech., Sect. B. 37:314–335.
12. French, R. C. 1933. Polish on metals. Roy. Soc. (London) Proc. A. 140:637.
13. Gibb, J. G., P. D. Ritchie, and J. W. Sharp. 1953. Physicochemical studies on dusts. VI. Electron-optical examination of finely ground silica. J. Appl. Chem. 3:213–218.
14. Gordon, R. L., and G. W. Harris. 1955. Effect of particle-size on the quantitative determination of quartz by X-ray diffraction. Nature 175:1135.
15. Mitchell, B. D., V. C. Farmer, and W. J. McHardy. 1964. Amorphous inorganic materials in soils. Advan. Agron. 16:327–383.
16. Nagelschmidt, G., R. L. Gordon, and O. G. Griffin. 1952. Surface of finely ground silicon. Nature 169:539–540.
17. Noake, Hirota, and Mizushima. 1956. On the contamination of MgO smoke crystals in the electron microscope. J. Electron Microscopy 4:50–51. (Soc. of E. M., Japan).
18. Rieck, G. D., and K. Koopmans. 1964. Investigations of the disturbed layer of ground quartz. Brit. J. Appl. Phys. 15:419–425.
19. Thrush, Paul W., ed. 1968. A dictionary of mining, mineral, and related terms. Bureau of Mines, US Dep. of the Interior.
20. Van der Marel, H. W. 1966. Quantitative analysis of clay minerals and their admixtures. Contr. Mineral. Petrol. 12:96–138.
21. Watson, J. H. L. 1947. An effect of electron bombardment upon carbon black. J. Appl. Phys. 18:153–161.

Identification of Ferrihydrite in Soils by Dissolution Kinetics, Differential X-ray Diffraction, and Mössbauer Spectroscopy[1]

U. SCHWERTMANN, D. G. SCHULZE, AND E. MURAD[2]

ABSTRACT

Ferrihydrite is a poorly crystalline, natural Fe^{3+} oxide which occurs in ochreous spring precipitates and hydromorphic soils of humid temperate climates. The identification of ferrihydrite in soils is complicated by its association with goethite, quartz, and layer silicates.

The following criteria were used to identify ferrihydrite in Fe-oxide accumulations from soils: high solubility in acid oxalate, five to six broad x-ray diffraction lines, and the existence of a typical magnetic hyperfine field distribution at 4K in Mössbauer spectra rather than a discrete field value. Identification of low concentrations of ferrihydrite ($\leq$ 20% oxalate-soluble Fe) by x-ray diffraction was made possible by subtracting diffraction data obtained after oxalate treatment from data obtained before such a treatment (differential XRD). Oxalate treatment preferentially dissolves ferrihydrite over goethite. This led to an increase in the quadrupole splitting observed in Mössbauer spectra from ~0.10 to ~0.23 mm s^{-1}, resulted in a significantly narrower field distribution, and intensified the goethite DTA peak.

Additional Index Words: Fe oxides, goethite, concretions, hydromorphic soils, oxalate method, DTA, silica.

Schwertmann, U., D. G. Schulze, and E. Murad. 1982. Identification of ferrihydrite in soils by dissolution kinetics, differential x-ray diffraction, and Mössbauer spectroscopy. Soil Sci. Soc. Am. J. 46:869-875.

F ERRIHYDRITE is a poorly crystalline natural Fe^{3+} oxide named by Chukhrov et al. (1973). Its bulk composition is 5 $Fe_2O_3 \cdot 9H_2O$. Various formulae have been proposed: $Fe_5HO_8 \cdot 4H_2O$ by Towe and Bradley (1967) for a synthetic product, $Fe_5(O_4H_3)_3$ by Chukhrov et al. (1973) for a natural product, both on the basis of x-ray and electron diffraction data, and, more recently, $Fe_2O_3 \cdot 2FeOOH \cdot 2.6H_2O$ by Russell (1979) using additional information from infrared (IR) spectroscopy.

The structure of ferrihydrite is made up of layers of octahedra with Fe in the center and O, OH, and OH_2 as ligands. The octahedra are arranged similarly to those in hematite. The a-dimension of the hexagonal unit cell is the same as for hematite (5.08Å), but c is two-thirds that of hematite (9.4 instead of 13.8Å) because the unit cell has only four instead of six octahedral layers in the z-direction. The crystals of natural ferrihydrite are usually very small (3 to 7 nm in diameter) and possibly defect; Fe positions can be vacant, and O and OH are partly replaced by OH_2. This leads to strong broadening of the five to six x-ray diffraction lines (see, e.g., Fig. 2).

Ferrihydrite has frequently been identified in brown, gel-like precipitates from Fe-bearing waters (Schwertmann and Fischer, 1973; Henmi et al., 1980; Carlson and Schwertmann, 1981). In these, Fe is quickly oxidized and precipitated when the water becomes oxygenated upon its appearance at the earth's surface. The rapid formation may be partly responsible for the poor crystallinity of the product.

Furthermore, soluble constituents in the water with a high affinity towards the Fe-oxide surface such as silicate and organic compounds will be bound, thereby inhibiting crystallization and stabilizing the ferrihydrite.

Because of its poor crystallinity ferrihydrite can only be recognized easily by routine x-ray diffraction (XRD) if it is reasonably pure, such as in the surface precipitates described above. To identify it in soils at lower concentrations and in mixtures with other minerals (e.g., quartz, mica, goethite), special procedures such as differential x-ray diffraction and Mössbauer spectroscopy need to be applied. Examples of this are described in this paper.

MATERIALS AND METHODS

Because of the difficulty in detecting ferrihydrite, samples which have a fairly high Fe-oxide content together with a high oxalate-soluble portion were most suitable for this study. Such samples occur in hydromorphic (gley) soils of humid temperate areas in the form of nodules, pipe stems, or ferricretes and are the types of samples used for this study. Sample FE51 is a vesicular ferricrete from a gleysolic soil in Canada which was recently described by Evans et al. (1978). Samples K1, K2, and K3 are iron-rich pipe stems from three different gleys from the Netherlands. Sample R28c consists of nodules from Go1 horizon of a peaty gley from northern Germany. Four almost pure ferrihydrites with different crystallinity from spring deposits from Finland (LC2, LC40A, LC31; Carlson and Schwertmann, 1981) and northwestern Germany (N162; Schwertmann and Fischer, 1973) were used for comparison. The hard samples were ground in an agate mortar before further analysis. Sample FE51f is a fine fraction ($\leq$ 2 μm) of FE51 obtained by repeated gravity sedimentation in water. This fraction was used throughout this study because it gave a better differential x-ray diffraction pattern than the original samples because of less interference from coarse quartz grains. Samples K1, K2, and K3 contained less quartz, and fractionation was not necessary. Samples K1 through K3 had $Fe_{o/d}$ ratios between 0.06 and 0.72, thereby covering a wide range of this ratio. The $Fe_{o/d}$ ratio of sample FE51 varied with different pieces of the ferricrete between 0.8 and 1.0, indicating the dominance of poorly ordered Fe oxides. The "pure" ferrihydrites from spring deposits had $Fe_{o/d}$ ratios of 0.89 to 0.97 (Table 1).

Oxalate-extractable Fe (Fe_o) was determined after Schwertmann (1959, 1964) and DCB-extractable Fe (Fe_d) after Mehra and Jackson (1960). Iron in the extract was measured photometrically by the sulfosalicylic acid method (Koutler–Anderson, 1953) and Si by the heteropoly blue method (Boltz and Mellon, 1947). Dissolution kinetics were studied by shaking approximately 500 mg of the sample in 250 ml of the oxalate solution at 25°C and taking 1-ml subsamples at appropriate time intervals.

X-ray diffraction patterns were obtained using CoKα radiation and a Philips PW 1050 vertical goniometer equipped with a 1-degree divergence slit, a 0.2-mm receiving slit, a

---

[1] Contribution from the Institut für Bodenkunde der Technischen Universität München, 8050 Freising-Weihenstephan, F.G.R. Paper presented in part before Div. S-9, Soil Science Society of America, Fort Collins, Colo., August 1979. Received 21 July 1981. Approved 12 Feb. 1982.

[2] Professor, Graduate Research Assistant, and Research Associate, respectively.

Table 1—Color, mineralogy, and some chemical properties of the samples.

Sample	Moist color (Munsell)	Fe-oxide mineralogy†	Fe_d‡, %	Fe_o‡, %	$Fe_{o/d}$	Si_o, %
Soil samples						
FE51f (<2 μm)	5YR 3/4	Fh > Gt	30.7	25.3	0.82	1.12
FE51f ox (1 × 2h)§		Gt	3.21	0.50	0.16	
K2	5YR 3/3	Fh > Gt	45.4	35.1	0.77	1.31
K2 ox (1 × 2h)		Gt > Fh	43.8	20.9	0.48	
K2 ox (2 × 2h)		Gt > Fh	41.0	9.7	0.24	
K2 ox (24h)	7.5YR 3/4	Gt	35.1	2.58	0.07	
K3	5YR 3/4	Gt > Fh	49.3	13.3	0.27	0.63
K3 ox (2 × 2h)	7.5YR 4/6	Gt	51.2	4.40	0.09	
K1	7.5YR 4/6	Gt	52.7	2.91	0.06	0.04
R28c		Gt > Fh	29.1	8.6	0.30	0.86
Samples from cold springs						
LC2	10YR 6/6	Fh	31.7	30.8	0.97	6.84
LC40A	7.5YR 6/8	Fh	41.6	38.7	0.93	2.40
LC31	7.5YR 6/9	Fh ⪢ Gt	48.0	42.7	0.89	2.35
N162	5YR 4/8	Fh ⪢ Gt	51.3	46.8	0.90	3.11

† Fh = ferrihydrite, Gt = goethite.
‡ Subscripts d and o refer to dithionite- and oxalate-soluble Fe, respectively.
§ ox = Oxalate treated.

1-degree scatter slit, and a diffracted beam graphite monochromator. A scanning speed of 1-degree/min and a time constant of 2 s were used. Self-supporting powder mounts were prepared by back-filling the samples into aluminum frames and then gently pressing the material against filter paper to minimize preferred orientation.

To facilitate the identification of ferrihydrite by XRD, a subtraction technique, Differential XRD (DXRD), was used (Schulze, 1981). Powder mounts of untreated and oxalate-treated preparations of each sample were step-scanned at intervals of 0.02° 2θ from 18 to 80° 2θ and counted for 10 s/increment. Instrumental conditions were otherwise the same as above. By trial and error subtraction of the patterns of the oxalate-treated samples from those of the untreated samples, diffractograms of the ferrihydrite were obtained.

Mössbauer spectra were taken using a 5 mCi ^{57}Co/Rh source mounted on a loudspeaker-type drive system. Sample quantities of 40 mg spread uniformly over an area of 2 cm^2 in a Plexiglas holder served as absorbers. Spectra were taken at room temperature and after cooling both source and absorber to about 120 and 4°K in a cryostat. The transmitted radiation was registered with a proportional counter and fed into a multichannel analyser using either 512 or 1,024 channels. Counting proceeded until at least $4 \cdot 10^5$ counts had been accumulated per channel. The data were plotted, and Lorentzian curve fits were carried out by a computer procedure. Pure metallic iron served as a standard for velocity calibration.

Samples for electron microscopy were ultrasonically dispersed in ethanol and a drop of the suspension was dried on a carbon-coated copper grid. The samples were examined with a Zeiss EM 10 transmission electron microscope operated at 80 kV. Differential thermograms were run with a Linseis instrument using 50-mg samples, hematite as an inert material, and a heating rate of 10°C/min. The temperature was calibrated with KNO_3 and quartz. Infrared spectrograms were recorded with a Beckman IR-4250 instrument using KBr pellets.

RESULTS AND DISCUSSION

Color and Oxalate Solubility

The three soil samples with $Fe_{o/d}$ ratios > 0.27, samples FE51f, K2, and K3, are redder (5YR) than sample K1 (7.5YR) which has an $Fe_{o/d}$ ratio of 0.06 (Table 1). After a longer or a double treatment with oxalate the color of samples K2 and K3 shifted from 5YR to 7.5YR. Among 15 samples of bog iron ores and nodules containing goethite and ferrihydrite ($Fe_{o/d}$:0.2–0.9) 12 became yellower after oxalate treatment (Schwertmann and Lentze, 1966). These results show that the material dissolved in oxalate is redder than the material which is not dissolved.

The oxalate method (Tamm's reagent) was proposed by Schwertmann in 1959 (1959, 1964) and later by MacKeague and Day (1967) to separate a more "active" part of the Fe oxides from a less active one. The active part was frequently called amorphous, partly in agreement with the fact that none of the crystalline Fe oxides such as goethite, hematite, or lepidocrocite could be identified in samples with a high proportion of oxalate-soluble Fe. The oxalate solubility of goethite, hematite, and lepidocrocite is low (goethite ≃ hematite < lepidocrocite) and it can there-

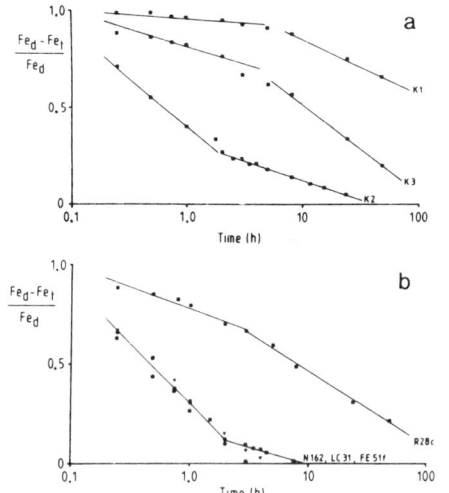

Fig. 1—Dissolution time curves of various samples containing ferrihydrite and goethite in 0.2M oxalate at pH 3.0 and 25°C.

fore be expected that what is dissolved by oxalate in 2 hours is mainly ferrihydrite.

Dissolution time curves for oxalate give further indication of the existence of iron oxides with significantly different dissolution rates. A plot of $(Fe_d - Fe_t)/Fe_d$ (Fe_t = Fe dissolved at time t) against log t results in two straight lines for all samples (Fig. 1a, 1b). Only the two very poorly crystalline ferrihydrites of samples LC2 and LC40A were completely dissolved in < 15 min so that no time curve could be recorded. The low crystallinity of these two samples is indicated by their XRD spectra (Carlson and Schwertmann, 1981) which consist of only one (sample LC2) or three (sample LC40A) very broad lines instead of five to six for a well-crystallized ferrihydrite. The curves consisting of two linear parts were therefore interpreted as due to the dissolution of ferrihydrite followed by that of goethite. The inflection point occurs between 2 and 4 h, and it may be concluded from this that a somewhat better separation of ferrihydrite from goethite may be achieved by using a slightly longer time of extraction than the 2 h presently used.

In three parallel oxalate extractions with sample FE51f, between 61 and 66% of the original sample weight was lost. Between 81 and 84% of this loss would correspond to ferrihydrite on the basis of the Fe dissolved and a bulk composition of $5Fe_2O_3 \cdot 9H_2O$ for ferrihydrite.

Oxalate-Soluble Silicon

Oxalate-soluble Si (Si_o) varied between 0.04 and 6.84% (Table 1) with a tendency to increase with increasing ferrihydrite in the soil samples (FE51 ≃ K2 > K3 > K1). Among the "pure" ferrihydrites, the poorly crystalline one (LC2) contained much more Si_o than the better crystalline ones (LC31, N162). It was shown by IR adsorption that considerable amounts of Si are associated with ferrihydrite, leading to a shift of the Si-O-vibration to lower wave numbers (Carlson and Schwertmann, 1981). It can be concluded, therefore, that Si is also involved in the formation and stabilization of ferrihydrite in soils.

X-ray Diffraction

Of the three samples K1, K2, and K3, K2 had the largest $Fe_{o/d}$ ratio (Fig. 2a). The diffraction pattern of the untreated sample consists of peaks due to goethite and quartz with the broad asymmetric peak between 30 and 43° 2θ, suggesting the presence of ferrihydrite. The pattern from the oxalate treated sample is dominated by goethite and quartz peaks with almost all goethite peaks being well resolved. The intensity of the undissolved minerals (e.g., goethite) was about four times higher than the pattern from the untreated sample. The DXRD pattern clearly shows the peaks diagnostic for ferrihydrite at 2.5, 2.2, 1.97, 1.71, and 1.5Å (Chukhrov et al., 1973; Schwertmann and Fischer, 1973; Carlson and Schwertmann, 1981) and shows beyond a doubt that the material dissolved by the oxalate treatment is ferrihydrite. The peak at 1.5Å is actually a poorly resolved doublet.

Sample K3 (Fig. 2b) shows a pattern dominated by goethite along with some quartz. After the oxalate treatment, which dissolved 27% of the DCB-extractable iron, the XRD pattern appears almost identical to the untreated pattern. Only the DXRD diagram clearly shows the pattern of ferrihydrite. The reason why DXRD is able to detect the ferrihydrite in this sample becomes clearer when one examines the scale used to plot the diagrams. The untreated and oxalate-treated patterns are plotted using 600 counts full scale, whereas the DXRD diagram is plotted at 50 counts full scale, a 12-fold expansion. In other words, the contribution of the ferrihydrite to the XRD pattern of the untreated sample is only to make the "background" slightly wavey.

Sample K1, with an $Fe_{o/d}$ ratio of 0.06, shows the XRD pattern (Fig. 2c) of goethite along with a small amount of quartz. As one would expect with so little material dissolved by the oxalate treatment, the pattern from the oxalate-treated sample is almost identical with that from the untreated sample. The DXRD pattern is essentially a straight line and ferrihydrite can not be clearly identified, although the weak, broad peak at 2.5Å could be due to the strongest ferrihydrite peak.

The presence of mica and quartz makes it difficult to identify the iron oxide minerals in sample FE51f. The DXRD diagram (Fig. 2d) shows ferrihydrite to be the major iron oxide mineral. The very broad peak at 4.2Å is at the same position as the strongest goethite peak and could be due to poorly crystalline goethite in agreement with a $Fe_{o/d}$ ratio of <1(0.82). The "negative" peaks in the DXRD pattern of FE51f are caused by an increased degree of preferred orientation of mica after the oxalate treatment. Consequently the basal reflections at 5.0 and 3.3Å (coincident with quartz) are increased relative to the nonbasal reflection at 4.5Å. The mica peaks common to both patterns do not match exactly, and negative peaks result in the DXRD diagram. The negative peaks do not, however, interfere with the identification of ferrihydrite in this sample.

The fact that ferrihydrite can still be easily identified in sample K3 (13.3% Fe_o) but not positively identified in sample K1 (2.9% Fe_o) indicates that the lower limit of detection lies somewhere between these two Fe_o values. Ten percent Fe_o as ferrihydrite (roughly 15% by weight) is probably about the lower limit of detection using DXRD. This might seem to be a rather large amount, but it is unlikely that smaller amounts can be detected, simply due to the very broad, weak peaks of ferrihydrite. If one considers that using routine XRD procedures ferrihydrite must be almost pure for positive identification, DXRD represents an improvement in the detection limit by a factor of six or seven.

Mössbauer Spectroscopy

Mössbauer spectra of oxalate-treated and untreated samples K1, K2, K3, and FE51f were taken at 4K and, of selected samples, at 120K and room temperature. The room temperature and 120K spectra consist of superimposed superparamagnetic doublets and sextets of different intensity ratios. The 120K spectrum of sample K1 consists almost exclusively of a magnetically split sextet, whereas K2 has only a subor-

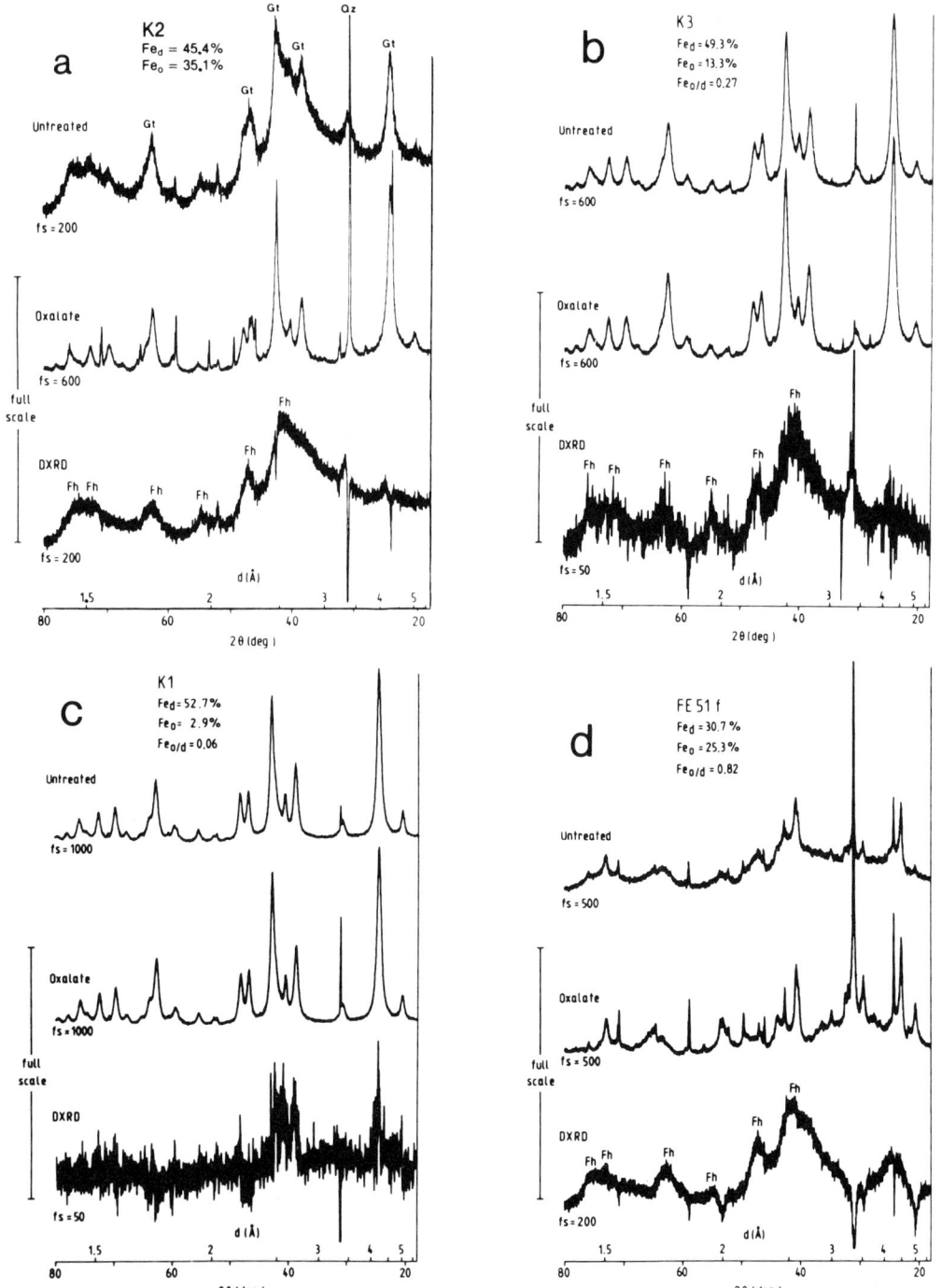

Fig. 2a–d—X-ray powder diffraction patterns of the soil samples before and after oxalate treatment and the differential x-ray diffraction (DXRD) pattern obtained by subtracting the oxalate pattern from the untreated pattern (Fh: ferrihydrite, Gt: goethite, Q: Quartz). Oxalate extraction times for the patterns shown: K1, 1 × 2h; K2, 1 × 20h; K3, 2 × 2h; FE51f, 1 × 6h.

dinate magnetically ordered component. Hyperfine fields of 484 and 465 KOe, respectively, indicate that the magnetically ordered constituents of these spectra result from goethites of different particle size and/or Al substitution.

At 4K all iron oxides, including ferrihydrite, exhibit magnetically ordered Mössbauer spectra (Murad and Schwertmann, 1980). Therefore at this temperature all samples show magnetically split spectra which, however, differ significantly in their line widths and quadrupole splittings. Previous studies on pure natural and synthetic ferrihydrites showed these to possess not a discrete magnetic hyperfine field at 4K, but rather a distribution of hyperfine fields. The widths and maxima of the distributions give an indication of the degree of crystallinity of the ferrihydrites (Murad and Schwertmann, 1980).

Analyses of the 4K spectra showed the ferrihydrite-rich samples FE51f and K2 to have similar broad hyperfine field distributions with maxima at 493 and 499 kOe, respectively, and half-widths of 58 kOe and low quadrupole splittings of 0.10 mm s^{-1}. Oxalate treatments of 6-hours' duration reduced the half-widths of hyperfine field distributions (HWHD) of these samples to 40 and 36 kOe and increased the quadrupole splittings to 0.12 and 0.18 mms^{-1}, respectively (Table 2). A prolonged oxalate treatment of 24 hours reduced the HWHD of K2 to 25 kOe and increased the quadrupole splitting to 0.23 mm s^{-1} (Fig. 3). The maximum of the hyperfine field distribution of FE51f increased

Table 2—Mössbauer parameters of the samples at 4K.

Sample	δ(Fe)	ΔE_Q	$H_{i_{max}}$†	HWHD‡
	mm s^{-1}		kOe	
FE51f	0.37	0.09	493	58
FE51f (3 × 2h ox)	0.36	0.12	497	40
K2	0.34	0.10	499	58
K2 (3 × 2h ox)	0.34	0.18	499	36
K2 (24h ox)	0.37	0.23	499	25
K3	0.35	0.22	502	24
K3 (2 × 2h ox)	0.36	0.24	502	22
K1	0.36	0.23	504	21
K1 (2h ox)	0.36	0.24	504	21

† $H_{i_{max}}$ = magnetic hyperfine field with maximum probability.
‡ HWHD = half-width of hyperfine field distribution.

slightly to 497 kOe, whereas that of K2 remained unchanged at 499 kOe following oxalate treatment. In contrast, the ferrihydrite-poor samples K3 and K1 had narrower initial hyperfine field distributions (HWHD 24 and 21 kOe) and high quadrupole splittings (0.22 and 0.23 mm s^{-1}, respectively). Oxalate treatment had only a minor effect upon the Mössbauer spectra of these samples, giving HWHD values of 22 and 21 kOe, respectively, a quadrupole splitting of 0.24 mm s^{-1} for both samples, and maxima of the hyperfine field distributions at 502 and 504 kOe, respectively. The isomer shifts for all samples averaged 0.36 ± 0.01 mm s^{-1} with respect to metallic iron and showed no systematic variations either between the individual samples, or as a result of oxalate treatment.

Room temperature and 120K spectra consist of

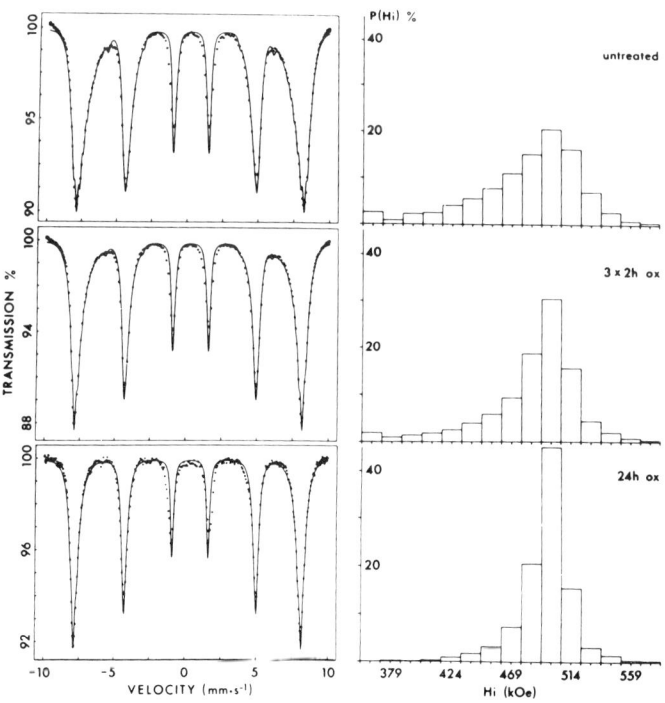

Fig. 3—Mössbauer spectra and hyperfine field distributions of sample K2 at 4K: untreated sample (top), 3 × 2h oxalate-treated sample (middle), and 24h oxalate-treated sample (bottom).

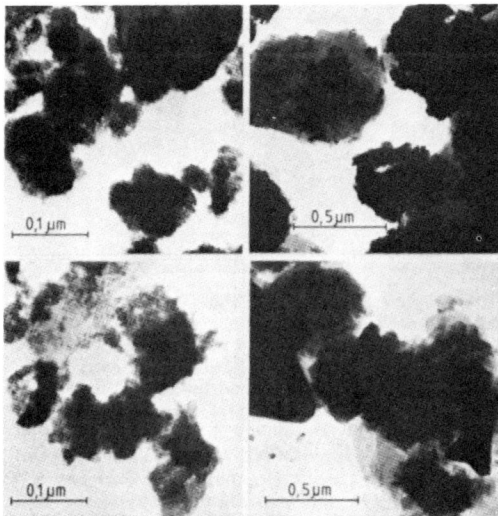

Fig. 4—Electron micrographs of sample K2 before (upper left) and after (lower left) 24h oxalate treatment and of sample FE51 before (upper right) and after (lower right) DCB treatment.

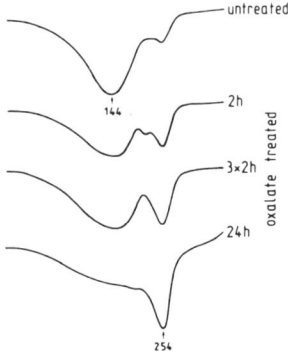

Fig. 5—DTA curves of sample K2 before and after various oxalate treatments.

broad, unspecific Fe^{3+} doublets and are not suited for the identification of ferrihydrite. The isomer shift and hyperfine field maximum at 4K do not differ significantly from those of goethite. Wide hyperfine field distributions (HWHD about 60 kOe) and low quadrupole splittings (≤ 0.1 mm s^{-1}), however, are indicative of ferrihydrite. The resonant lines should be significantly narrowed (HWHD about 25 kOe) after treating the sample with oxalate, and the quadrupole splitting should approach that of the remaining magnetically split phases.

Electron Microscopy, DTA, and IR

The difference in particle morphology between the untreated and 24h-oxalate-treated sample K2 is surprisingly small in view of the fact that ferrihydrite was completely extracted and only goethite was left (Fe$_{o/d}$: 0.77 → 0.07, Table 1). Figure 4 (upper left) shows opaque aggregates with a fine granular structure before oxalate treatment, whereas after 95% of the Fe$_d$ of the original sample was extracted, some transparent flaky material, probably layer silicates, are visible. Occasionally, laminar structures in the opaque material particles occur which are typical for poorly crystalline soil goethites (Schwertmann and Taylor, 1977).

Sample FE51f (after H$_2$O$_2$ treatment) shows large silicate flakes again covered partly with opaque, highly aggregated material with irregular edges probably consisting of ferrihydrite. This material disappeared after DCB treatment (Fig. 4), and thin clean clay flakes become visible.

The gradual concentration of goethite at the expense of ferrihydrite is also shown by an increase in the endothermic DTA peak at about 250°C (Fig. 5). The two OH-bending vibrations of goethite at 790 and 885 cm^{-1} are present in IR spectrograms (not shown) after treatment but are hardly visible before treatment. In contrast, the Si-O feature at 1,030 cm^{-1} weakens on oxalate extraction in accordance with the considerable amounts of Si extracted (Table 1).

ACKNOWLEDGMENT

The authors are grateful to Miss B. Schönauer for performing the chemical analyses, Mr. O. Herrmann for preparing the differential thermograms, and Dr. H.-Ch. Bartscherer for preparing the electronmicrographs. Drs. L. C. Carlson, Helsinki, L. J. Evans, Guelph, W. R. Fischer, Freising, H. W. Van der Marel, the Netherlands, and E. Schlichting, Stuttgart, kindly supplied samples.

REFERENCES

1. Boltz, D. F., and M. G. Mellon. 1947. Determination of P, Ge, Si and As by the heteropoly blue method. Anal. Chem. 19:873-878.
2. Carlson, L., and U. Schwertmann. 1981. Natural ferrihydrites in surface deposits from Finland and their association with silica. Geochim. Cosmochim. Acta 45:421-429.
3. Chukhrov, F. V., B. B. Zvyagin, L. P. Ermilova, and A. I. Gorshkov. 1973. New data on iron oxides in the weathering zone. p. 333-341. In J. M. Serratosa and Sanchez (ed.) Proc. of the Int. Clay Conf. 1972 (Madrid). Division de Ciencias C.S.I.C., Madrid.
4. Evans, L. J., J. G. Rowsell, and J. D. Aspinall. 1978. Massive iron formations in some gleysolic soils cf southwestern Ontario. Can. J. Soil Sci. 58:391-395.
5. Henmi, T., N. Wells, C. W. Childs, and R. L. Parfitt. 1980. Poorly-ordered iron rich precipitates from springs and streams on andesitic volcanoes. Geochim. Cosmochim. Acta 44:365-372.
6. Koutler-Anderson, E. 1953. The sulfosalicylic method for iron determination and its use in certain soil analysis. Ann. Roy. Agric. Coll. Sweden 20:297-308.
7. McKeague, J. A., and J. H. Day. 1967. Dithionite- and oxalate-extractable Fe and Al as aids in differentiating various classes of soils. Can. J. Soil Sci. 46:13-22.
8. Mehra, O. P., and M. L. Jackson. 1960. Iron oxide removal from soils and clays by a dithionite-citrate system buffered with sodium bicarbonate. Clays Clay Miner. 7:317-327.
9. Murad, E., and U. Schwertmann. 1980. The Mössbauer spectrum of ferrihydrite and its relations to those of other iron oxides. Am. Mineral. 65:1044-1049.
10. Russell, J. D. 1979. Infrared spectroscopy of ferrihydrites; evidence for the presence of structural hydroxyl groups. Clay Miner. 14:109-114.
11. Schulze, D. G. 1981. Identification of soil iron oxide minerals by differential x-ray diffraction. Soil Sci. Soc. Am. J. 45:437-440.
12. Schwertmann, U. 1959. Die fraktionierte Extraktion der freien Eisenoxyde in Böden, ihre mineralogischen Formen und ihre Entstehungsweisen (in German). Z. Pflanzenernähr., Düng., Bodenkunde 84:194-204.
13. Schwertmann, U. 1964. Differenzierung der Eisenoxide des

Bodens durch photochemische Extraktion mit saurer Ammoniumoxalat-Lösung (in German). Z. Pflanzenern., Düng., Bodenkunde 105:194–202.

14. Schwertmann, U., and W. R. Fischer. 1973. Natural "amorphous" ferric hydroxide. Geoderma 10:237–247.

15. Schwertmann, U., und W. Lentze. 1966. Bodenfarbe und Eisenoxidform (in German). Z. Pflanzenern., Düng., Bodenkunde 115:209–214.

16. Schwertmann, U., and R. M. Taylor. 1977. Iron oxides. p. 145–180. *In* J. B. Dixon and S. B. Weed (ed.) Minerals in soil environments. Soil Science Society of America, Madison, Wis.

17. Towe, K. M., and W. F. Bradley. 1967. Mineralogical constitution of colloidal "hydrous ferric oxides". J. Colloid. Interface Sci. 24:384–392.

A Study of a Deep Weathering Profile on Granite in Peninsular Malaysia: III. Alteration of Feldspars[1]

H. ESWARAN AND WONG CHAW BIN[2]

ABSTRACT

Feldspar grains or their pseudomorphs from the different weathering zones are studied with the SEM. Weathering commences with dissolution to form voids in the grains. Close to the rock, the first product is allophanic material present as globules adhering to the voids in the grains. In the δ zone the product is halloysite with some amorphous silica spherules. The surface of the pseudomorph has a hummocky appearance suggesting an amorphous intermediary phase prior to halloysite formation. Kaolinite formation commences in the γ_p zone and gibbsite in the γ_m zone. Admixtures of kaolinite and halloysite may be found on the same pseudomorph, but admixture of gibbsite with another secondary mineral is not encountered.

Biotite follows a similar trend of alteration. It is concluded that irrespective of the primary weatherable mineral, the type of secondary mineral found is a function of the microenvironment.

Additional Index Words: amorphous alumino-silicates, amorphous silica spherules, biotite, feldspar, gibbsite, halloysite, kaolinite, micro-environment, SEM, weathering.

STUDYING the physico-chemical properties of the soils or the mineralogy of the clay, silt, or sand fractions (6, 7) does not indicate the origin of the secondary minerals. This has been a major handicap in previous studies, though in some cases the source mineral is inferred. This is frequently overcome by hand-picking grains, when it is possible, and monitoring their changes in different depths in the profile. Some minerals like micas, are amenable to separation and these have been studied in detail. Feldspars are more difficult to separate and there have been correspondingly less studies. The scanning electron microscope (SEM) is an ideal instrument for such studies and if complimented with electron microprobe investigations, the study would be more complete.

The SEM study is undertaken with two objectives in mind. First is to monitor the products of alteration of plagioclases in the deep weathering profile described previously (6, 7) and secondly is to try to establish mineral weathering sequences. The mineralogical zonations in the deep profile on granite are similar to some previous work and in older publications or textbooks on soils, it is frequently implied that the halloysite is formed from feldspars, that the kaolinite is derived from the halloysite, and later, that the gibbsite is from the kaolinite. However, by the use of TEM and replica techniques, Wilson et al. (16) could not find any intermediate crystalline or well defined amorphous phases in their mineral transformation studies.

SEM micrographs of alteration of feldspars to kaolinite, halloysite, and montmorillonite have been shown by Eswaran et al. (9). Under certain conditions, weathering results only in dissolution, producing etching on the grains as shown by Wilson (15). These are only incidental observations and no explanations regarding the differential alterations are given. In a similar study, Eswaran et al. (5) have shown the alteration of feldspars to allophane or imogolite. Laboratory simulated weathering experiments have been attempted to reproduce natural weathering. Parham (13) showed, by a soxhlet extraction technique, that one of the initial products of intensive leaching is halloysite. The small spicules that he observed on the replica of the feldspar were interpreted as halloysite. Bondham (2) obtained perfectly spherical bodies of amorphous silica gels in his soxhlet. This experiment has been reproduced, with granite and microcline in our laboratories (unpublished), with similar results. Such spherules have not been reported in soil clays and are distinctly different from allophane, both in morphology and in composition.

METHODS

The soil and saprolitic material were studied by other techniques (6, 7) prior to the present. Soil micromorphological and mineralogical studies indicated the absence of feldspars in the solum of the profile.

Fracture surfaces of the saprolitic material were prepared using an incident light microscope. Feldspars (pseudomorphs) were identified and mounted on aluminum stubs using normal glue. The material is coated with a thin film of gold using a splutterer and is then ready for observation with the SEM. After SEM study, if necessary, the material is ground for XRD analysis (the peak for gold is ignored). If the grain is too small, a single crystal diffraction is made using a Debije-Scherrer powder camera. When amorphous materials are suspected, a DTA using a Du Pont Thermoanalyzer is run. More than one hundred photographs were taken and some of these are presented later. The alteration of biotite is also studied but as the results obtained are similar to a previous study (10), only some micrographs are included.

RESULTS AND DISCUSSION

SEM micrographs (Fig. 1a, b, c) show a very initial state in the alteration of the plagioclase as has been previously observed by Wilson (15). In thin sections, the feldspars show transgranic voids (Part I, Fig. 2h) and under plain light, material with high relief is seen adhering to the void walls. Delvigne (3) considers such features as amorphous alumino-silicate gels. Fig. 1a shows the transgranic voids and in (b) and (c), the accumulation of amorphous materials on the void walls and fracture surface is seen. Such features are present only in the δ zone, decreasing in amount from the rock to the surface. The other observation here is that the feldspars do not show any characteristic etch features as could be seen on quartz (6). Weathering and dissolution produces the transgranic voids which generally have a parallel alignment which is related to the twinning planes of the original mineral (4).

In the upper δ and in γ_p zones, the plagioclase pseudomorphs have a speckled appearance in thin sections [Part I(6), Fig. 2g] and the isotic plasmic fabric is attributed to the halloysite and the birefringent spots to the few silt-sized, unweathered fragments of feldspars in the pseudomorph. Micrographs (Fig. 1 *d, e,* and *f*) show the fracture

---

[1]Contribution from the Geological Institute, Krijgslaan, 271, Gent, Belgium. Work supported in part by a grant from the National Science Foundation, Belgium. Received 1 July 1977. Approved 12 Oct. 1977.

[2]Assistant and Graduate Student, respectively.

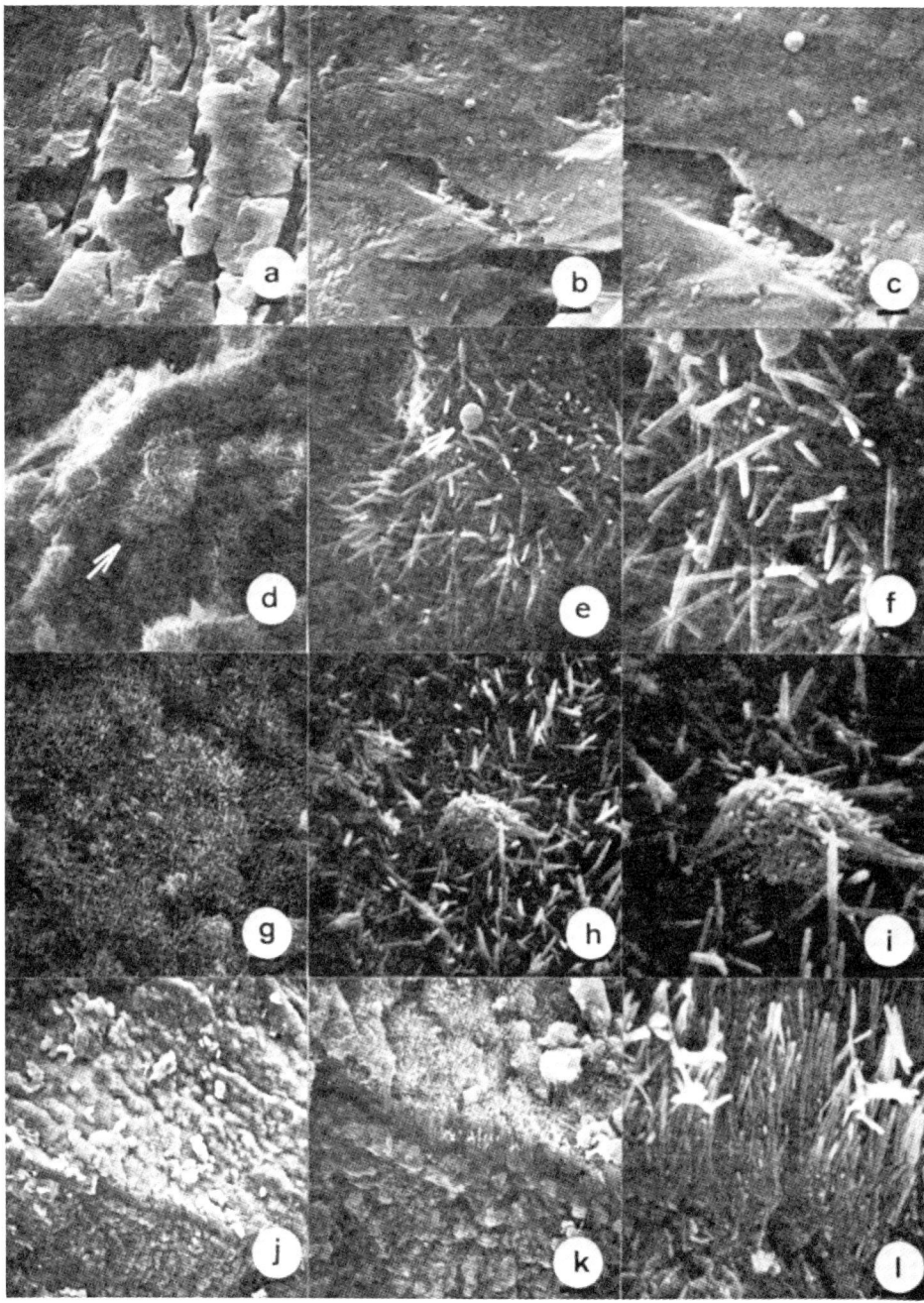

Fig. 1—SEM micrographs of handpicked feldspar grains or their pseudomorphs.
(a), (b), (c)—Feldspar grain from the δ zone showing formation of dissolution pits and accumulation of amorphous materials. (a) 4×500; (b) = $\times 2,000$; (c) = $\times 10,000$.
(d), (e), (f) Feldspar grain from the δ_p zone. Alteration to halloysite. Some spherules of probably amorphous silica are present. (d) = $\times 2,000$; (e) = $\times 10,000$; (f) = $\times 20,000$.
(g), (h), (i)—Alteration of feldspar in γ_p zone. The halloysite crystals are present as clumped aggregates. (g) = $\times 1,000$; (h) = $\times 7,500$; (i) = $\times 20,000$.
(j), (k), (l)—A partially altered feldspar showing topotactic growth of halloysite in the γ_p horizn. (j) = $\times 500$; (k) = $\times 5,000$; (l) = $\times 20,000$.

of the pseudomorph of the γ_p zone. At the low magnification, (d), the halloysite tubes appear to be clumped together and at very low magnification ($\times 200$), the surface of the pseudomorph appears hummocky. This suggests that the clumps may be originally amorphous globules. On higher magnifications, the halloysite tubes are distinct but in addition, there are a few spherules (arrow in Fig. 1e, f) which resemble those reported by Bondham (2).

The hummocky nature of the surface, seen to be common in all the cases of the pseudomorphs examined, is further observed in the γ_p zone (Fig. 1g, h, i). At higher magnifications (h, i) there is halloysite aggregated with a net globular form. These features could indicate that the halloysite crystallised via an amorphous gel phase.

The alteration of feldspars to halloysite is topotactic and when space permits, long tubes are formed. Figures 1j, k, and l show a side view of a partially altered feldspar. At low magnification the hummocky surface is evident. The halloysite tubes are about 20-μm long. Beneath the tubes the material is composed of small fragments of the original, unaltered feldspar. The intragranic voids in the unaltered parts is also coated with the amorphous gels similar to those in Fig. 1c. The unaltered fragments, with the halloysite and some amorphous gels, give the speckled appearance in thin sections.

In the rest of the γ_p zone, similar alteration of the feldspar to halloysite is observed, but proceeding to the γ_m zone, the frequency of such alteration decreases. Alteration to kaolinite commences in the upper part of the γ_p zone. It is not uncommon to find kaolinite flakes in between halloysite tubes on the surface of the pseudomorph here. This suggests a transitional case, possibly indicating that conditions were favorable for alterations to both types.

Figure 2a, b, c, shows a large grain of feldspar pseudomorphically altered to kaolinite in the upper part of the γ_p zone. The kaolinite crystals are rounded with smooth edges and do not resemble the hexagonal forms in TEM. This could be an effect of the magnification employed or because several crystals are adhering together. They do not resemble the macrokaolinite or dickite crystals reported by Bohor et al. (1). The pretreatments for TEM, especially the ultrasonic dispersion technique, may break up the booklets. Alteration to kaolinite appears to be confined to the smaller grains of feldspars in the upper part of the γ_p zone. The larger grains alter to gibbsite. This observation is not followed up due to the inherent difficulties in making more detailed studies.

The biotites alter to halloysite in the δ and γ_p zones, and in the γ_m zone they alter to a mixture of kaolinite and goethite. Figure 2d, e, f, show the accumulation of goethite on the edges and surfaces of a biotite flake. The morphology of the goethite aggregates is similar to those observed in petroplinthite (5) or in other biotites (10). SEM micrographs of gibbsite crystals adhering to the interlamellar surfaces are also available, but as it is uncertain if it is a product of alteration of the biotite or if the alumina is an external addition; they are not presented here. However, gibbsite is present in the interlamellar spaces in the weathering zone where the feldspars pseudomorphically alter to gibbsite. When the alteration product of feldspar is halloysite or kaolinite, biotites show a similar alteration. This common alteration phenomena indicates that the nature of the secondary mineral is determined purely by the microenvironment, irrespective of the composition of the primary mineral. The only exception to this is when the primary mineral cannot supply the major element of the secondary as in the case of iron for goethite.

In the γ_p zone, the feldspar also alters to gibbsite and the frequency of occurrence of such pseudomorphs increases to the surface. Figure 1, in Part I(6), shows the pseudomorphic alteration of the feldspars. The alignment of the gibbsite crystals is very distinct. Fig. 2g, h, i, shows this in the C_3 horizon and (Fig. 2j, k, l) in the C_2 horizon. The gibbsite crystals are euhedral and the crystal habit does not differ from those which are precipitated from the soil solution (13). There is, however, one difference between the two types of gibbsite aggregates. In the case of aggregates formed from feldspars, the crystals are relatively uniform in size, while in gibbsite aggregates formed from soil solution, the size of the crystals varies with the position in the matrix of the aggregates. Those adjoining voids are large and tend to develop a tabular habit while those within the matrix are fine. Examples of gibbsite formed from soil solution is given by Eswaran et al. (13).

Alterations by the same feldspar to give admixtures of gibbsite and halloysite or gibbsite and kaolinite are not encountered. However, in the same horizon, one feldspar grain may alter to kaolinite while another to gibbsite. This indicates that the presence of a void beside a feldspar grain may determine the course of alteration.

CONCLUSION

In order to evaluate the mineralogical changes in this profile, the following postulations are made:

1) The transformations of the feldspars to kaolinite or gibbsite is direct, without intermediary crystalline or amorphous phases. The transformation of feldspar to halloysite appears to be via an amorphous gel phase.
2) The formation of each secondary mineral requires a specific microenvironmental condition.
3) Irrespective of the primary weatherable mineral, the type of secondary mineral formed is a function of the microenvironment.
4) Once the secondary mineral is reduced to extinction size, further weathering results in its destruction; silica is lost in the groundwater and alumina precipitates as gibbsite.
5) The mineralogical changes with the depth of the weathering front are:
    a) During the formation of the present day solum, weathering produced kaolinite and some halloysite.
    b) As the weathering front reached the γ_m zone, supply of organic acids was sufficient to result in the alteration of feldspars to gibbsite with some associated kaolinisation of the feldspar.
    c) Deepening of the weathering front was not accompanied by a similar supply of organic acids and with the moisture saturated conditions and the slightly higher pH conditions prevailing, alteration to halloysite and amorphous materials took place.

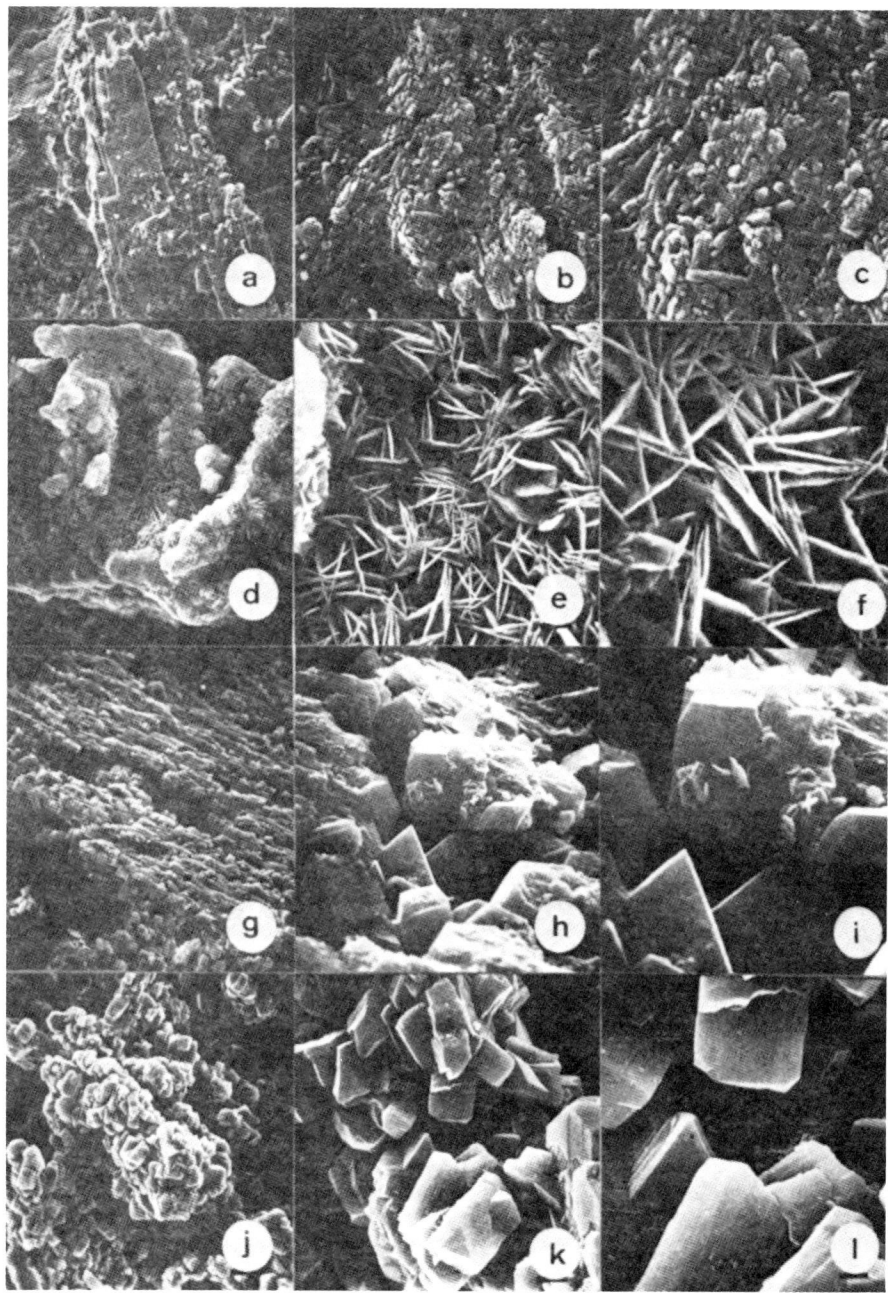

Fig. 2—SEM micrographs of handpicked feldspar and biotite grains or their pseudomorphs.
(a), (b), (c)—Alteration of feldspar to kaolinite. (a) = × 500; (b) = × 5,000; (c) = × 20,000.
(d), (e), (f)—Formation of goethite from biotite in the μ_m zone. (d) = × 1,000; (e) = × 10,000; (f) = × 20,000.
(g), (h), (i)—Gibbsite pseudomorphs after feldspar in γ_m zone. (g) = × 500; (h) = × 2,000; (i) = × 5,000.
(j), (k), (l)—Gibbsite pseudomorphs in C_2 horizon. (j) = × 800; (k) = × 2,000; (l) = × 5,000.

LITERATURE CITED

1. Bohor, B. F., and R. E. Hughes. 1971. Scanning electron microscopy of clays and clay minerals. Clays and Clay Minerals 19:49–54.
2. Bondham, J. 1969. Soxhlet extraction of albite. Proc. Int. Clay Conf. Jerusalem. 475–492 pp.
3. Delvigne, J. 1965. Pédogenése en zone tropicale. La formation des mineraux secondaires en milieu ferralitique. ORSTOM, Paris.
4. Delvigne, J., and Boulange. 1973. Micromorphologia des hydroxides d'aluminium dans la niveaux d'alteration et dans les bauxites. In: Soil Microscopy (Ed. Rutherford) Publ. Limestone Press, Kingston, Ontario. 665–681.
5. Eswaran, H. 1972. Morphology of allophane, imogolite and halloysite. Clay Minerals 9:281–285.
6. Eswaran, H., and Wong Chaw Bin. 1977. A study of a deep weathering profile on granite in Peninsular Malaysia: I. Physicochemical and micromorphological properties. Soil Sci. Soc. Amer. J. 42:144–149 (this issue).
7. Eswaran, H., and Wong Chaw Bin. 1977. A study of a deep weathering profile on granite in Peninsular Malaysia: II. Mineralogy of the clay, silt and sand. Soil Sci. Soc. Amer. J. 42:149–153 (this issue).
8. Eswaran, H., and Wong Chaw Bin. 1977. A study of deep weathering profile on granite in peninsular Malaysia: III. Alteration of feldspars. Soil Sci. Soc. Am. J. 42:154–158 (this issue).
9. Eswaran, H., and F. De Coninck. 1972a. Clay mineral formation and transformations in basaltic soils in tropical environments. Pedologie, XXI:181–210.
10. Eswaran, H., and Yeow Yew Heng. 1976. The weathering of biotite in a profile on gneiss in Malaysia. Geoderma 16:9–20.
11. Eswaran, H., and N. G. Raghu Mohan. 1973a. The microfabric of Petroplinthite. Soil Sci. Soil Sci. Soc. Amer. Proc. 37:79–82.
12. Eswaran, H., G. Stoops and P. D. Paepe. 1973b. A contribution to the study of soil formation on Isla Santa Cruz, Galapagos. Pedologie XXIII:100–122.
13. Eswaran, H., G. Stoops and C. Sys. 1977. The micromorphology of gibbsite forms in soils. J. Soil Sci. 28:136–143.
14. Parham, W. E. 1969. Formation of halloysite from feldspar. Clays and Clay Minerals. 17:13–22.
15. Setlow, L. W. and R. P. Karpourch. 1972. "Glacial" microtextures on quartz and heavy mineral sand grains from the littoral environment. J. Sedi. Petro. 42:864–875.
16. Wilson, M. J. 1975. Chemical weathering of some primary rock-forming minerals. Soil Sci. 119:349–355.
17. Wilson, M. J., D. C. Bain, and W. J. McHardy. 1971. Clay mineral formation in a deeply weathered boulder conglomerate in N. E. Scotland. Clays and Clay Minerals 19:345–352.

Part II
DISSOLUTION AND PRECIPITATION

Editor's Comments
on Papers 8 Through 12

8   **RAUSELL-COLOM et al.**
       Studies in the Artificial Weathering of Mica

9   **MORTLAND, LAWTON, and UEHARA**
       Alteration of Biotite to Vermiculite by Plant Growth

10   **KITTRICK**
       Mica-Derived Vermiculites as Unstable Intermediates

11   **VAN BREEMEN, MULDER and DRISCOLL**
       Acidification and Alkalinization of Soils

12   **EGGLETON and BUSECK**
       High Resolution Electron Microscopy of Feldspar Weathering

      The papers in Part I deal mainly with the major minerals of a soil profile, or some larger unit, where the mineralogy is previously unknown. While this approach certainly provides a good general view of soil mineral weathering, Part II emphasizes a smaller scale of weathering. In Part II, researchers started with known minerals and attempted to determine what happens to them during weathering, either in nature or in the laboratory. Although the weathering of primary minerals is essentially the process of dissolving them, this simple concept may involve much intriguing detail. For example, in the case of the mica minerals, the details of vermiculite formation have been emphasized to the point of obscuring the fact that vermiculite formation is merely a step in the process of dissolving mica.
      The mica minerals are important in soil fertility and produce an indentifiable weathering product in the laboratory within a convenient amount of time. For these reasons, the weathering of mica has been studied more than all other primary minerals combined. While Rausell-Colom, Sweatman, Wells, and Norrish (Paper 8) do not apply any new techniques, they do systematically apply appropriate techniques to a series of mica minerals with a suitable range of chemical properties. This approach permits the authors a view of mica conversion to vermiculite that is both relatively broad and detailed. They

have been particularly successful in relating solution conditions and mica type to the critical K concentration in solution (which if exceeded, stops K replacement in mica).

Laboratory studies like those in Paper 8, are necessary for a detailed understanding of the conversion of mica to vermiculite. The question remains as to what will happen when conditions more closely approximate those in the soil, especially when plants are present. Mortland, Lawton and Uehara (Paper 9) were the first to show that wheat plants can function as an effective K sink, resulting in an appreciable conversion of biotite to vermiculite over a period of four growing seasons. Their straightforward procedure and clear-cut results apparently have left little for subsequent investigators to expand upon.

In recent years it has become increasingly evident that, although vermiculite, smectite, and illite are all secondary 2:1 layer silicates, they are incorporated into the soil through different routes. The enthusiasm for investigating the conversion of some micas to vermiculite tended to endow vermiculite with a stability it does not possess. Therefore, it came as a surprise when Kittrick (Paper 10) showed that vermiculite derived from trioctahedral micas is merely an unstable intermediate in the dissolution process. After all, unstable intermediates are usually fast-forming precipitates that are small particles and are also poorly crystalline. Vermiculite occurs as relatively large crystalline particles, and it probably cannot precipitate from solution at all. Paper 10 shows that trioctahedral vermiculites have no solution stability area and are doomed to dissolve. Perhaps this finding will help to balance research efforts by focusing more attention on the mica dissolving process.

With few exceptions, the dissolution of soil minerals consumes a proton while the precipitation of soil minerals releases a proton. Individual processes that involve protons are studied intensively, whereas the proton cycle itself is not. Paper 11 provides the best place to discover the sources and sinks of protons. In it, Van Breemen, Mulder, and Driscoll show how the proton cycle can be understood in terms of a proton intensity factor (pH) and capacity factor (acid neutralizing capacity) and the coupling and decoupling of various reactions. Because protons are the single most important constituent in weathering, Paper 11 is an important contribution.

The work described in Paper 11 helps to establish the major chemical processes at the dissolving mineral surface, which sets the stage for a more detailed look at the surface mineralogy. The examination of small-scale mineralogy at mineral surfaces requires sophisticated instruments and training. An excellent example of this work is given in Paper 12 by Eggleton and Buseck. Using a high-resolution

Transmission electron microscope (TEM), they found that feldspar dissolution during natural weathering begins at structural defects and gradually invades the feldspar crystal. Amorphous ring-shaped structures, about 250 Å in diameter, precipitate initially, changing later into a crystalline phase with a 10 Å basal spacing. Eggleton and Buseck (Paper 12) show that minerals change with time in microenvironments as small as etchpits, reflecting mineral formation kinetics or changing equilibria due to changing chemical microenvironments. Just as Eswaran and Wong (Paper 7) set a high standard for investigating weathering in a soil profile, the authors of Paper 12 set a high standard for investigating the weathering of soil mineral surfaces. Both papers were successful by going to a finer scale than their predecessors. This change in scale permits an important observation. Notice that as we go from the particle size fractions given in Paper 2 by Jackson et al. in 1948 to the mineral surfaces given in Paper 7 by Eswaran and Wong in 1978, the variety of secondary minerals present decreases sharply, reflecting a sharp decrease in the variety of chemical microenvironments.

STUDIES IN THE ARTIFICIAL WEATHERING OF MICA

J. A. RAUSELL-COLOM, T. R. SWEATMAN,
C. B. WELLS and K. NORRISH

C.S.I.R.O., Division of Soils, Adelaide, South Australia

INTRODUCTION

THE MICA minerals are such a reservoir for potassium that a knowledge of the factors controlling their rate of release of potassium into the soil solution is basic to other work in fertility, pedogenesis and weathering.

Micas weather by loss of potassium and uptake of water. Biotite in some Scottish soils weathered to vermiculite by this process (Walker, 1949). The potassium was replaced by hydrated magnesium ions entering the interlayer regions which caused the 10 Å mica spacing to expand to the 14 Å of vermiculite. At the same time, ferrous iron in octahedral sites in the silicate layers was oxidized to ferric, causing a reduction in lattice charge.

Experimentally, the same processes have been reproduced by Barshad (1954), Mortland (1958) and Bassett (1959) using solutions of inorganic salts to provide cations (Mg, Ca, Na) to replace the potassium from micas. In all cases the potassium was readily removed from biotite and vermiculite but much less readily from muscovite and illite which had to be of clay size to yield significant results. Kinetic studies indicate that the reaction with inorganic salts may be a diffusion-controlled one (Mortland, 1958; Bassett, 1959; Sumner and Bolt, 1962); sodium tetraphenylboron, which forms a precipitate with potassium, has been used to maintain low potassium concentrations in the extracting solution (Scott, Hunziker and Hanway, 1960; Scott and Reed, 1962).

Organic alkylammonium ions have been used for replacing potassium from micas and are able to extract it from large single crystals of muscovite (Weiss, Mehler and Hofmann, 1956).

The progress of changes taking place in the interlamellar region can be seen and measured by the appearance of a boundary that moves inwards parallel to the edges of the crystal (Weiss and Hofmann, 1951; Weiss, Mehler and Hofmann, 1956). Hydration of vermiculites can cause it and so can ion exchange, in which case it coincides with the furthest limit of penetration of the substituting cation (Walker 1956, 1959, 1963). Walker also found that the amount of inwards movement of the boundary line was proportional to the square root of the duration of treatment: a relationship characteristic of diffusion-controlled processes.

The present detailed study of the factors involved seeks to relate the behaviour of each mica to its chemistry and structure. This preliminary paper reports findings on the general conditions governing the exchange of K with other cations and the changes that accompany it.

MATERIALS AND METHODS

Reagents

Laboratory Reagent and Analytical Reagent grade chemicals, usually chlorides, caused consistent and expected responses from the micas used in the early stages of the work. Other micas used later behaved inconsistently even between different batches of the same grade and manufacturer. Hence Merck* Guaranteed Reagent or Electronic Grade reagents have since been used wherever possible.

Specimens

Pure specimens of biotite, phlogopite, lepidolite and muscovite micas were used. Their chemical analyses and structural formulae are given in *Tables 1* and *2*.

Specimen preparation and treatment

Blocks guillotined from samples of micas were cleaved under water with a needle into thin (0·02–0·05 mm), uniform, crack-free sheets. Subsamples 1–2 × 10 mm or 2 × 2 mm were cut with sharp, stout razor blades. There was no evidence in the present studies that crystallographic direction influenced diffusion rate, and diffusion into natural crystal edges took place at the same rate as into cut edges. A flake and 5 ml. of salt solution were sealed into a soda glass ampoule and maintained at constant temperature. At appropriate intervals the ampoules were opened, the flake removed for measurement and non-destructive analysis, the solution renewed (except in equilibrium studies), the same flake returned and sealed once more into the ampoule to undergo further treatment.

For specific purposes some treatments were kept as low as 22°C, but most of the work was done at 100°C or 120°C to achieve acceptable reaction rates.

* Supplied by E. Merck, Darmstadt

Table 1

Chemical analyses and localities of mica specimens

	1	2	3	4	5	6	7	8	9	11	12	13	14
SiO_2	34·64	39·4	41·3	38·9	39·17	42·2	40·6	39·3	39·9	52·0	48·9	45·5	46·2
Al_2O_3	20·31	17·7	11·6	17·0	11·24	10·3	13·4	14·2	15·4	21·9	21·8	31·8	34·28
Fe_2O_3	1·39	—	0·98	1·26	1·86	1·43	0·84	1·20	—	0·11	0·76	3·90	2·29
FeO	20·76	8·7	8·3	4·70	16·58	8·0	2·65	2·15	3·57	0·00	1·20	0·61	—
TiO_2	3·01	3·25	0·61	0·98	2·33	1·59	0·91	0·62	0·71	0·09	0·20	0·48	—
MnO	0·24	0·08	0·05	0·06	0·89	0·37	0·04	0·03	0·04	0·56	2·90	0·03	—
MgO	5·95	19·1	21·8	21·1	13·51	20·9	26·7	25·5	25·0	0·03	0·04	1·00	0·60
Li_2O	0·05	0·04	0·08	0·04	0·18	0·20	0·04	—	0·02	5·80	5·10	—	—
CaO	0·04	0·02	0·12	0·23	0·20	0·29	0·18	0·19	0·02	0·08	0·15	0·10	—
Na_2O	0·25	0·36	0·12	0·72	0·62	0·50	0·35	0·22	0·20	0·39	0·42	0·56	0·44
K_2O	9·37	9·64	9·70	9·90	9·29	9·40	10·2	9·65	10·03	9·95	9·95	10·00	10·91
Rb_2O	0·10	—	—	—	0·05	—	—	—	—	—	—	—	—
BaO	0·00	—	—	—	0·05	—	—	1·05	—	—	—	—	—
F	0·24	0·98	2·7	1·7	3·46	4·4	3·9	4·15	3·72	8·75	8·45	0·20	—
Cl	0·06	—	0·68	0·03	0·12	0·05	0·07	—	—	—	—	—	—
P_2O_5	—	—	—	—	—	—	—	0·01	—	0·01	0·01	0·04	—
H_2O^+	3·44	—	—	—	1·64	—	—	3·25	—	1·50	1·50	4·85	5·0
H_2O^-	0·14	—	—	—	0·09	—	—	0·35	—	0·52	0·47	0·57	—
Total	99·99				101·28			101·87		101·69	101·85	99·6	99·72
less $O \equiv F$	0·11				1·48			1·77		3·68	3·55	0·08	
	99·88				99·80			100·1		98·0	98·3	99·5	

Sample 1 Brown biotite from Quebec, Canada; AUGD 18059 (Analyst: C. O. Ingermalls, Rock Laboratory, University of Minnesota)

2 Metallic lustred phlogopite, locality unknown (Analyst: K. Norrish by x-ray spectrography; all Fe expressed as FeO)

3 Olive-green phlogopite from Pierrepont, New York (Analyst: H. W. Sears, Australian Mineral Development Laboratories, Adelaide)

4 Brown phlogopite from North Burgess, Ontario, Canada; AUGD 18060 (analyst as sample 3)

5 Dark green biotite from North Burgess, Ontario, Canada; AUGD 18058 (Analyst: E. H. Oslund, Rock Laboratory, University of Minnesota)

6 Light brown phlogopite from Mannum, South Australia; AUGD 15457 (analyst as sample 3)

7 Light brown phlogopite from Lanark County, Ontario; AUGD 114 (analyst as sample 3)

8 Clear pale brown phlogopite, near Hart's Range, Northern Territory (analyst as sample 3)

9 Light brown phlogopite. Locality unknown (analyst as sample 2)

10 Clear, colourless, synthetic fluor-phlogopite

11 Lilac-coloured lepidolite, Grosmont, Western Australia (analyst as sample 3)

12 Lepidolite, Londonderry, Western Australia (analyst as sample 3)

13 Muscovite, Barrier Range, New South Wales (analyst as sample 3)

14 Muscovite from Spotted Tiger Mine, Central Australia (Analysis, given by Radoslovich (1960), by R. D. Bond, C.S.I.R.O., Division of Soils, Adelaide)

Note: AUGD numbers refer to the minerals collection in the Adelaide University Geology Department
Analyses of micas 1 and 5 quoted by courtesy of the Geological Society of America

Table 2
Structural formulae of micas
Calculated on a water-free basis (O = 11)

Sample	Tetrahedral			Octahedral								Interlayer						
	Si	Al	Fe³⁺	Al	Fe³⁺	Fe²⁺	Ti	Mn²⁺	Mg	Li	Σ	Ca	Na	K	Rb	Ba	Σ	F	Cl	Σ

Sample	Si	Al	Fe³⁺	Al	Fe³⁺	Fe²⁺	Ti	Mn²⁺	Mg	Li	Σ	Ca	Na	K	Rb	Ba	Σ	F	Cl	Σ
1	2·665	1·335		0·506	0·080	1·335	0·174	0·016	0·682	0·016	2·809	0·003	0·037	0·920	0·005		0·964	0·058	0·008	0·066
2	2·760	1·240		0·221		0·510	0·171	0·003	1·990	0·011	2·906	0·001	0·049	0·860			0·909	0·217		0·217
3	3·038	0·962		0·200	0·054	0·498	0·033	0·003	2·332	0·023	2·960	0·009	0·016	0·889			0·914	0·612	0·083	0·695
4	2·789	1·211		0·225	0·068	0·282	0·053	0·004	2·253	0·011	2·896	0·018	0·100	0·898			1·016	0·385	0·008	0·393
5	2·960	1·040		0·002	0·106	1·047	0·133	0·057	1·521	0·055	2·919	0·016	0·091	0·896			1·003	0·826	0·015	0·841
6	3·055	0·878	0·067		0·011	0·484	0·087	0·023	2·254	0·058	2·917	0·016	0·070	0·868			0·961	1·006	0·006	1·012
7	2·856	1·111	0·033		0·012	0·156	0·048	0·003	2·798	0·011	3·027	0·023	0·048	0·915			0·977	0·867	0·008	0·875
8	2·830	1·170		0·034	0·065	0·129	0·034	0·002	2·735		2·999	0·014	0·030	0·887		0·030	0·962	0·945		0·945
9	2·835	1·165		0·122		0·213	0·038		2·640	0·004	3·017	0·015	0·027	0·908			0·935	0·835		0·835
10*	3·00	1·00							3·00		3·00			1·00			1·00	2·00		2·00
11	3·532	0·468		1·285	0·006		0·005	0·032	0·003	1·584	2·915	0·006	0·051	0·862			0·919	1·880		1·88
12	3·403	0·597		1·190	0·040	0·070	0·010	0·171	0·004	1·426	2·911	0·011	0·057	0·883			0·951	1·857		1·857
13	3·110	0·890		1·670	0·200	0·035	0·025	0·002	1·018		3·950	0·007	0·074	0·872			0·953	0·043		0·043
14	3·110	0·890		1·885	0·118				0·061		2·014		0·058	0·936			0·994	0	0	0

* Theoretical formula

Optical observations

Visual evidence of the potassium-cation replacement front was seen through a suitably adjusted petrographic microscope (Walker, 1959). In some cases it appeared as a clearly defined line at an accurately

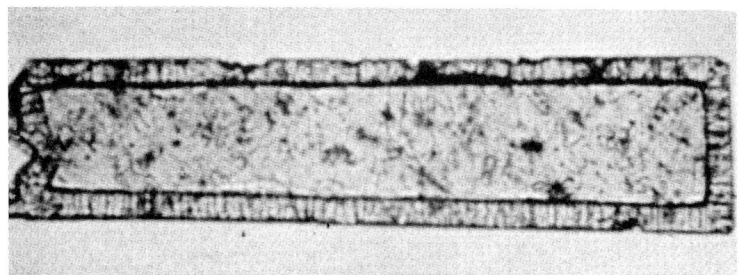

Plate 1. Replacement boundary line parallel to, and striations normal to, the edge of a flake of mica 4 after treatment with 1 N $SrCl_2$ at 120°C; (a) transmitted light

measurable distance from a scratch or edge of the flake (*Plates 1 and 2*) or as a uniform peripheral band of accurately measurable width (*Plates 3a–c*). In other cases there were multiple or diffuse boundaries (*Plates 4 and 5*) that could not be accurately located. The observed line boundary did not have the characteristics of a Becke line (*Plates 1a and 1b*). Although the position of the optical boundary could often be measured accurately, it could be altered a little

Plate 1(b). Reflected light

STUDIES IN THE ARTIFICIAL WEATHERING OF MICA 45

Plate 2. Replacement boundary lines parallel to, and striations normal to, the edge and to surface scratches of a flake of mica 1 after treatment with $1 \text{ N CaCl}_2 + 1 \times 10^{-4} \text{ N KCl}$; *transmitted light*

Plate 3. Replacement band around the periphery of a flake of mica 1 after treatment in $1 \text{ N SrCl}_2 + 1 \times 10^{-3} \text{ N KCl}$ *at 120°C, compared with no visible change after treatment in* $1 \text{ N SrCl}_2 + 1 \times 10^{-2} \text{ N KCl}$ *at 120°C;* (a) *transmitted light*

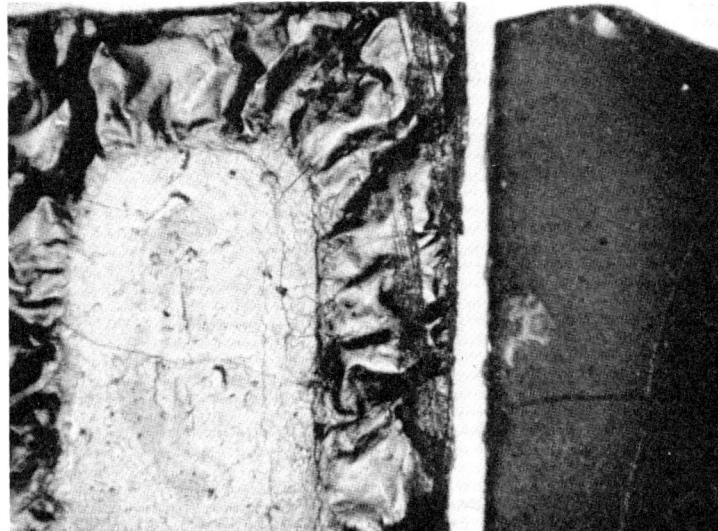

Plate 3b. Reflected light

($\leqslant 20$ μm) by using different conditions of viewing. It is therefore necessary, having found the most suitable system for each mica, to use the same optics at each measurement. Any other visible microscopic changes in the flakes were recorded as they were seen (*Plate 6*).

Plate 3c. Transverse section across a flake from Plate 3a showing expansion of replacement band; reflected light

X-ray diffraction measurements

X-ray diffraction measurements with CoKα radiation were made on flakes immersed in a drop of their replacing salt solution and covered by a thin polythene sheet to prevent evaporation. Diffraction traces recorded the presence and intensity of the 10 Å mica peak and of other spacings ascribable to the entry of potassium-replacing cations into the interlayer regions. The intensities at these spacings

STUDIES IN THE ARTIFICIAL WEATHERING OF MICA 47

Plate 4. Multiple replacement boundary lines of a flake of mica 7 after treatment in 1 N $CaCl_2$ at 100°C; transmitted light (through epiobjective)

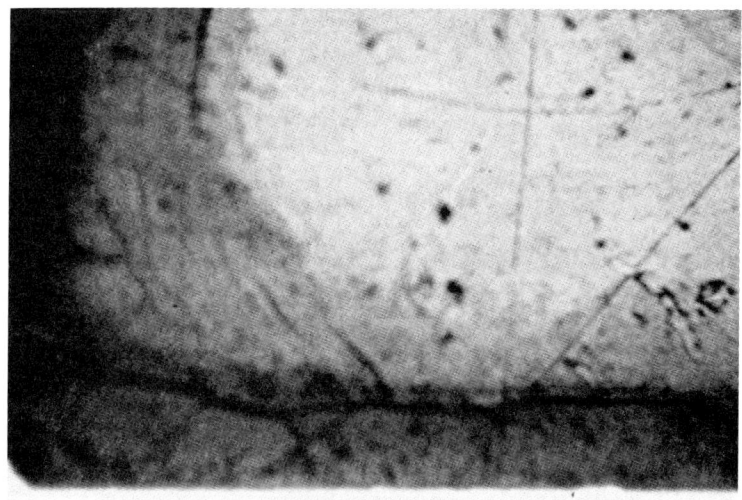

Plate 5. Diffuse replacement band around the periphery of a flake of mica 1 after treatment with 1×10^{-4} N $SrCl_2$ at 120°C; (a) transmitted light (through epiobjective)

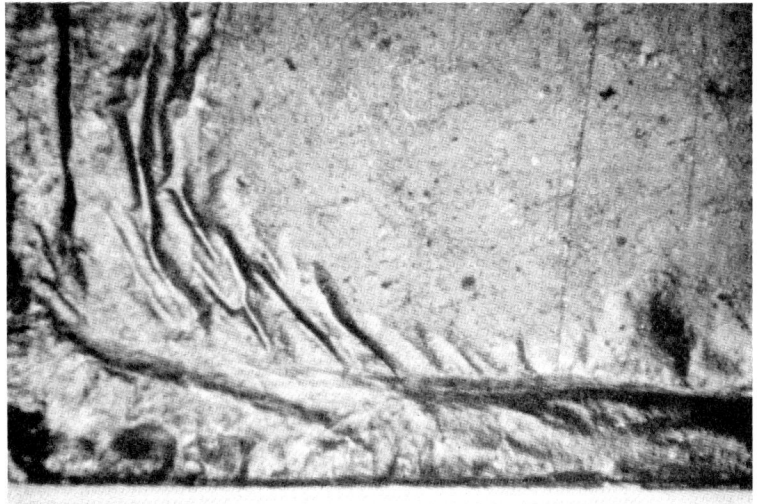

Plate 5(b). *Reflected light*

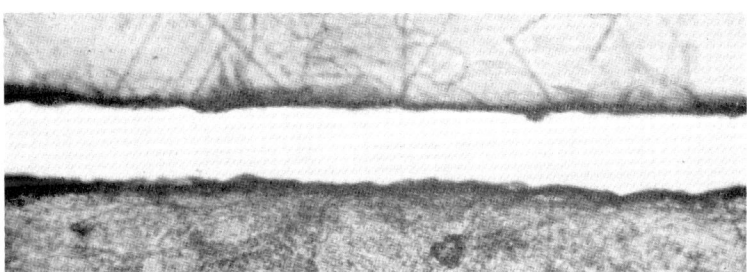

Plate 6. Surface changes in a flake of mica 11 after treatment in 5×10^{-1} N $MgCl_2$ at 100°C, *compared with no visible change in a control flake in distilled water; transmitted light*

relative to the 10 Å intensity were a satisfactory semi-quantitative measure of the extent of replacement. The method is particularly useful for measurements in the very early stages of the reaction, before the visible boundary line has had time to move beyond the optical disturbances associated with the cut edges of the flakes. The b axes of treated and untreated micas were measured using a 19 cm diam. powder camera.

X-ray spectrographic analyses

The amounts of potassium and of substituting cation (Ca or Sr) and associated anion (Cl) in a flake at different times during the reaction were followed by using x-ray spectrographic analysis which, being a non-destructive method, preserved the flake for other kinds of analyses and for a further period of treatment. Mica flakes 0·02 mm thick are effectively infinitely thick with respect to the $K\alpha$ radiation of potassium, calcium and chlorine, and for small flakes carefully positioned in the sample holder, the fluorescent intensity from these elements was established to be proportional to the product of their *concentration* in the flake and the *area* of the flake. For the $K\alpha$ radiation of strontium, the flakes are infinitely thin and the fluorescent intensity is proportional to the *total mass* of strontium in the sample.

The amounts of potassium appearing in the salt solutions as a result of the reaction were measured by x-ray spectrography. Discs of hydrogen-saturated cation exchange resin paper* were soaked in 2 ml. of the solution for an hour, blotted between filter papers and dried in a stream of warm air. The method was calibrated using similar salt solutions containing known additions of potassium. Potassium concentration in the solution is linearly related to the fluorescent intensity from the resin paper. In dilute solutions ($\sim 0\cdot 01$ N) all the salts used gave very similar calibrations, but in strong solutions the slope of the line differed for each salt.

Flame spectrographic analyses

In some cases the solutions were analysed for potassium using the 7,680 Å emission line in an E.E.L. flame photometer. All the 1 N salt solutions were diluted 1:10, those of Na, Li, Mg with H_2O, Ba with $H_2O + H_2SO_4$ and of Sr and Ca with 0·1 M $AlCl_3$. Sulphuric acid reduces the concentration of barium in solution; aluminium chloride suppresses interference from the calcium and strontium oxide bands.

Results obtained with this method agreed well with the x-ray spectrographic determination.

Electron probe x-ray microanalyses

The concentration gradients of potassium and the replacing ion across the mica flakes were studied by making electron microprobe traverses†. Rough surface topography (*Plates 1b, 3b, 5b*) could not be entirely eliminated by polishing or cleaving and caused such

---

* 'Amberlite' ion-exchange papers, grade SA–2, supplied by Rohm & Haas, Philadelphia 5, Pa.

† This work was done in conjunction with Dr. J. Lovering and Mr. K. Williams, Australian National University, Canberra.

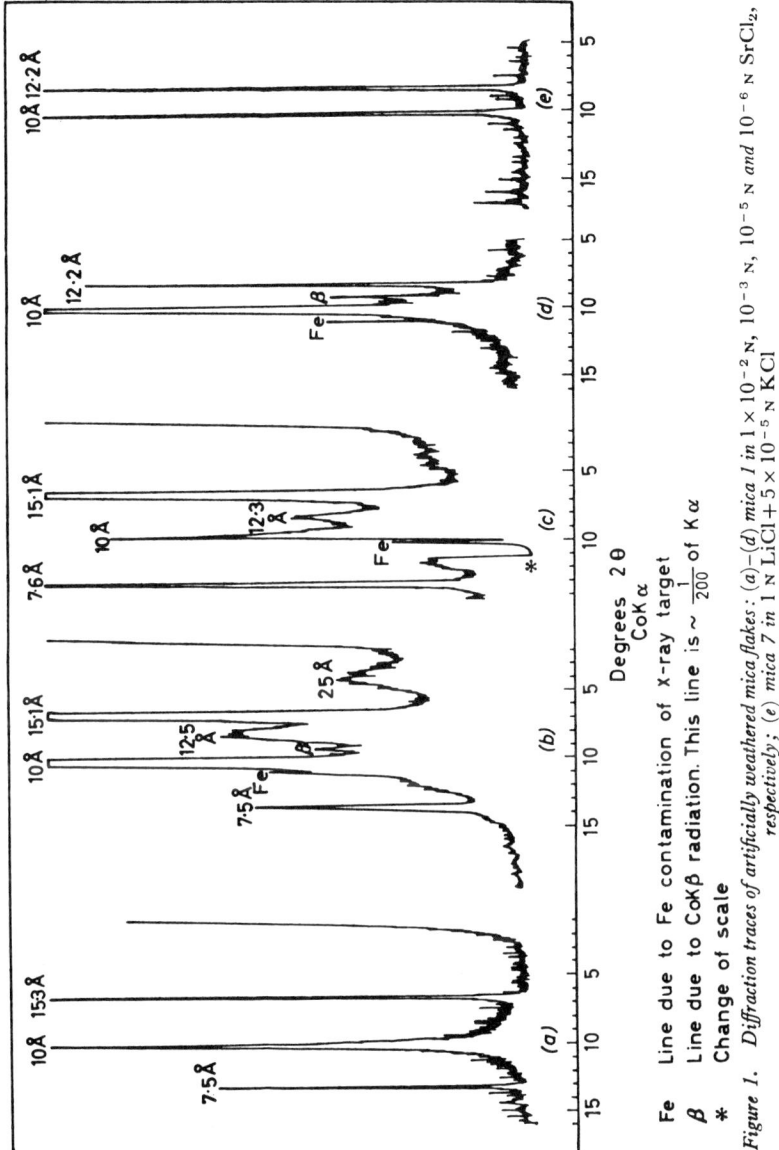

Figure 1. Diffraction traces of artificially weathered mica flakes: (a)–(d) mica 1 in 1×10^{-2} N, 10^{-3} N, 10^{-5} N and 10^{-6} N $SrCl_2$, respectively; (e) mica 7 in 1 N LiCl + 5×10^{-5} N KCl

erratic results for the area outside the optical boundary that quantitative results could not be obtained for this region. However, the method provided satisfactory results in the immediate vicinity of the replacement front.

RESULTS

Dependence of diffusion on salt concentration

In all the salt solutions, i.e. chlorides of Li, Na, Mg, Ca, Sr and Ba, the cations exchanged readily with the K of natural phlogopites and biotites, but there was no optical or diffraction evidence of replacement in any experiment with lepidolites, muscovites or synthetic fluorphlogopite, even after six months at 120°C. The salt solutions usually had a pH within the range 5–6, and there were no appreciable changes as a result of the reaction with mica.

Replacement of K in micas causes an additional sharp line at 12–15 Å to appear on the diffraction traces (*Figure 1a*). The precise spacing depends on the cation involved (*Table 3*) but not on solution

Table 3

Interlayer spacings (Å) in mica 1 after replacement of K by other ions

Temperature °C	1 N solution of chloride of					
	Li	Na	Mg	Ca	Sr	Ba
20	15·42	14·65	14·30	14·97	15·40	12·40
100		14·80		14·95	15·13	12·33

concentrations over a wide range nor on mica species. With small amounts of K in solution or with very low salt concentrations, the diffraction traces are much less predictable (*Figure 1b, c, e*) and are often accompanied by diffuseness and irregularity of the optical boundary. In some cases an ion causing a 15 Å spacing in strong solutions produces a 12 Å spacing when exchange takes place in very dilute solutions (Figure *1d*). In other cases a tailing of the 10 Å peak towards higher spacings is the only diffraction evidence of exchange. The diffraction trace of mica 1 in reaction with 1×10^{-3} N $SrCl_2$ (*Figure 1b*) shows the normal sharp lines at 10 Å and 14 Å and, in addition, peaks at 25 Å and 12 Å, indicating the presence of a mixed layer phase with partial ordering.

52 EXPERIMENTAL PEDOLOGY

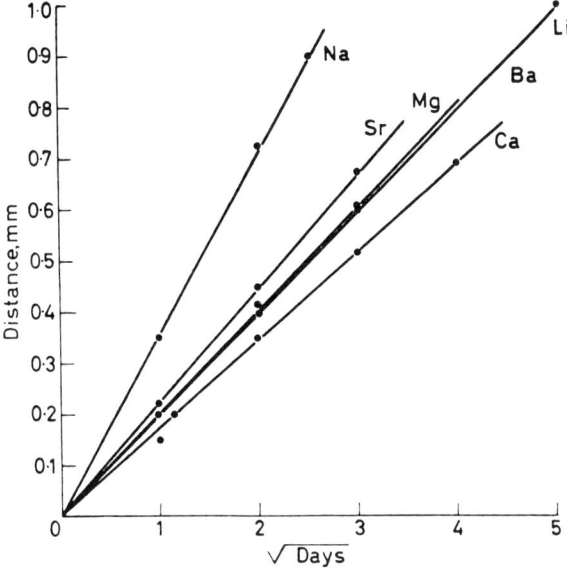

Figure 2. Diffusion of cations into mica 1 at 122°C

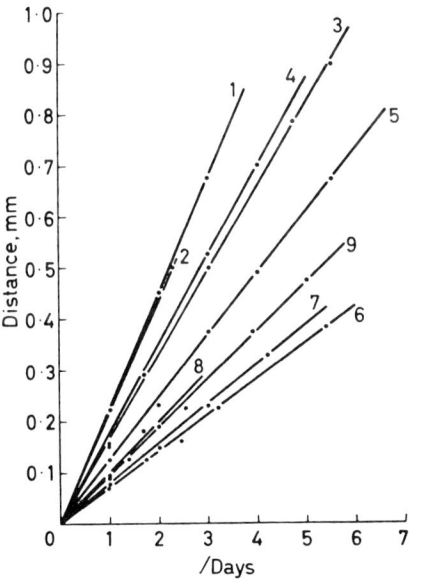

Figure 3. Diffusion of Sr into different micas at 122°C

103

STUDIES IN THE ARTIFICIAL WEATHERING OF MICA

Table 4

Effect of solution concentration on diffusion coefficient
($cm^2/sec \times 10^{10}$)

Mica	SrCl$_2$, 122°C					MgCl$_2$, 100°C			
	1	10^{-1}	10^{-2}	10^{-3}	10^{-5}	5×10^{-1}	5×10^{-2}	5×10^{-3}	5×10^{-4}
9	3·3	—	—	—	—	1·0	D*	0	0
8	4·0	3·1	0·9	0	0	1·2	D*	0	0
7	4·4	2·6	2·2	—	0	1·2	D*	0	0
6	4·2	2·1	0·9	0	0	6·3	0·2	—	0
5	12·5	7·5	4·8	0	0	2·5	0·6	D*	0
4	2·9	1·9	0·8	0·3	—	20·7	10·0	1·2	D*
2	51	35	12	0·8	0·5	14·4	3·5	1·2	0·6
1†	51	—	51	—	—	31	13	D*	D*

D* diffusion detectable but too slow to measure
† diffusion detectable in mica 1 using 10^{-5} and 10^{-6} N SrCl$_2$

The distances traversed by the optical boundary with the various salts and micas are linearly related to the square root of time (*Figures 2* and *3*). *Figure 4* shows nevertheless that salt concentration affects the *rate* of diffusion. This is expressed in a different way in *Table 4* which gives the diffusion coefficients (distance^2/time) of most

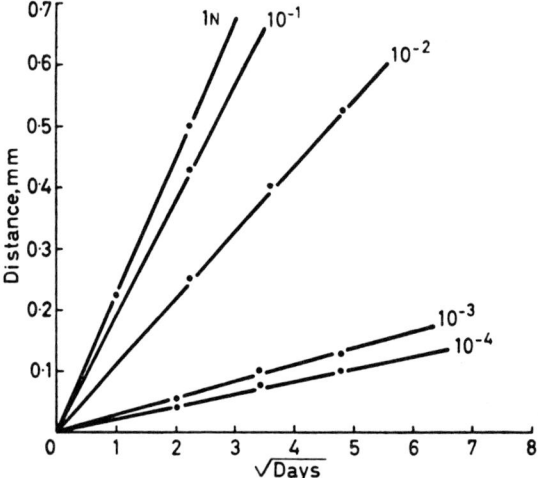

Figure 4. Diffusion of Sr into mica 2 as a function of SrCl$_2$ concentration

of the micas for several concentrations of Mg and SrCl$_2$. In most cases a tenfold decrease in solution concentration causes about a twofold reduction of the diffusion coefficient, but only down to a certain threshold dilution beyond which diffusion ceases. Different micas stop at different dilutions, and the value is further affected by the cation used and by temperature: at higher temperatures, exchange will continue in more dilute solutions.

The diffusion rate also falls away in the stronger solutions. Between 1 N and 3 N it changes little, but at higher concentrations

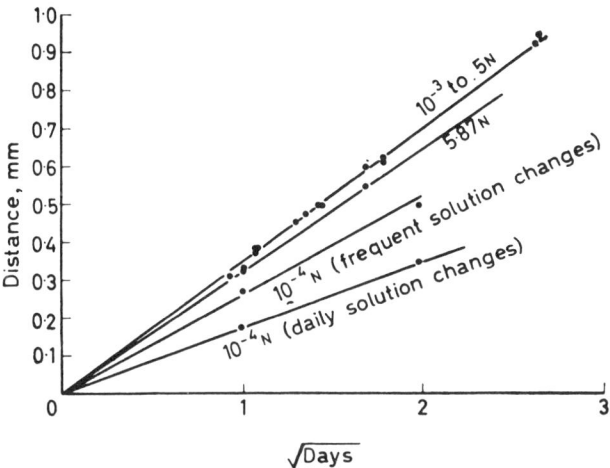

Figure 5. Diffusion of Sr into mica 1 as a function of SrCl$_2$ concentration

the rate is reduced, and at the same time appreciable amounts of chlorine are detected within the mica. Mica 1, for instance, after exchange in 3 N SrCl$_2$ contained 45 m.e. Cl/100 g compared with the 2 m.e./100 g found in the untreated mica and after exchange in 1×10^{-3} N and 1 N SrCl$_2$.

In most cases mica 1 behaved just as others did, but its reaction to strontium treatment was completely different: there was no appreciable change in diffusion rate for Sr concentrations from 5 N to 1×10^{-3} N (*Figure 5*).

Dependence of diffusion on potassium concentration of the external solution

Small amounts of potassium in the solution retard or prevent the exchange reaction, and it was therefore necessary to test whether K released from the flakes was limiting diffusion, and whether K contributed from the glass might do so. Experiments at 122°C with different-sized mica 4 flakes (0·1 and 2·5 mg) and 1 N $SrCl_2$ in continuously agitated soda and Pyrex glass and polypropylene-lined tubes all gave diffusion coefficients of $3·1 \times 10^{-9}$ cm²/sec. Under the same conditions in 0·01 N $SrCl_2$, the value in all cases was of the order of $8·4 \times 10^{-10}$ cm²/sec, and the K concentration remained less than 2×10^{-5} N; a lower limit of K in solution is set by release from the glass giving concentrations of 2×10^{-5} N.

Very small amounts of potassium deliberately added to solutions had no measurable effect, but larger additions caused the diffusion

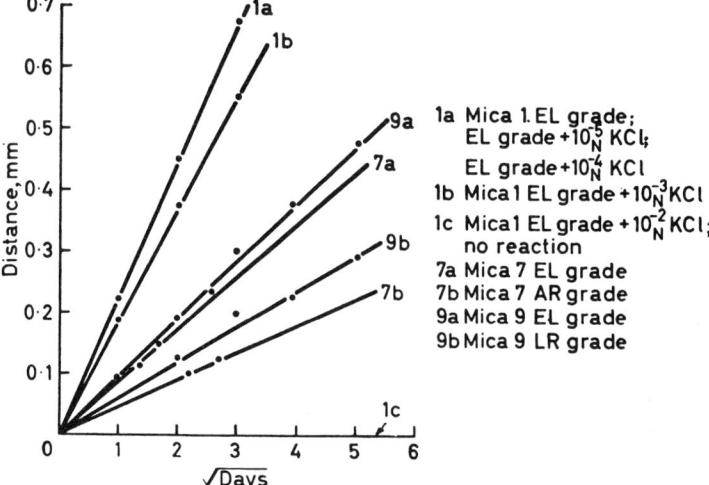

Figure 6. *Effect of K additions and reagent quality on diffusion rate in 1 N $SrCl_2$ at 122°C*

rate to become lower until, at a critical K concentration, exchange was inhibited (*Figure 6*). A first estimate of the critical concentration was obtained by inspecting micas optically after reaction with solutions containing serial double dilutions of K. It was more critically assessed (*Table 5*) by reacting an excess of chopped-up mica for long periods in salt solutions, until K release from the mica had brought the concentration in solution up to the critical value. Mortland (1958) has shown that K content under the above conditions increases slowly for a long time, so that equilibrium is unlikely to be attained over short periods. However, experiments incorporating a number of times and weights of micas show that the data in *Table 5* correspond to the critical K values. They also fall within the limits determined by the potassium addition method mentioned above. The results

Table 5
Concentration (m.e./l.) of K which stops diffusion at 122°C

Mica	1 N Chloride solutions					
	Li	Na	Mg	Ca	Sr	Ba
9	0·14	0·35	0·60	0·18	0·21	2·4
8	0·16	0·37	0·75	0·42	0·23	2·4
7	0·11	0·40	0·78	0·23	0·40	2·9
6	0·26	0·62	0·86	0·42	0·48	2·3
5	0·34	0·82	1·3	0·57	0·60	3·2
4	0·60	0·95	2·03	1·5	1·6	4·6
3	0·80	1·45	2·6	1·8	1·6	5·0
1	0·92	1·48	3·0	1·80	1·4	6·6
	0·01 N chloride solutions					
8		0·06		0·07		0·20
4		0·08		0·26		0·47
1		0·09		0·43		0·97

show that in some cases 4 p.p.m. of potassium (1×10^{-4} N) prevents exchange in strong salt solutions. In more dilute solutions the critical K concentration also drops, but not as fast as the dilution rate. Thus the 2×10^{-5} N K contribution from the glass ampoules probably assumes importance as a critical concentration for salt solutions of 0·01 N and lower. This was shown to be so using sodium tetraphenylboron, NaTB, to maintain very low K concentrations around the flake. Because the reagent decomposed at the temperatures used, the experiment was not continued beyond the first day's reading. Diffraction traces of the micas treated with 0·1 N and 0·01 N Na-tetraphenylboron showed 12·4 Å spacings, whereas those treated with the same concentrations of NaCl had sharp 14 Å spacing peaks. Heating the flakes to 100°C brought the NaCl-treated ones back to a sharp 10 Å line whereas the NaTB gave a 10 Å line that had a tail or band extending to higher spacings.

Table 5 shows that the K concentration required to stop replacement varies with both the mica and the replacing cation. Li was the ion, and sample 7 the mica most sensitive to small amounts of K in solution. Experiments with mica 7 in 1 N LiCl initially showed no reaction. Repetition, with higher-quality LiCl, gave no observable replacement at 90°C and only very slow replacement at 122°C ($D = 6 \times 10^{-11}$ cm²/sec). An x-ray diffraction trace of the latter specimen showed a sharp 12·1 Å spacing (see *Figure 1e*), in contrast to the normal 15·4 Å spacing observed with Li as replacing cation. This behaviour suggested that the K impurity in the salt (5×10^{-5} N) was retarding diffusion in 1 N solutions.

Mica 7 treated with a Laboratory Reagent grade 1 N $SrCl_2$ showed reduced diffusion rates at 122°C, and diffusion ceased at 90°C. These and the results with LiCl show that the K required to prevent diffusion depends on temperature, the concentration decreasing with temperature.

Dependence of diffusion on temperature

Under any particular conditions diffusion coefficients change little in the range 1–3 N. Measurements of these values at different temperatures in 1 N solutions (*Figure 7*) show that the rate increases

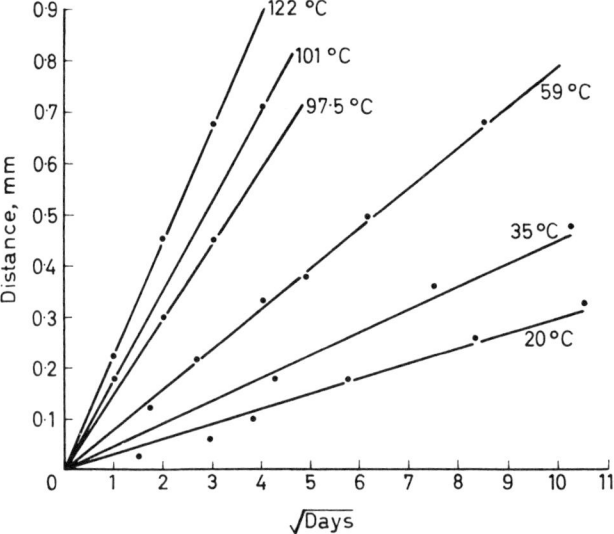

Figure 7. *Effect of temperature on the diffusion rate of Sr into mica 1 from* 1 N $SrCl_2$

with temperature. Plots of these data after transformation to ln D and $1/T$ (temp. °K) give a linear relationship from which it is possible to calculate the constant A and the activation energy, E, in the Arrhenius equation

$$D = Ae^{-E/RT}$$

A summary of these values for a range of ions and micas is presented in *Table 6*.

Table 6

Effect of replacing cation and mica species on diffusion

Mica	Replacing ion	E kcal/mole	A cm^2/sec	ΔS e.u./mole	ΔF kcal/mole
1	Li*	24·9	$2·6 \times 10^5$	30·3	15·1
	Li†	9·3	$6·1 \times 10^{-4}$	$-9·2$	12·3
	Li‡	>40	—	—	—
	Na	10·4	$7·8 \times 10^{-3}$	$-4·1$	11·8
	Mg*	9·7	$9·3 \times 10^{-4}$	$-8·4$	12·4
	Ca	9·9	$9·4 \times 10^{-4}$	$-8·5$	12·6
	Sr	9·5	$1·1 \times 10^{-3}$	$-8·6$	12·3
	Ba*	10·7	$4·4 \times 10^{-3}$	$-5·4$	12·4
3	Sr	11·4	$5·9 \times 10^{-3}$	$-4·8$	13·0
4	Ca*	11·9	$2·7 \times 10^{-2}$	$-3·0$	12·9
	Sr	12·5	$2·6 \times 10^{-2}$	$-1·8$	13·1
	Ba*	10·9	$4·9 \times 10^{-3}$	$-5·1$	12·6
5	Ca	9·1	$3·0 \times 10^{-4}$	$-10·6$	12·6
	Sr	10·8	$1·5 \times 10^{-3}$	$-7·6$	13·2
	Ba*	11·1	$1·2 \times 10^{-3}$	$-7·9$	13·6
6	Sr	26·3	$1·9 \times 10^5$	29·7	16·7
7	Sr	22·1	$9·6 \times 10^2$	19·1	15·9
	Sr†	10·8	$1·1 \times 10^{-3}$	$-8·1$	13·4

* Errors in these cases are possibly large since the plot of ln D versus $1/T$ did not yield a straight line
† Preliminary re-determination using higher quality salts
‡ Determination with a salt of poorer than normal quality

Dependence of diffusion on oxidation-reduction conditions

Cation exchange measurements (see later) suggest considerable oxidation of ferrous iron during K replacement but give no indication whether it has a controlling influence on the replacement rate. Mica 1, the sample highest in ferrous iron, reacted with the same diffusion coefficient of $1·30 \times 10^{-8}$ cm^2/sec towards (a) 1N NaCl at 122°C, (b) the same solution with H_2O_2 added, (c) the same solution with air removed and replaced by nitrogen, (d) as for (c) but with hydrogen-saturated sponge palladium added.

Concentration gradients

The concentration gradients of K and replacing cation from the edge of an altered flake inwards towards its unaltered centre were found by analysing, by x-ray fluorescence, small strips (about 0·2 mm) serially sectioned from a flake of mica treated with 1 N $SrCl_2$ (*Figure 8*). The sum of the K and Sr, as m.e. per cent, is found to be less than the original K content of the mica. Sr is detectable in the unaltered area well in from the optical boundary. These two observations are considered to be due to surface alterations not

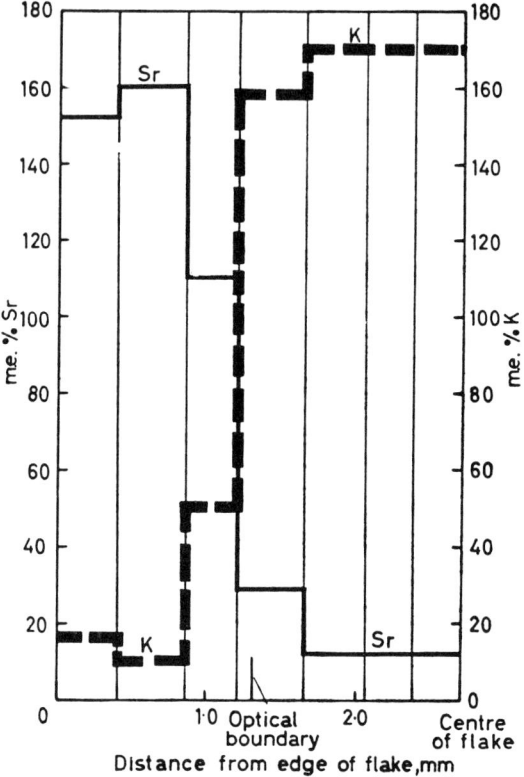

Figure 8. Concentration gradients of K and Sr in mica 1 treated for 32 days in 1 N $SrCl_2$ at 122°C; results obtained by analysing serial sections of a mica flake cut parallel to its edge

associated directly with the diffusion process. The slight increase in K, and corresponding decrease in Sr, at the edge of the flake was unexpected. In magnitude it is greater than experimental error, and a similar reversal of gradient appeared in results obtained from duplicate samples of mica 1 and from micas 4 and 5. The same increase in K at the flake edge was evident in some analyses made with an electron probe micro-analyser. It was not practical to cut and

60　EXPERIMENTAL PEDOLOGY

analyse sections of flakes smaller than those shown in *Figure 8*. Attempts were made to alter much larger flakes for a longer period so that sections would give the concentration gradient in relatively more detail. However, the large flakes began to disintegrate physically at the edges after long periods of reaction.

Figure 9 shows the result of an electron probe traverse across the optical boundary of mica 5 which had been exchanged with Ca. Attempts to continue the analysis beyond the boundary into the

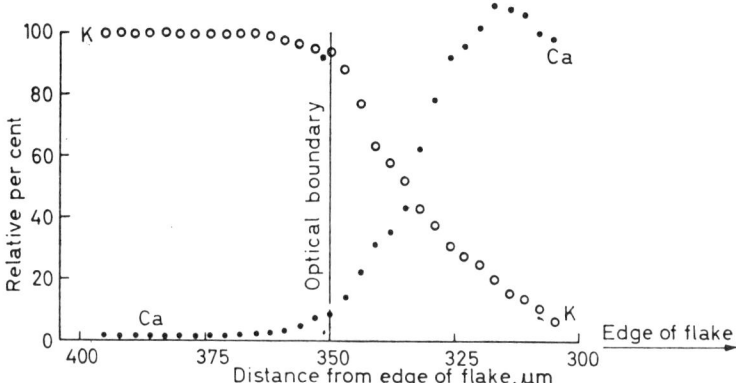

Figure 9. Traverse with an electron probe microanalyser across the optical boundary (3 μm steps) of mica 5 exchanged with Ca

altered mica were not successful owing to rough surface topography. The slight inflections in both curves at ~340 μm are considered significant experimentally since they occurred in traverses made on several different mica flakes.

Surface charge and structural changes

Surface alteration of the micas after long reaction periods imparts a dull lustre to biotite and phlogopite flakes and a white opacity to lepidolite and muscovite flakes. Microscopically they become pitted (*Plate 6*). The evidence which follows concerns the nature of the alterations causing these effects.

The progress of the exchange was generally followed, using x-ray fluorescence analyses, by measuring the decreasing amounts of K in each flake. This, as an estimate of the fraction of the flake remaining unaltered, was usually lower than was indicated by optical observations of the area inside the boundary (*Table 7*). That the discrepancy was not due to penetration of the replacing ion beyond the boundary is shown by x-ray fluorescence and electron microprobe results;

there was an insufficient amount of the replacing ion in the area inside the optical boundary. The discrepancy was due to the method of analysis adopted, for the potassium Kα radiation used in making the analyses is highly absorbed by mica, so that the method measures K in only the surface 5 µm or so. Such a discrepancy could thus arise from a surface film 1–2 µm thick completely free of K, or a slightly thicker film partially depleted of K. When K remaining in the flake was measured on a surface freshly exposed by cleaving, the estimates of replacement agreed well with those obtained from optical measurements.

Table 7

Apparent disagreement between estimates of weathering based on optical observations and K content of mica

Mica	Distance of penetration of optical boundary	Relative to untreated flake	
	mm	Per cent area within boundary	Per cent K remaining
1 (Na)	0·11	86	90
	0·32	66	57
	0·52	48	43
	0·67	39	29
4 (Sr)	0·15	86	82
	0·25	78	75
	0·27	75	70
	0·52	55	43
	0·70	42	28

X-ray diffraction results also indicate the nature of the surface alterations. The visible changes in the surface were accompanied by the emergence of a sharp 7·1 Å spacing line that persisted when the 14–15 Å spacing of the expanded material was shifted by ion exchange or heat treatment; it also appeared on the diffraction patterns of lepidolite and muscovite flakes which had at no time expanded beyond 10 Å (*Figure 10*). The 7 Å spacing line was much reduced after heating the flake to 400°C and disappeared at 500°C. The mineral responsible had a very high degree of orientation parallel to the mica surface. Attempts to concentrate the material for positive identification failed, and powder diffraction patterns from whole flakes failed to reveal any additional lines. However, the evidence available is all indicative of kaolinite.

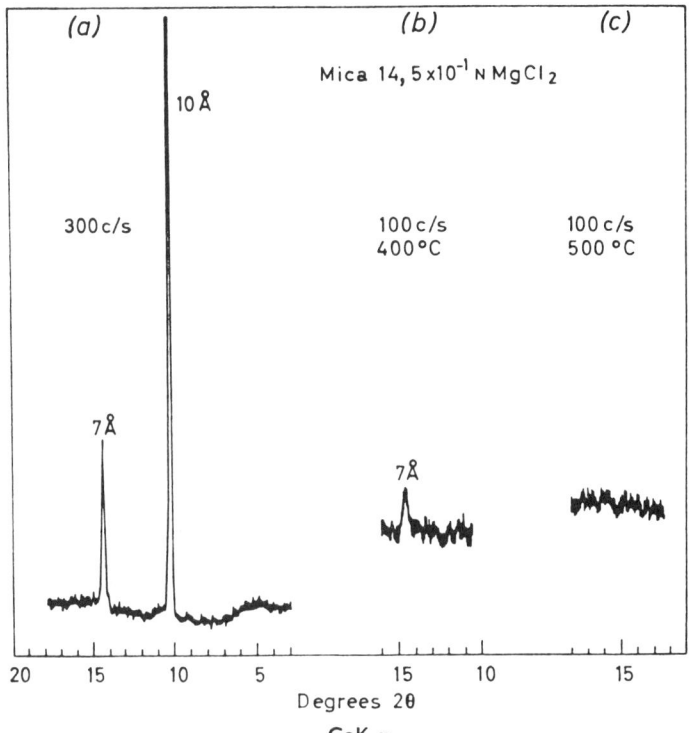

Figure 10. Diffraction trace of artificially weathered flake of mica 14 after treatment with 5×10^{-1} N $MgCl_2$ at 100°C
(a) *after removal from* 100°C *oven* (b) *after heating to* 400°C *and* (c) *to* 500°C

Table 8

Reduction in charge of silicate sheets during weathering

Mica	Charge on silicate sheet m.e./100 g of original mica			FeO *before reaction* millimoles/100 g*
	Before reaction *	*After reaction* †	*Reduction in charge*	
7	234	201	33	37
4	241	203	38	65
6	226	174	52	111
5	226	158	68	225
1	208	160	48	295

* From complete chemical analyses
† From analysis for Sr and K after reaction with $SrCl_2$

Measurements of total interlayer cations in completely exchanged mica flakes indicated that the silicate layer charge had been reduced during K replacement. There was not sufficient material available for direct analysis, but the results given in *Table 8* support the hypothesis that the effect is caused by the oxidation of ferrous iron.

Considerations of the structure by Radoslovich (1962) and Franzini and Schiaffino (1963) suggest that, because interlayer potassium exerts a small influence on the b axis of micas, potassium replacement should cause a small change in b axis. It was not possible to confirm this in the present work because the treatments caused such a broadening of the (060) diffraction line as to prevent its accurate measurement.

DISCUSSION

Mechanism of diffusion as the rate-controlling process

The most rapid and most obvious alterations in micas treated with salts solutions are due to the replacement of interlayer potassium: the lattice is forced to expand to such an extent as to visibly change the optical properties of the crystal; the replacing ion is present in the expanded part but virtually absent from the optically unchanged part, and potassium is low in the altered part. The data recorded for these alterations, especially the distance–time relationships (*Figures 2–7*), confirm the conclusions of earlier workers by establishing diffusion as the rate-controlling process but are not sufficient to establish a satisfactory model. This requires more precise data on the effect of salt concentration, K concentration and their interaction and an understanding of the anomalous concentration changes at the flake edges (*Figure 8*). Accurate concentration gradients across the weathered zone of flakes subjected to a variety of treatments are also necessary to propose a mathematically precise model. However, some qualitative aspects of the diffusion mechanism can be discussed from the results.

With the exception of Ba, all the ions normally give a 14–15 Å spacing, but in very dilute solutions (*Figure 1d*) or with a small amount of K in solution (Figure *1e*), a 12 Å phase appears instead. This suggests that when the supply of substituting ions is limited, an expanded phase of 12 Å forms which contains more K than the normal 14–15 Å phase. Near the boundary there would always be only a limited supply of ions, so a 12 Å phase might be expected between the 10 and 14–15 Å phases. In strong salt solutions, x-ray diffraction patterns gave no evidence of such a spacing, but in the curves of concentration gradient (*Figure 9*) the slight inflection in both the K and Ca curves near the boundary is consistent with a very narrow region of a third phase. If the diffusion coefficient of the 12 Å phase is much less than that of the 14–15 Å phase, it would normally be restricted to a very narrow region near the diffusion front, and would only increase in area when diffusion in the 14–15 Å phase was stopped or greatly retarded.

Barshad (1954) has suggested that for replacement to occur at all, the replacing ions must first enter the unexpanded 10 Å phase without their hydration shells. Expansion due to hydration of these ions would occur when they reached a particular concentration. The experimental results are not inconsistent with a slight penetration of the replacing ions into the 10 Å phase but they do not establish such a process.

The presence of replacing ions beyond the optical boundary (*Figure 8*) appears to indicate some diffusion of the unhydrated ions, but the location of the optical boundary with respect to the 10 to 14–15 Å phase transition, and to the concentration gradient, cannot be made with sufficient accuracy to establish the presence of replacing cations in the 10 Å phase. If diffusion occurs at all in this phase, its diffusion coefficient is much smaller than in the expanded phase(s) so that, under normal conditions, the unhydrated ions do not penetrate very far beyond the phase boundary. In principle, it is not essential that unhydrated ions enter the 10 Å phase before expansion and diffusion can proceed; alkylammonium ions have been used to replace the K of micas (Weiss *et al.*, 1956) and these ions could hardly enter the 10 Å phase prior to expansion. Their entry must be coincident with expansion. From the above it is seen that the potassium replacement probably involves diffusion in two and possibly three different phases, each of which will have a different coefficient. Crank (1956) gives a mathematical solution to the processes controlled by two diffusion coefficients and shows that the movement of the boundary separating the two phases will be linearly related to the square root of time; that is, the movement of the boundary will bear the same relationship to time whether the overall process is controlled by one or two diffusion coefficients.

Energy relationships of diffusion

The experimental results show that the different micas have different temperature coefficients for diffusion. Therefore, comparisons between them must be in terms of the parameters involved in the kinetics of diffusion. *Table 6* gives the values of E, the activation energy, and A, the constant, of the Arrhenius equation, derived from optical observations of micas treated with different ions. Several boundaries were visible in many barium-treated micas, and occasionally in others (*Plate 4*): under these circumstances plots of the types shown in *Figure 7* were not straight lines, and the E and A values derived from them are therefore less reliable.

It can be seen from *Table 6* that E is usually 9–12 kcal/mole. It has a much higher value if potassium is present in solution. The high values of E are for those combinations of ion and mica which are most sensitive to K (*Table 5*), that is micas 6 and 7 in Sr and mica 1 in Li. Re-determinations of mica 1 with Li and of mica 7 with Sr, using higher-quality salts, give values of E in the range 9–12 kcal/mole.

The activation energies measured for ion exchange in vermiculite (Walker, 1963; Keay and Wild, 1961) are normally in the range 9–13 kcal/mole, although Keay and Wild obtained values up to 22 kcal/mole, where NH_4 was present in appreciable amounts in the interlayer positions. This behaviour is analogous to the increases in E due to K in solution.

In spite of high values for E, diffusion in the micas has not been greatly retarded because of a sympathetic variation of A with E (*Table 6*). The large variation in A implies an entropy change and excludes activation energy as a reliable parameter for comparing rates of diffusion; instead it is necessary to use the free energy of activation, ΔF. The diffusion coefficient may be expressed as

$$D = e\lambda^2 \cdot \frac{kT}{h} \cdot e^{\Delta S/R} \cdot e^{-E/RT}$$

where ΔS is the entropy of activation, the other symbols have their usual meanings (Glasstone *et al.*, 1941), and λ is the distance from one equilibrium position to the next in diffusion. The water structure governs the positions ions can occupy in vermiculite (Mathieson and Walker, 1954) and these ion positions are approximately the same distance (6 Å) apart as the potassium ions in the interlayer region of mica. In calculating ΔS, λ was taken as 6 Å and the temperature as 50°C. ΔF was found from the equation*

$$\Delta F = E - T \cdot \Delta S$$

From *Table 6* it can be seen that ΔF shows less fluctuation than E. It also shows, as Keay and Wild (1961) found for E when sodium was replaced from vermiculite by divalent cations, no significant differences in E or ΔF for Ca, Sr, Ba and, in one case, Mg treatments of any one mica. Hence the most reliable procedure for comparing the micas is by averaging their ΔF values for divalent ions. *Table 9* shows the micas arranged in this order, with certain of their chemical and

* Any error in the assumed value of λ, though it affects the absolute values of ΔS and ΔF, has no effect on their relative values.

Table 9

Relationship between the structural features of the micas and their diffusion behaviour

Mica	ΔF^* kcal/mole	Critical K concentration (m.e./l.)†	Number per 12 anions (from Table 2)				
			fluorine and chlorine	octahedral positions not occupied by divalent cations	octahedral tri- and tetra-valent cations	octahedral Fe^{2+} and Mn^{2+}	tetrahedral charge (Al and Fe)^{IV}
1	12·4	1·5	0·066	0·97	0·76	1·35	1·33
4	12·9	1·5	0·393	0·46	0·35	0·29	1·21
3	13·0	1·7	0·695	0·17	0·11	0·50	0·96
5	13·1	0·6	0·841	0·38	0·23	1·10	1·04
7	13·4	0·31	0·875	0·04	0·06	0·16	1·14
6	13·4	0·45	1·01	0·24	0·10	0·51	0·95
8	—	0·33	0·945	0·13	0·13	0·13	1·17
9	—	0·20	0·835	0·15	0·16	0·21	1·16

* Average of the values obtained with divalent cations
† Average of values for Ca and Sr

structural features added for comparison. The total structural charge, expressed as charge per unit cell, can be taken as approximately constant for all specimens (*Table 2*).

Although the energies of activation for mica and vermiculite are similar, the micas have appreciably higher free energies of activation. ΔF calculated for vermiculites from the data of Walker (1963) and Keay and Wild (1961) with the same assumptions as for the micas has values of 8·3 and 7·0 kcal/mole. This difference from micas shows up in the diffusion rates.

The effect of potassium in solution on rate of diffusion is still reflected in the ΔF values, although less markedly than by E (*Table 6*). There is obviously a correlation in *Table 9* between ΔF and the critical potassium concentration that stops diffusion. It is in the direction which would be expected: low ΔF and high critical K concentrations, both reflecting a low specific bonding of the potassium to the mica surface.

Influence of mica structure on weathering

The chemical and structural features which might be expected to influence the bonding of K, and hence diffusion are fluorine, octahedral cations and tetrahedral charge.

Fluorine, by replacing hydroxyl ions, affects potassium bonding. Hydroxyl orientation in trioctahedral micas is mainly perpendicular to the silicate layers, whereas in muscovites it is inclined (Serratosa and Bradley, 1958; Bassett, 1960). In perpendicular orientation, protons would be so much closer to the interlayer potassium that they might weaken its bonding to the silicate sheet. Conversely, where the hydroxyl is inclined, bonding of the potassium will be strong.

Replacement of the hydroxyl by fluorine removes the protons, and potassium remains firmly bonded. Bassett offered this model to explain his experimental results which are confirmed in the present work; that is, potassium is displaced relatively easily from biotites and phlogopites but not at all from synthetic fluorphlogopites or muscovite. The correlation of fluorine content with ΔF and critical potassium concentration (*Table 9*) is as would be expected from the model.

Many biotites and phlogopites have hydroxyls with the two orientations. In most trioctahedral micas the octahedral cations are not all the same, nor all divalent. Therefore, the three octahedral positions associated with a particular hydroxyl may be occupied differently (Mg, Mg, Li; Mg, Fe, Al, etc.) and the asymmetry probably results in an inclination of the hydroxyl orientation away from the perpendicular. The probability of having asymmetrical occupation around one hydroxyl should be related to the number of octahedral positions not occupied by divalent cations, or the number occupied by tri- and tetravalent cations. They should, therefore, both be positively correlated with ΔF. In *Table 9* the correlation, though probably not significant, tends instead to be negative.

Reduction in lattice charge due to oxidation that normally takes place in the field and in laboratory experiments, ought to aid the replacement process, even though experiments conducted under controlled oxidation conditions indicated that oxidation itself was not a necessary part of the diffusion process. The oxidizable ion content (Fe^{++}, Mn^{++}) in the lattice should therefore show a negative correlation with ΔF in *Table 9*. It may do so, but its significance is obviously low.

The tetrahedral charge, because it originates near the silicate surface, is considered to favour the specific bonding of potassium (Schwertmann, 1962). In the present instance, the magnitude of the tetrahedral charge does not vary greatly (*Table 9*), but any correlation with ΔF is in the reverse direction to that expected.

An examination of *Table 9* shows that although the fluorine content need not in principle be related to the other figures, between which a partial interdependence would be expected, it seems to show some correlation with them. This makes it difficult to nominate which structural features are pertinent in the diffusion process, but fluorine content does seem to be an important factor. It is worth noting that vermiculites generally have a low fluorine content, e.g. 0·1–0·5 per cent (Deer, Howie and Zussman, 1962; Barnes and Clabaugh, 1961), which suggests that only micas low in fluorine weather to vermiculite.

Influence of potassium in solution on weathering

Experiments with constant shaking or frequent changes of solution established that the amounts of potassium released by the mica during treatment were not normally so high as to depress the rate of diffusion. However, potassium contributed from the glass reaction vessels was sufficient to do so at low salt concentrations. This and the distinct drop in diffusion coefficient on passing from 0·01 N solutions to 0·001 N solutions suggests that the ratio of the concentrations of displacing ion to potassium reaches a threshold value below which the rate of diffusion is much reduced and finally stopped. The results of the experiment with NaTB did not clarify the role of K in solution, owing to the decomposition of the salt and possible entry of decomposition products into the interlayer positions.

The sensitivity of the micas to potassium in solution varies with the replacing cation (*Table 5, Figure 6*). Barium is unusual in that it will continue to enter the micas in the presence of comparatively high concentrations of potassium. Furthermore, it produces a 12·4 Å spacing even in strong salt solutions, that is, it is the only ion which does not enter with two water layers. Because they are not surrounded by a hydration shell, barium ions will be in contact with one silicate sheet and, being the same size as potassium ions, they may have a higher specific adsorption than other ions surrounded by water.

With biotite, Mortland (1958) showed that the concentration which prevented further exchange, using 0·1 N NaCl solutions, increased with temperature from $1·5 \times 10^{-4}$ N at 20°C to 3×10^{-4} N at 50°C. It also increased with salt concentration from 9×10^{-5} N in 0·01 N NaCl to 3×10^{-4} N in 0·25 N NaCl. The present results are in accord with both these findings.

Sodium tetraphenylboron which reduces the potassium concentration in solution to ca. 5×10^{-7} N (or ~0·02 p.p.m.) has been used successfully to extract potassium from illites and muscovites (Scott, Hunziker and Hanway, 1960; Scott and Reed, 1962). In the present study no potassium replacement was observed in lepidolites or muscovites, presumably because of potassium levels up to almost 1 p.p.m. derived from the glass reaction vessels. This indicates the exceeding sensitivity of muscovites to traces of potassium, and makes it evident that muscovites will not exchange their potassium in most natural waters.

In strong salt solutions the replacement of potassium is orderly and appears to take place between every silicate sheet at the same rate, so that a well-defined boundary is observed. In natural weathering the salt solutions will usually be dilute and the replacement very slow against the back reaction of potassium in solution. In these circumstances, the replacement is not so orderly, and once a particular interlayer region has been opened up, continued replacement in that region will be favoured over the opening up of other interlayer regions. Therefore, interstratification would be expected during natural weathering, and is commonly observed. *Figure 11* shows diffraction traces of weathered mica particles from soils; there is obviously a

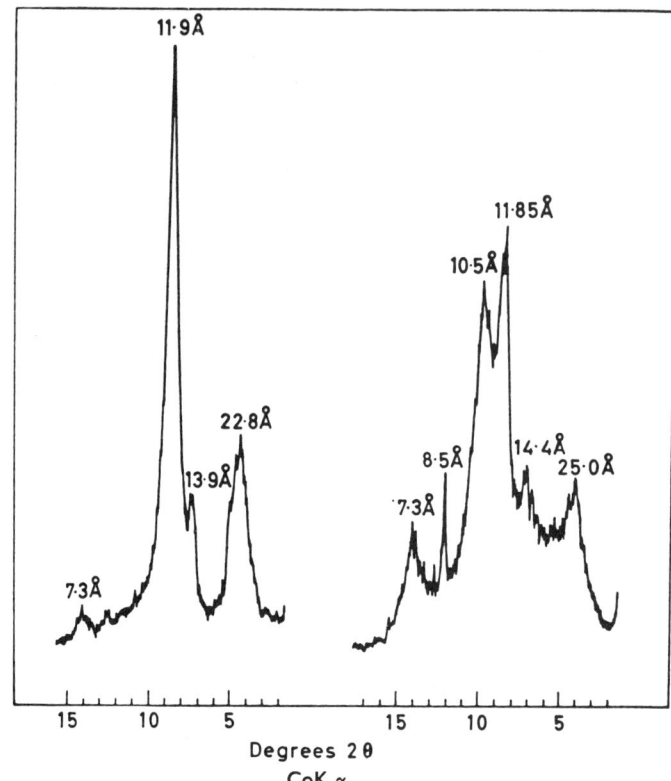

Figure 11. Diffraction traces of two naturally weathered mica flakes from soil

basic similarity to *Figure 1b* which is for mica reacted with a dilute solution. Vermiculites produced experimentally from micas in strong salt solutions differ from natural vermiculites. Strains associated with all the layers expanding together from 10 to 15 Å cause large flakes to break up physically when artificially weathered. X-ray diffraction patterns of these samples show sharp (001) lines but broadened (060) lines probably also caused by the strains. Vermiculites formed naturally do not show evidence of these strains; they are not always fragile, and their diffraction patterns do not show broadening of the (060) line. These comparisons confirm that under natural conditions only a few layers expand at a time, and physical strains are limited.

Calculations from the free energies of activation, presented in *Table 10*, give some idea of the times that might be involved in exchange in small micaceous mineral grains. These figures show that where diffusion is not influenced by potassium in solution, there is comparatively little difference, only a factor of 10, in the rates of weathering

Table 10

Time required for cation replacement to 1 mm depth in vermiculite and mica specimens at 20°C

Specimen	Time
Vermiculite*	89 h
Vermiculite†	14 h
Mica 1 (Sr only)	3·6 years
Mica 7 (Sr only)	33 years
Mica 7 (1 N $SrCl_2$ + 2 p.p.m. K)	10,000 years

* Calculated from Walker (1963)
† Calculated from Keay and Wild (1961)

of the slowest and fastest micas studied. They also show that under these conditions relatively few years are required to alter millimetre-sized crystals. In the field, micas rarely weather this fast, so the figures suggest that the rate of weathering will not generally be independent of the back reaction due to potassium in solution. The addition of several parts per million of potassium to the 1 N solution causes the calculated time for 1 mm penetration into mica 7 to increase from 33 years to 10,000 years.

Summing up from the foregoing paragraphs, it is clear that the most important property of mica in weathering is its sensitivity to potassium in solution (*Table 5*). It is the difference between the figures in this Table and the potassium present in the surrounding solution that governs whether exchange will take place, and if it does, what its rate will be.

Surface alterations

Concurrently with the replacement of interlayer potassium, but much more slowly, phlogopites and biotites undergo surface changes. These also occurred in lepidolites and muscovites and were, in fact, the only changes observable in these micas. Potassium is lost from the surface and a superficial layer of another mineral phase, possibly kaolin, appears, but the experimental data would suggest that the two changes are not necessarily coupled; more potassium is generally lost than could be explained by the amount of kaolin formed, and the replacing cation is present on the mica surface inside the optical boundary (*Figure 8*). The potassium loss probably takes place as exchange via dislocations and other defects in the mica surfaces. Rosenqvist (1963) has observed a similar surface replacement in muscovites. The kaolin formation could represent a topotactic change of parts of the mica surface or might be an epitaxial crystallization onto the surface. The mineral has yet to be positively identified, its relationship to the mica surface established, and a source found for its silica and alumina. It is possible the silica may originate from the glass ampoules.

CONCLUSIONS

Interlayer potassium in mica can be replaced by exchange with hydrated cations from neutral salt solutions. The ions move principally between the lattice sheets and, in strong solutions, cause expansion from 10 Å to 15 Å. The replacement front is clearly visible optically as a well defined boundary line parallel to the edges of the flake.

The rate of replacement is controlled by diffusion and depends on the concentration of the salt solution.

Potassium concentration in the solution determines whether replacement will take place. For each case there is a critical potassium concentration which, if exceeded, will stop the replacement. The absence of any observable replacement in muscovites and lepidolites is probably due to an extreme sensitivity to potassium in solution.

The diffusion rates and sensitivities to potassium of the biotites and phlogopites are closely related to their fluorine contents.

The slow rates of natural weathering of micas are due to the presence of potassium in solution causing a reduction in the rate of diffusion.

REFERENCES

Barnes, V. E. and Clabaugh, S. E. (1961). *10th Nat. Clay Conf. Guide Book No. 3*, p. 45, Bureau of Economic Geology, University of Texas

Barshad, I. (1954). *Soil. Sci.* **78**, 57

Bassett, W. A. (1959). *Am. Miner.* **44**, 282; (1960) *Bull. geol. Soc. Am.* **71**, 449

Crank, J. (1956). *The Mathematics of Diffusion*, Clarendon Press, Oxford

Deer, W. A., Howie, R. A. and Zussman, J. (1962). *Rock-forming Minerals*, Vol. 3, *Sheet silicates*, Longmans, Green, London
Franzini, M. and Schiaffino, L. (1963). *Z. Kristallogr.* **119,** 297
Glasstone, S., Laidler, K. J. and Eyring, H. (1941). *The Theory of Rate Processes*, McGraw-Hill, New York
Keay, J. and Wild, A. (1961). *Soil Sci.* **92,** 54
Mathieson, A. McL. and Walker, G. F. (1954). *Am. Miner.* **39,** 231
Mortland, M. M. (1958). *Proc. Soil Sci. Soc. Amer.* **22,** 503
Radoslovich, E. W. (1960). *Acta crystallogr.* **13,** 919; (1962) *Am. Miner.* **47,** 617
Rosenqvist, I. Th. (1963). *Int. Clay Conf. 1963* (Stockholm) **2** (in press)
Schwertmann, U. (1962). *Beitr. Miner. Petrogr.* **8,** 199
Scott, A. D., Hunziker, R. R. and Hanway, J. J. (1960). *Proc. Soil Sci. Soc. Amer.* **24,** 191
— and Reed, M. G. (1962). *Proc. Soil Sci. Soc. Amer.* **26,** 41, 45
Serratosa, J. M. and Bradley, W. F. (1958). *J. phys. Chem.* **62,** 1164
Sumner, M. E. and Bolt, G. H. (1962). *Proc. Soil Sci. Soc. Amer.* **26,** 541
Walker, G. F. (1949). *Miner. Mag.* **28,** 693; (1956) *Nature, Lond.* **177,** 239; (1959) *ibid.* **184,** 1392; (1963) *Int. Clay Conf. 1963* (Stockholm) **1,** 177
Weiss, A. and Hofmann, U. (1951). *Z. Naturf.* **6b,** 405
— Mehler, A. and Hofmann, U. (1956). *Z. Naturf.* **11b,** 435

ALTERATION OF BIOTITE TO VERMICULITE BY PLANT GROWTH

M. M. MORTLAND, K. LAWTON, AND G. UEHARA

Michigan Agricultural Experiment Station[1]

Received for publication May 28, 1956

Vermiculite was proposed by Walker (5) as the immediate weathering product of biotite in soils the parent materials of which were neutral or acidic in nature. Walker suggests that as potassium is removed and a double layer of water molecules forms in some zones, a mixed-layer structure arises. As weathering proceeds the number of the layers of the vermiculite type increases.

Jackson *et al.* (3) suggest that the persistence of the 110 diffraction line and the decrease in the 00L diffraction intensity as weathering proceeds, is the result of disorder of the 00L sequence with preservation of the inherent layer silicate structure responsible for the 110 diffraction maxima. This they propose as the principal change accompanying weathering of micas.

Barshad (1) studied the relationship of biotite and vermiculite and found that vermiculite forms a biotite-like structure upon saturation with potassium and drying.

The present investigation was concerned with direct evidence of the weathering products of biotite. The effect of plant growth on the mineral transformations of biotite was of interest since this approaches soil conditions to a greater degree than empirical laboratory methods. This approach also would have a bearing on the availability of the potassium in biotite to plants.

EXPERIMENTAL METHODS

Large biotite crystals[2] were broken down by ball-milling to pass a 60-mesh screen. Quantities of the biotite in the amounts of 25, 50, and 100 g. were mixed with enough acid-washed quartz sand (ASTM-C-109) to give a total weight of 5 kg. These materials were placed in closed, glazed, 1-gallon ceramic pots and four successive crops of wheat were grown over a 1-year period. All nutrients other than potassium were supplied in a nutrient solution similar to that of Shive. Biotite was the only source of potassium supplied to the plants. After the top growth of the last crop was harvested, the quartz sand and biotite mixture was agitated in water. The minerals in suspension were immediately decanted, air dried, passed through a 60-mesh screen, and stored in a desiccator with calcium chloride.

Cation-exchange capacity data were obtained on the samples using the sodium acetate method (4). Total potassium was determined with a Perkin-Elmer flame-photometer after 0.5-g. samples had been treated in platinum crucibles with hydrofluoric acid and the residue dissolved in 0.1 N HCl. Ethylene glycol retention data were obtained using the method of Dyal and Hendricks (2). Electro-

[1] Journal Article No. 1920, Michigan Agricultural Experiment Station, East Lansing.
[2] Obtained from Ward's Natural Science Establishment, Inc., Rochester, New York.

dialysis was carried out in continuous operation for 2 weeks in a Mattson cell at a voltage of 210. The x-ray diffraction patterns were obtained with a Norelco unit using copper radiation, goniometer, and Brown recorder. The samples x-rayed were oriented aggregates, natural, and glycerol-solvated. The amount of mineral in the oriented aggregate was 40 mg. per square inch. A part of the intensity of the 3.3 A maxima is due to quartz in the sample holder. The 4.2 A quartz line was deleted from the patterns after it was established that it came from the sample holder.

DISCUSSION OF RESULTS

The x-ray diffraction data obtained on the unaltered biotite, on the electrodialyzed biotite, and on the biotite segregated from the pots which produced

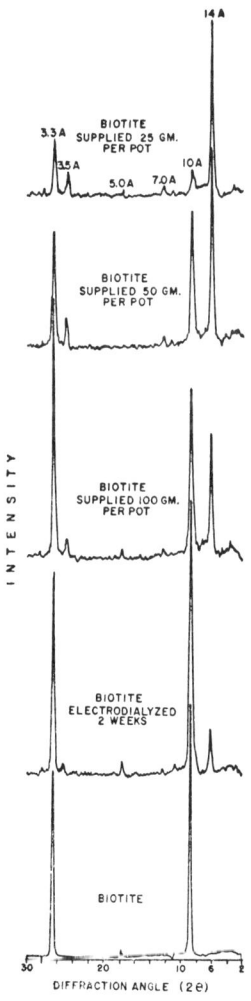

FIG. 1. X-RAY DIFFRACTION PATTERNS OF GLYCEROL-SOLVATED BIOTITE AND ITS WEATHERING PRODUCTS

four crops of wheat, are presented in figure 1. The samples in this figure were glycerol-solvated. The intensity of the diffraction maxima may be compared directly with the exception of the pure biotite which must be multiplied by a factor of 2. The data show that when biotite was used as the sole source of potassium for plant growth, vermiculite was a direct weathering product. This conclusion is established when the glycerol-solvated samples gave a diffraction maximum at 14A., characteristic of the 002 plane of vermiculite. This is also characteristic of chlorite but the fact that the 7, 4.7, and 3.5A. peaks are relatively weak compared with the 14A. peak indicates vermiculite. The fact that when the samples were potassium-saturated and heated to 110°C. the 14A. maximum moved to 10A. again demonstrates this mineral to be vermiculite and not chlorite. The data show also that as the amount of biotite supplied to the wheat was reduced the proportion of biotite altered to vermiculite increased because of the increase in potassium stress applied by the growing crop.

When the diffraction pattern of the biotite subjected to electrodialysis at 210 volts for 2 weeks is compared with the other samples, it is evident that a small amount of vermiculite was produced as a result of this treatment, and very evident that the four crops of wheat were much more effective in the alteration of biotite to vermiculite than was the electrodialysis.

Glycerol treatment seemed to have less effect on the intensity of the 10A. and 14A. maxima than the diagrams of Walker (5) would suggest.

Table 1 gives total potassium content, cation-exchange capacity, ethylene glycol retention, and total charge of the samples (cation-exchange capacity plus potassium content). These values tend to substantiate x-ray diffraction results, showing a progressive decrease in total potassium content and a corresponding increase in cation-exchange capacity as the proportion of vermiculite increased. There was a linear increase in cation-exchange capacity as potassium content decreased. Ethylene glycol retention increased as a function of the increase in vermiculite content.

The general observation can be made that the total x-ray diffraction intensity decreased as potassium was extracted from the biotite during its alteration to vermiculite. This phenomenon may be due to the production of single interplanes or zones of too few layers to give diffraction maxima, as suggested by Jackson et al. (3). These investigators suggest that the 10A. diffraction maxima

TABLE 1

Effect of four crops of wheat on the properties of biotite

Biotite sample	Biotite Added	Total K	C.E.C.	Ethylene-Glycol Retention	Total Charge*
	g./pot	%	me./100 g.	mg./g.	me./100 g.
Original	—	5.8	14	4.9	162
Cropped	100	4.5	30	12.6	145
Cropped	50	3.6	38	24.2	130
Cropped	25	2.4	54	28.5	114

* Sum of the potassium content and cation-exchange capacity.

result from the yet unweathered (X-crystalline) zones. The diffraction maxima at 14A. result when weathered and expanded 14A. layers are repeated with sufficient regularity to give X-crystalline materials. It is noteworthy that the sample which was supplied to the plant at the 25-g. level still contained 2.4 per cent potassium, but the 10A. diffraction maximum was quite small, indicating a very small quantity of crystalline biotite. Most of the potassium may exist in zones amorphous to x-ray, too few layers being present to give diffraction maxima. Emphasis should be placed on the observation that the diffraction maxima were either 10A. or 14A. indicating no intermediate hydrous mica or partially expanded material or at least in enough quantity or periodicity to be crystalline to x-ray.

Wheat plants were effective weathering agents of biotite in this investigation, which suggests an interaction between root and biotite particles. It may be that the root aids or is an agent in the oxidation of octahedral ferrous iron in the biotite which would compensate the negative tetrahedral charge and allow rapid interlayer expansion and potassium release. One can observe (table 1) that as the potassium content decreased, there was a reduction in total charge which could result from the oxidation of octahedral ferrous iron.

Since the biotite was so readily altered to vermiculite by cropping, it is possible that under soil weathering conditions including vegetative effects, finely divided biotite may be rapidly altered to vermiculite and probably would not exist to any great extent in the clay fraction. This point has some basis in that most clay micas in soils reported in the literature are of the dioctahedral type. The total quantity of biotite present and its particle size as well as the presence of other potassium minerals are assumed to influence the rate at which biotite weathered to vermiculite.

SUMMARY

Biotite when supplied as the only source of potassium to growing wheat plants was readily altered to vermiculite. This alteration is direct evidence that vermiculite is a weathering product of biotite. The growing plant seemed to be very effective in promoting the alteration of biotite to vermiculite. A decrease in potassium content and a corresponding increase in cation-exchange capacity and ethylene-glycol retention were observed as biotite was altered to vermiculite. A decrease in total charge was observed as a function of the release of potassium. No hydrous micas of intermediate expansion were observed in the transition of biotite to vermiculite. A general decrease in the basal spacing diffraction intensity was observed as weathering proceeded, which indicates a development of some layers amorphous to x-ray. Since biotite was so easily altered to vermiculite by plant growth, apparently biotite can not usually exist to any great extent in the clay fraction of soil horizons subjected to extensive biological influence.

REFERENCES

(1) BARSHAD, I. 1948 Vermiculite and its relation to biotite as revealed by base exchange reactions, x-ray analyses, differential thermal curves, and water content. *Am. Mineralogist* 33: 65.

(2) DYAL, R. S., AND HENDRICKS, S. B. 1950 Total surface of clays in polar liquid as a characteristic index. *Soil Sci.* 69: 421.
(3) JACKSON, M. L., et al. 1952 Weathering sequence of clay-size minerals in soils and sediments: II. *Soil Sci. Soc. Amer. Proc.* 16: 3–6.
(4) RICHARDS, L. A. 1954 *Agricultural Handbook No. 60.* U. S. Dept. Agr., Washington, D. C.
(5) WALKER, G. J. 1949 Trioctahedral minerals in the soil-clays of northeast Scotland. *Mineralog. Mag.* 29(208): 72–84.

10

Copyright © 1973 by the Clay Mineral Society
Reprinted from Clays Clay Miner. **21**:479-488 (1973)

MICA-DERIVED VERMICULITES AS UNSTABLE INTERMEDIATES*

J. A. KITTRICK†

Department of Agronomy and Soils, Washington State University, Pullman, Washington 99163, U.S.A.

(Received 30 November 1972)

Abstract—Stability determinations were made by solubility methods on two trioctahedral mica-derived vermiculites. The phlogopite-derived vermiculite was found to be unstable under acid solution conditions, where stabilities of montmorillonite, kaolinite and gibbsite had previously been determined. An attempt was next made to locate a possible montmorillonite–vermiculite–amorphous silica triple point. This triple point involved conditions of alkaline pH, high pH_4SiO_4 and high Mg^{2+}. These are conditions where phlogopite and biotite-derived vermiculites are most likely to control equilibria *if* they are stable minerals. The montmorillonite–vermiculite–amorphous silica samples went to the montmorillonite–magnesite–amorphous silica triple point, leaving no stability area whatsoever for the vermiculites. These large particle-size, trioctahedral, mica-derived vermiculites appear to be unstable under all conditions of room T and P.

Arguments are presented indicating that micas are unstable in almost all weathering environments. A hypothesis is proposed that mica-derived vermiculites result from the unique way in which unstable micas degrade in these environments. It is proposed that vermiculite derives from a series of reactions whose relative rates often result in an abundance of vermiculite. These relative reaction rates are slow for mica dissolution, rapid for K removal and other reactions pursuant to vermiculite formation, and slow for vermiculite dissolution. In chemical terms, mica-derived vermiculites may be considered fast-forming unstable intermediates.

INTRODUCTION

THE MAJOR clay minerals of soils and sediments compete for a relatively small group of elements during their formation. In theory at least, each clay mineral forms only under solution conditions where it is the least soluble of the clay minerals competing for that group of elements. Thus, because of the stability of other minerals, the stability field of a particular mineral represents a much more restricted chemical environment than does a solution saturated with respect to that mineral alone. The objective of this investigation was to provide some initial stability information for vermiculite. It was anticipated that this information would help define the stability fields of vermiculite and of other clay minerals, and thereby contribute to a better understanding of the formation and weathering of clay minerals in soils and sediments.

Vermiculite is a 2:1 layer silicate with a negative charge of about 0.6 to 0.9 per formula unit. Operationally, the term 'vermiculite' is usually applied to a 2:1 layer silicate that has a dehydrated c-repeat distance of about 10 Å, that expands to about 14 Å when solvated with a polar liquid, provided the exchange sites are saturated with one of the more highly hydrated cations. This behavior is exhibited by materials that are not necessarily identical in their other properties. Vermiculites used in this investigation are those most readily obtained in relatively pure form; i.e. they were large particle size, trioctahedral mica-derived vermiculites, with mica impurities artifically converted to vermiculite in the laboratory. It is from similar materials that most of the properties of vermiculites are inferred.

MATERIALS AND METHODS

The vermiculite derived from phlogopite was from Palabora, South Africa, obtained from C. H. Kingsland, Lahabralite Co. The vermiculite derived from biotite was from Libby, Montana, obtained from J. D. Hayes, Zonolite Co. Both vermiculites were wet ballmilled to pass a 300 mesh sieve ($<50\,\mu m$). Both vermiculites contained considerable micaceous impurities which were converted to vermiculite by repeatedly

---

* This investigation was supported in part by grant 16060 DGK from the Federal Water Pollution Control Administration and from the U.S. Department of the Interior in support of the State of Washington Water Research Center project A-042. Published as Scientific Paper No. 3965, College of Agriculture, Washington State University, Pullman, Washington 99163, U.S.A. Project No. 1885.

† Professor of Soils. Appreciation is expressed to Mr. E. W. Hope for his help in the experimental work.

extracting the K with hot 1 N NaCl. Samples were then Mg-saturated by 4 centrifuge washes with 1·0 M MgCl$_2$, followed by four water washes. The NaCl-treated Palabora sample contained no detectable K, with the calculated (Jackson, 1956) unit cell formula,

$$[(Si_{6.29}Al_{1.71})(Mg_{5.38}Fe^{3+}_{0.47}Fe^{2+}_{0.15})O_{20}(OH)_4]^{1.24-}$$

X-ray diffraction analysis indicated no detectable impurities (Fig. 1). The NaCl-treated Libby sample contained 0·34 per cent K corresponding to 5·8 per cent biotite, with the calculated vermiculite unit cell formula,

$$[(Si_{6.38}Al_{1.62})(Mg_{5.22}Al_{0.16}Fe^{3+}_{0.59}Fe^{2+}_{0.03})\,O_{20}(OH)_4]^{0.87}$$

X-ray diffraction analysis indicated a small interstratified biotite impurity remaining (Fig. 1). X-ray traces of the vermiculites did not appear to change during the course of the experiments.

Total chemical analysis of the vermiculites (with Ba as the exchangeable ion) for Mg, Al, Si, Fe, K and Ba was done with the X-ray spectrograph. Standard rock and mineral samples were used as references. The proportion of Fe^{2+} and Fe^{3+} was determined with ortho-phenanthroline (Roth et al., 1968). Composition of the Palabora vermiculite was similar but not identical to that determined previously by wet chemical methods on a separate NaCl-treated batch (Kittrick, 1969a).

Vermiculite samples were suspended in solutions whose compositions are described in detail later. After suitable equilibration times, aliquots of these solutions were removed from the supernatant liquid after centrifuging to the absence of suspended material in a Tyndall beam. Al was determined colorimetrically with aluminon (Hsu, 1963), and Si with molybdate and aminonaphthol sulfonic acid (APHA, 1960). Na and K were determined with a flame photometer and Mg with an atomic absorption spectrophotometer. The pH measurements were made with a Corning Model 12 unit calibrated to within 0·01 pH units of two or sometimes three buffers. Precision of the analyses for pAl, pSi and pH was approximately ±0·02.

The Al^{3+} component of total Al was calculated from the first hydrolysis constant where necessary (Schofield and Taylor, 1954). Ion activities were calculated from the extended form of the Debye–Hückel equation (Klotz, 1964). Room temperature during analysis periods was between 23 and 25°C. The samples were agitated almost continuously on a shaker. X-ray diffraction analysis at high sensitivity before and after equilibration did not reveal the formation of any new phases, although solution analyses indicated precipitation must be occurring in some cases, as described later.

For equilibration under low pH conditions, four 1·0 g samples of Palabora vermiculite were given three washes with 0·010 N HCl, suspended in 200 ml of 0·0010 N HCl and then acidified to approximately pH 3 with HCl. In addition to the vermiculite, sample three contained 1·0 g of kaolinite (No. 3 Colloidal, Hammil and Gillespie, Inc., from England) and sample 4 contained 5·0 g of gibbsite (synthetic C-730, from Aluminum Co. of America).

For equilibration under high pH conditions, samples contained a mixture of 5·0 g vermiculite, 5·0 g montmorillonite (Belle Fourche, South Dakota, from A. G. Clem, American Colloid Co., <150 mesh), 5·0 g hematite (Mapico Red 347, from Columbia Carbon Co.), and 5 g amorphous SiO_2 (the amorphous SiO_2 was precipitated from 500 ml of solution containing 47·3 g $Na_2SiO_3\cdot 9H_2O$, by decreasing the pH with HCl. The precipitate was centrifuge-washed with water, twice with the solution in which the vermiculite samples were equilibrated and then divided). To ensure removal of possible carbonates, soluble hydroxides and other soluble substances, the vermiculite–montmorillonite–amorphous silica mixtures were given pH 4·0 HOAc treatments until the pH of the solution rose

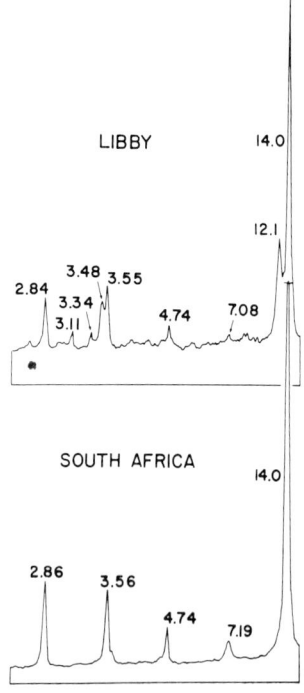

Fig. 1. Oriented X-ray diffraction patterns of <50 μm Palabora South Africa and Libby vermiculites after K-removal by hot 1 N NaCl extractions. Glycerated samples, CuKα, Ni filter, 1·2θ/min.

no higher than pH 6·0 after overnight contact. They were then given three centrifuge washes with a solution $1 \times 10^{-2 \cdot 0}$ M in Mg and $1 \times 10^{-2 \cdot 52}$ M in Si, pH 7·90. Finally, the samples were equilibrated in 30 ml of this same solution.

RESULTS

Acid pH

If vermiculite is stable under soil solution conditions similar (not necessarily identical) to those of montmorillonite, kaolinite and gibbsite, a reasonable approach would be to equilibrate vermiculite under the acid solution conditions already used for stability determinations on these minerals. Under acid conditions, the pH_4SiO_4 of the four Palabora vermiculite samples decreased with time (Table 1) until most of them became supersaturated with respect to amorphous silica ($pH_4SiO_4 < 2·72$). The $pH-1/3pAl^{3+}$ of the samples increased with time, approaching or exceeding 2·66 (Table 1), which is the $pH-1/3pAl^{3+}$ of gibbsite. These compositions are also supersaturated with respect to kaolinite and montmorillonite. Most of the preceding changes took place in the first 2 months, with the samples remaining supersaturated for over 2 yr. Apparently under the low pH conditions of this experiment, vermiculite dissolves faster than more stable minerals can precipitate.

The worldwide abundance of vermiculite suggests a rather large stability area in terms of solution composition, yet it is evident that the phlogopite-derived vermiculite is unstable relative to gibbsite, kaolinite and montmorillonite under conditions that previously permitted stability determinations on these minerals. To confirm the surprising contrast between vermiculite abundance and stability and to delimit a vermiculite stability area even if it were very small, a search was made for conditions of maximum vermiculite stability.

Alkaline pH

Strategy for equilibrating vermiculite. In order to determine the solution conditions under which vermiculite is most likely to be stable, consider the dissolution of Palabora vermiculite

$$[(Si_{6·29}Al_{1·71})(Mg_{5·38}Fe^{3+}_{0·47}Fe^{2+}_{0·15})O_{20}(OH)_4]^{1·24-}$$
$$+ 1·16 H_2O + 18·84 H^+$$
$$= 1·71 Al^{3+} + 5·38 Mg^{2+} + 0·47 Fe^{3+}$$
$$+ 0·15 Fe^{2+} + 6·29 H_4SiO_4.$$

Then $pK = 1·71 pAl^{3+} + 5·38 pMg^{2+}$
$$+ 0·47 pFe^{3+} + 0·15 pFe^{2+}$$
$$+ 6·29 pH_4SiO_4$$
$$- 18·84 pH$$

where K is the equilibrium constant, and the activity of vermiculite and water are assumed to be unity. By appropriate grouping of terms and division to reduce the coefficient of $pH-1/3pAl^{3+}$ to unity, we have

$$pH-1/3 pAl^{3+} = 1·23 pH_4SiO_4 - [2·10$$
$$(pH-1/2 pMg^{2+})$$
$$+ 0·27 (pH-1/3 pFe^{3+})$$
$$+ 0·04 (pH-1/2 pFe^{2+})$$
$$+ 0·25 pH - 0·19 pK]. \quad (1)$$

On $pH-1/3pAl^{3+}$ vs pH_4SiO_4 coordinates, this equation represents a straight line of slope 1·23 and an intercept of minus the quantity inside the brackets.

Selecting $pH-1/3pAl^{3+}$ as the variable controlling the relative stability of Al-containing clay minerals (Kittrick, 1969b), it can be shown from (1) that the level of $pH-1/3pAl^{3+}$ supported by Palabora vermiculite will be lowest (the vermiculite will be least soluble) when pH_4SiO_4 is low and pH is high. When the pH is high, $pH-1/2pMg^{2+}$, $pH-1/3pFe^{3+}$ and $pH-1/2pFe^{2+}$ are also high, which also makes the vermiculite more stable. There are no experimental difficulties in equilibrating vermiculite in low pH_4SiO_4 solutions (up to slight supersaturation with respect to

Table 1. Composition of the solution in contact with Palabora vermiculite samples after 918 days equilibration. Initial solution was pH 3·0 HCl

Sample	Composition	pMg^{2+}	pH	pAl^{3+}	$pH-1/3pAl^{3+}$	pH_4SiO_4
1	Vermiculite	2·92	4·05	4·58	2·52	2·52
2	Vermiculite	2·92	4·05	4·64	2·50	2·51
3	Vermiculite-kaolinite	2·93	4·08	4·78	2·49	2·51
4	Vermiculite-gibbsite	2·78	4·48	5·15	2·76	2·77

The negative logarithm of the molar activity is denoted by p. The Al^{3+} is corrected for hydrolysis. The ionic strength, essentially determined by KCl released from the reference electrode during pH measurements, was 0·047 for sample 4 and 0·036 for the rest.

amorphous silica), but alkaline pH prevents the measurement of Al^{3+} and Fe^{3+} and usually Fe^{2+} (depending upon the Eh).

The strategy adopted for equilibrating vermiculite under alkaline pH and high silica conditions, where it is most likely to be stable, was first to add hematite to the sample and assume control of pH-1/3pFe^{3+} at the hematite level. Fe^{2+} can be calculated from Fe^{3+} and an Eh measurement, when the sample reaches equilibrium. Next, amorphous SiO_2 was added to control H_4SiO_4. The H_4SiO_4 was initially raised to the level of slight supersaturation with respect to amorphous silica by the addition of Na_2SiO_3. No assumptions were necessary with respect to control of H_4SiO_4 by amorphous silica, since the H_4SiO_4 could be measured. The level of pH-1/3pAl^{3+} was assumed to be fixed by the addition of Belle Fourche montmorillonite of known stability (Kittrick, 1971a). The initial Mg^{2+} level was set at 0·0100 M by the addition of $MgCl_2$. The uncontrolled variable in the system was pH, permitting the course of equilibration to be followed by periodic pH measurements.

Montmorillonite and amorphous silica were added to the vermiculite samples so that at equilibrium the solution composition should be somewhere along the montmorillonite–amorphous silica join in Fig. 2. Specifically, if vermiculite is a stable mineral, then at equilibrium the solution composition should be at the montmorillonite–vermiculite–amorphous silica triple point, at some value of pH-1/2pMg^{2+} below that supported by magnesite. The stability plane of vermiculite resulting from this hypothetical value of pH-1/2pMg^{2+} would then be bounded by the dashed lines in Fig. 2. If vermiculite were unstable under the conditions of these experiments, the pH-1/2pMg^{2+} of the solution would increase until the montmorillonite–amorphous silica–magnesite triple point was reached. Further dissolution of vermiculite at this point should result in the precipitation of one or more of these triple-point minerals.

The pH-1/2pMg^{2+} of magnesite. Since solutions of pH-1/2pMg^{2+} greater than that supported by magnesite are supersaturated with respect to magnesite, the area encompassed by the vermiculite stability plane will then depend upon how much lower, if any, are the pH-1/2pMg^{2+} values of the solutions of Table 2 as compared to the pH-1/2pMg^{2+} supported by magnesite under similar conditions.

To determine the pH-1/2pMg^{2+} supported by magnesite in equilibrium with atmospheric CO_2, consider the following (where standard free energy of formation values, herein indicated for simplicity as ΔG, are taken from Robie and Waldbaum (1968) and ΔG_r is the standard free energy of the reaction):

$$MgCO_3 = Mg^{2+} + CO_3^{2-}$$

$$\Delta G_r = \Delta G_{Mg^{2+}} + \Delta G_{CO_3^{2-}} - \Delta G_{MgCO_3}$$

$$= 108·9 \pm 0·2 - 126·2 \pm 0·15$$

$$+ 246·1 \pm 0·3$$

$$= 11·0 \pm 0·65 \text{ kcal.}$$

Where K is the equilibrium constant, (K values taken from Garrels and Christ, 1965), and considering the activity of the solid phase to be unity,

$$pK = \Delta G_r/1·364 = 8·06 \pm 0·48 \text{ and}$$

$$pMg^{2+} + pCO_3^{2-} = 8·06 \pm 0·48. \quad (2)$$

When the CO_2 in the air at an average pressure of $10^{-3·5}$ atm dissolves in water,

$CO_2 + H_2O = H_2CO_3$
$pK = 1·47 = pH_2CO_3 - pCO_2$ and
$pH_2CO_3 = 1·47 + 3·50 = 4·97$.

Further,

$H_2CO_3 = H^+ + HCO_3^-$
$pK = 6·4 = pH^+ + pH_2CO_3^- - pH_2CO_3$ and
$pH^+ + pHCO_3^- = 6·4 + 4·97 = 11·37$.

Then,

$HCO_3^- = H^+ + CO_3^{2-}$
$pK = 10·3 = pH^+ + pCO_3^{2-} - pHCO_3^-$
$pHCO_3^- = -10·3 + pH^+ + pCO_3^{2-}$
$pHCO_3^- + pH^+ = -10·3 + 2pH^+ + pCO_3^{2-} = 11·37$
$pCO_3^{2-} = 21·67 - 2 \text{ pH}$.

Substituting in (2),

$pMg^{2+} + 21·7 - 2 pH^+ = 8·06 \pm 0·48$
and pH-1/2p$Mg^{2+} = 6·8 \pm 0·24 \simeq 6·6$ to $7·0$ (3)

where uncertainties in the ΔG values, but not in the K values are considered.

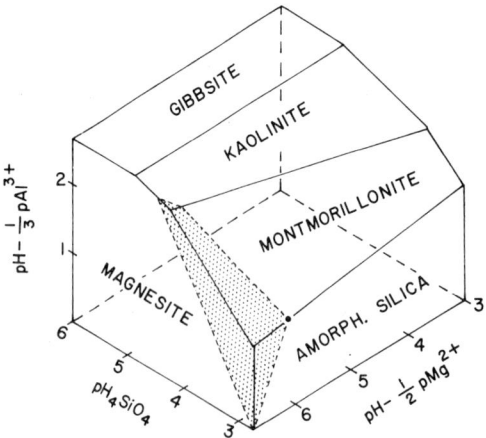

Fig. 2. Diagram indicating a possible vermiculite stability plane (shaded). The hypothetical montmorillonite–vermiculite–amorphous silica triple point is indicated by the dot.

Table 2. Composition of the solution in contact with Palabora and Libby vermiculites at alkaline pH. The samples also contained montmorillonite, amorphous silica and hematite

Equilibration time (days)	pH_4SiO_4	pMg^{2+}	pH	$pH-1/2pMg^{2+}$
Initial solution				
0	2.52	2.24	7.90	6.78
Palabora vermiculite				
8	2.72		7.55	
19	2.70	2.56	7.84	6.56
42	2.69	2.60	7.89	6.59
73	2.75	2.64	7.90	6.58
Libby vermiculite				
8	2.74		7.43	
19	2.73	2.63	7.87	6.55
42	2.73	2.73	7.89	6.53
73	2.75	2.73	7.97	6.60

The negative logarithm of the molar activity is denoted by p. The ionic strength was initially determined by the equilibrating solution (mainly 0.0100 M $MgCl_2$) at 0.030 for both samples. In spite of compositional changes during equilibration, the final ionic strength was still 0.030.

Vermiculite–montmorillonite–amorphous silica–hematite equilibria. A series of preliminary experiments (not shown) compared samples with two particle sizes of each of two montmorillonites, plus hematite and each of the two vermiculites. The samples did not contain amorphous silica, but initial solutions contained Si at the amorphous silica equilibrium level; that is, a pH_4SiO_4 of 2.72 (Kittrick, 1969b). The initial $pH-1/2pMg^{2+}$ of the solutions was 5.87 and after $4\frac{1}{2}$ hr of equilibration this value rose to 6.49. The difference between $pH-1/2pMg^{2+}$ of 6.49 and that of magnesite at 6.6–7.0 leaves only a very small possible stability area for the two vermiculites.

The results of the preliminary experiments were independent of the two particle-size fractions of each of the two montmorillonites so the final experiments reported in Table 2 used only a single type and particle size of montmorillonite. After 144 days the pH_4SiO_4 of the preliminary samples had increased to between 3.02 and 3.59, so amorphous silica was included in the final samples in an effort to maintain the samples on the amorphous silica plane and thereby permit clearcut interpretation of results.

Natural and experimental solutions often show some degree of silica supersaturation, but the solutions in Table 2 (which were supersaturated initially) apparently were controlled at the theoretical value by the amorphous silica added to the samples. The montmorillonite and hematite equilibria could not be measured, but both have been shown to come to equilibrium with similar solutions under similar conditions in about 2 weeks (Kittrick, 1971b). Provided that the $pH-1/3pAl^{3+}$ of the solutions was controlled by the montmorillonite added for that purpose, the solution compositions lie along the montmorillonite–amorphous silica join as originally anticipated.

When the initial solution of Table 2 was added to the acidified mineral mixture, the pH dropped. After 8 days the pH of the Palabora sample had risen to 7.55. Due to an oversight, no Mg^{2+} analyses were made at that time, but the pMg^{2+} must have been between the 2.24 of the initial solution and the 2.56 found at 19 days. The $pH-1/2pMg^{2+}$ at 8 days therefore must have been between 6.27 and 6.43. Values of $pH-1/2pMg^{2+}$ were probably lower than this prior to 8 days. Similar arguments apply to the Libby sample. It is evident that the samples of Table 2 were definitely undersaturated with respect to magnesite initially, and with time approached the lower range of values for magnesite saturation in equilibrium with atmospheric CO_2; that is, a $pH-1/2pMg^{2+}$ of 6.6.

Two factors would tend to make experimental values controlled by magnesite appear near the lower limit of the magnesite stability range. First and most important is the fact that the ΔG of a mineral corresponding to the solution conditions under which it precipitates may range from a few tenths to more than a kcal more negative than the ΔG determined for a particular mineral specimen (see Kittrick, 1970). For every kcal difference, the $pH-1/2pMg^{2+}$ calculated in (3) would be lowered 0.37 unit. Secondly, one must also consider that the air in a laboratory will ordinarily be higher in CO_2 than the average air content of 0.035 per cent used in arriving at the value of the $pH-1/2pMg^{2+}$ for magnesite in (3). A CO_2 content of 0.045 per cent would decrease this value by about 0.10 unit.

After 8 days, the pH of the samples increased with time as the pMg increased. This had the effect of maintaining the pH-1/2pMg^{2+} essentially constant and suggests control by some solid phase, presumably magnesite.* However, it was not possible to detect the formation of any magnesite in the samples by X-ray diffraction. It may be that the magnesite, if precipitated, was not crystalline. However, even crystalline magnesite in small amounts would have been difficult to detect, because the strongest X-ray diffraction peaks of magnesite are coincident with or close to the much stronger peaks of the vermiculite, montmorillonite and hematite which comprise the sample. It is evident from Table 2 that *some* Mg-containing phase is precipitating, because the Mg content of the solutions is decreasing with time.

It appears then that when the slightly acid mineral mixture was mixed with the initial solution, the samples approached the montmorillonite–amorphous silica join at a point well below the calculated position of the magnesite plane. Vermiculite dissolution moved the solution composition along the montmorillonite–amorphous silica join toward a vermiculite plane. Before the vermiculite plane could be reached, the solution composition intercepted the magnesite plane, remaining at the montmorillonite–magnesite–amorphous silica triple point. Thus, the two mica-derived vermiculites studied appear to have no stability area at room T and P.

When the stability of a mineral is determined by solubility methods, it is usually possible to show that the equilibrium was indeed controlled by the mineral in question. For example, this may be done by noting the behavior of solution compositions in relation to a theoretical stability line or plane whose slope is dependent upon the particular composition of the mineral. The negative assertion, that a mineral is not stable at all, cannot be proved with equal assurance because it is not possible to prove that non-equilibrium behavior is controlled by a particular mineral phase. It is possible that the dissolution of some unstable, undetected phase in each of the vermiculite samples controlled sample behavior. It seems much more likely that sample behavior was controlled by the large amounts of vermiculite present, and that the two vermiculites were unstable.

THE FORMATION OF MICA-DERIVED VERMICULITES: A HYPOTHESIS

A mineral is stable in a system if it can persist there at equilibrium, but in some cases a near-infinite

---

* The only other Mg-containing phase in the samples initially was Belle Fourche montmorillonite. Solubility determinations on this mineral in the absence of vermiculite did not encounter an equilibrium value of pH-1/2pMg^{2+} greater than 1·55 (Kittrick, 1971a).

amount of time may be required to reach equilibrium. Thus, persistence of a mineral on a geologic time scale does not necessarily mean thermodynamic stability. Muscovite mica can be stable relative to other minerals that form in soils and sediments if solution conditions are suitable (Routson, 1970). However, these conditions rarely obtain in river and upper ground waters, or even in ocean waters. The stabilities of biotite and ordinary phlogopite have not been determined, but weathering relationships indicate these minerals are even less stable than muscovite. With biotite and phlogopite rarely, if ever, stable in the soil solution, it is not surprising to find experimentally that the vermiculites derived from them are unstable in solution also. These considerations lead to the following hypothesis: *all mica-derived vermiculites are unstable as a result of the unique way in which micas degrade during weathering.*

It is the nature of hypotheses that they be continually tested by comparison with newly-observed facts. Thus, when more observations are available, it may be necessary to alter or even discard this hypothesis, especially since the hypothesis contains two absolutes; that is, *all* mica-derived vermiculites and the weathering environment (*all* naturally occurring solutions at essentially room T and P). Obviously, direct stability determinations such as those reported herein are needed on a wider variety of mica-derived vermiculites, particularly those that are dioctahedral and small in size. At the moment, it is instructive to examine the compatibility of this hypothesis with what is known of mica degradation.

Mica degradation reactions

Complete dissolution. One might expect biotite, for example, merely to dissolve in the soil solution as follows:

$$K_2(Mg, Fe^{2+})_6(Al_2Si_6)O_{20}(OH)_4 + 20H^+$$
$$= 2K^+ + 3Mg^{2+} + 3Fe^{2+} + 2Al^{3+} + 6H_4SiO_4. \quad (4)$$

This reaction (see Huang, Crosson and Rennie, 1968) apparently does occur as written to some extent (for example, Newman, 1969). However, dissolution of biotite more often takes place in a series of steps. Some of the Fe^{2+} is oxidized while still in the mica structure (Boettcher, 1966; Farmer and Wilson, 1970), some octahedral cations may go into solution before the tetrahedral cations (Sawhney and Voigt, 1969) and some of the OH$^-$ groups may be neutralized by the addition of H$^+$ (Newman and Brown, 1966).

K replacement. If the level of K$^+$ (or ions of lower hydration energy) in solution is low, the most striking mica degradation reaction of all is the rapid exchange of structural K$^+$ by ions of higher hydration energy.

K^+-replacement depends in part upon concurrent progress of reactions mentioned previously. For example, the replacement of K^+ is hastened by the incorporation of protons into the mica structure (Raman and Jackson, 1966; Newman, 1970b) and slowed by the oxidation of Fe^{2+} (Gilkes et al., 1973). Nevertheless, K^+-replacement can be illustrated as

$$K-mica + M^+ = M^+-vermiculite + K^+ \quad (5)$$

where M^+ is not restricted to monovalent cations. Reaction (5) is not strictly reversible since the original mica prior to K^+-removal is not exactly reconstituted following K^+ sorption (Newman, 1970a; Brown and Newman, 1970). Additional internal alterations in the vermiculite structure further diminish the likelihood of reversibility. The same is true for hydroxy interlayers, which persist on a geologic time scale even though they may be unstable. Although reactions (4) and (5) involve the same mica, to a large extent they appear to take place independently of each other. That is, (5) can be at equilibrium when (4) is proceeding to the right. However, if (4) is proceeding to the left or is at equilibrium, it seems reasonable to assume that (5) will be at equilibrium also.

The sensitivity of (5) to the *concentration* of K^+ and other weakly hydrated ions in solution appears to originate at the interface of unexpanded and expanded portions of the mica (Jackson, 1963; Wells and Norrish, 1968). The micas have a great selectivity for weakly hydrated ions, apparently because such ions may pass between partially expanded mica layers more readily by shedding much or all of their hydration shell. Thus, the concentration of Cs^+ may be greater than that of the more highly hydrated Na^+ at the site of K^+ exchange in micas even when the concentration of Na^+ is 10,000 times greater than Cs^+ in the external solution (see data of Wells and Norrish, 1968).

The *rate* limiting step in (5) is not the actual exchange of K^+ by M^+, but appears to involve the diffusion of M^+ into the mica structure and especially the outward diffusion of K^+. Experimental data have been fitted to several diffusion models (Mortland and Ellis, 1959; Reed and Scott, 1962; Quirk and Chute, 1968). There are numerous experiments demonstrating the more rapid replacement of K^+ in trioctahedral micas compared to dioctahedral micas. For example, Quirk and Chute (1968) find the diffusion coefficient for K^+ release in several illites to be nine to ten orders of magnitude smaller (10^{-19} vs 10^{-10} cm^2/sec) than those reported for biotite and phlogopite by Rausell-Colom et al. (1965). Thus, the K^+-replacement reaction has a strong structural bias.

Critical K^+ levels and the K^+ content of natural waters

The critical K^+ level is the level of K^+ in solution required to stop the replacement of K^+ in the mica structure by M^+ in solution. It is the equilibrium K^+ level of (5). The critical K^+ level of a particular mica-vermiculite reaction combines the contributions of many factors of mineral composition and structure (including particle size and strain), as well as solution composition. Some of these factors have been studied sufficiently to permit considerable understanding of the mechanism involved. Examples of such effects are tetrahedra rotation and OH^- tilt in dioctahedral micas and the F^- content of trioctahedral micas.

The critical K^+ level decreases with decreasing temperature, decreasing M^+ concentration, and varies with the nature of M^+. Critical K^+ levels have not been established for room temperature and the M^+ concentration in natural waters, but approximate levels can be estimated for these conditions. The level of about 0.02 ppm K^+ maintained by sodium tetraphenylboron permits the extraction of K^+ from illites and muscovites, whereas Mackintosh et al. (1971) found the critical K^+ level for a muscovite to be <3 ppm (lower still for a lepidolite). With the critical K^+ level for the muscovite thus apparently somewhere between 0.02 and 3 ppm, it is satisfying to note that Rausell-Colom et al. (1965) found a critical K level of almost 1 ppm K^+ for muscovite under high salt, high temperature conditions. Since Rausell-Colom et al. found the critical K^+ level to decrease with decrease in temperature and concentration of extracting salt, the critical K^+ level of muscovite is probably closer to 0.02 ppm than to 1 ppm $K+$.

The average K^+ content of river water is 2.3 ppm (Livingston, 1963), with many ground waters a little higher (see White et al., 1963). These average K^+ levels are probably higher than the critical K^+ levels of most dioctahedral micas, hence the conclusion by Rausell-Colom et al. (1965) that muscovites will not exchange their K^+ in most natural waters. However, the K^+ content of rain-water may often be low enough to exchange K^+ from dioctahedral micas at the soil surface.

Mackintosh et al. (1971) found the critical K^+ level for a biotite to be 290 ppm and Rausell-Colom et al. (1965) found that the critical K^+ level for a series of biotites and phlogopites ranged from almost that level down to about 2.4 ppm (depending upon temperature, salt concentration and F^- content). The critical K^+ levels of trioctahedral micas in natural waters are no doubt nearer the lower end of this range, as indicated by Hoda and Hood (1973). Since these critical K^+ levels still exceed the K^+ content of most natural waters, it seems evident that many trioctahedral micas should exchange their K^+ in natural fresh water, which they do.

Weathering and the formation of mica-derived vermiculites

If all mica-derived vermiculites are unstable intermediates, resulting from the unique way in which micas degrade, then in essence mica-derived vermiculites owe their existence to the rapidity of reactions corresponding to (5) as compared to reactions corresponding to (4). It seems likely that eventually the mica-derived vermiculites either dissolve to furnish essentially the same ions in solution that the parent mica does in equation (D) (except for ions already lost, such as K^+), or they may alter internally toward some mineral that is stable under the prevailing solution conditions, such as montmorillonite (for example see Sridhar et al., 1972). However, if dissolution of dioctahedral micas and dioctahedral vermiculites is slow, dioctahedral vermiculites could predominate after trioctahedral micas and vermiculites dissolve away. It is unfortunate that our present understanding of dioctahedral vermiculites is so rudimentary.

If future investigations confirm that mica-derived vermiculites are unstable, fast-forming intermediates in the degradation of mica, as suggested here, it does not necessarily follow that all 'vermiculites' are unstable. Non-mica-derived materials, that by present diagnostic criteria are called vermiculite, may also exist in soils and sediments. Unlike the mica-derived vermiculites, these vermiculites may precipitate from solution and have a stability field in the weathering environment. It has not yet been possible to obtain a sample of such material that is suitable for stability determinations.

REFERENCES

American Public Health Assoc. (1960) *Standard Methods for the Examination of Water and Waste Water*. 11th Edn., New York.

Boettcher, A. L. (1966) Vermiculite, hydrobiotite, and biotite in the Rainy Creek igneous complex near Libby, Montana: *Clay Minerals* **6**, 283–296.

Brown, G. and Newman, A. C. D. (1970) Cation exchange properties of micas—III. Release of potassium sorbed by potassium depleted micas: *Clay Minerals* **8**, 273–278.

Farmer, V. C. and Wilson, M. J. (1970) Experimental conversion of biotite to hydrobiotite: *Nature* **226**, 841–842.

Garrels, R. M. and Christ, C. L. (1965) *Solutions, Minerals and Equilibria*, 435 p. Harper and Row, New York.

Gilkes, R. J., Young, R. C. and Quirk, J. P. (1973) Artificial weathering of biotite—I. Potassium removal by sodium chloride and sodium tetraphenylboron solutions: *Soil Sci. Soc. Am. Proc.* **37**, 25–38.

Hoda, S. N. and Hood, W. C. (1972) Laboratory alteration of trioctahedral micas: *Clays and Clay Minerals* **20**, 343–358.

Huang, P. M., Crosson, L. S. and Rennie, D. A. (1968) Chemical dynamics of potassium release from potassium minerals common in soils: *Trans. 9th Int. Congr. Soil Sci.* **2**, 705–712.

Hsu, P. H. (1963) Effect of initial pH, phosphate, and silicate on the determination of aluminum with aluminon: *Soil Sci.* **96**, 230–235.

Jackson, M. L. (1956) *Soil Chemical Analysis—Advanced Course*. Published by the author. Univ. Wisconsin, Madison, WI.

Jackson, M. L. (1963) Interlayering of expansible layer silicates in soils by chemical weathering: *Clays and Clay Minerals* **13**, 29–46.

Kittrick, J. A. (1969a) Interlayer forces in montmorillonite and vermiculite: *Soil Sci. Soc. Am. Proc.* **33**, 217–222.

Kittrick, J. A. (1969b) Soil minerals in the Al_2O_3–SiO_2–H_2O system and a theory of their formation: *Clays and Clay Minerals* **17**, 157–167.

Kittrick, J. A. (1970) Synthesis of kaolinite at 25°C and 1 atm: *Clays and Clay Minerals* **18**, 261–267.

Kittrick, J. A. (1971a) Stability of montmorillonites—I. Belle Fourche and Clay Spur montmorillonites: *Soil Sci. Soc. Am. Proc.* **35**, 140–145.

Kittrick, J. A. (1971b) Montmorillonite equilibria and the weathering environment: *Soil Sci. Soc. Am. Proc.* **35**, 815–820.

Klotz, I. (1964) *Chemical Thermodynamics*, 468 p. Bemjamin, Inc. New York.

Livingstone, D. A. (1963) Chemical composition of rivers and lakes. In *Data of Geochemistry* (Edited by Fleischer, M.). U.S. Geol. Surey Prof. Paper 440–G.

Mackintosh, E. E., Lewis, D. G. and Greenland, D. J. (1971) Dodecylammonium–mica complexes—I. Factors affecting the exchange reaction: *Clays and Clay Minerals* **19**, 209–218.

Mortland, M. M. and Ellis, B. (1959) Release of potassium as a diffusion controlled process: *Soil Sci. Soc. Am. Proc.* **23**, 363–364.

Newman, A. C. D. (1969) Cation exchange properties of micas—I. The relation between mica composition and potassium exchange in solutions of different pH: *J. Soil Sci.* **20**, 357–373.

Newman, A. C. D. (1970a) Cation exchange properties of micas—II. Hysteresis and irreversibility during potassium exchange: *Clay Minerals* **8**, 267–272.

Newman, A. C. D. (1970b) The synergetic effect of hydrogen ions on the cation exchange of potassium in micas: *Clay Minerals* **8**, 361–372.

Newman, A. C. D. and Brown, G. (1966) Chemical changes during the alteration of micas: *Clay Minerals* **6**, 297–310.

Quirk, J. P. and Chute, J. H. (1968) Potassium release from mica-like clay minerals: *Trans. 9th Int. Congr. Soil Sci.* **2**, 671–681.

Raman, K. V. and Jackson, M. L. (1966) Layer charge relations in clay minerals and micaceous soils and sediments: *Clays and Clay Minerals* **14**, 53–68.

Rausell-Colom, J. A., Sweatman, T. R., Wells, C. B. and Norrish, K. (1965) Studies in the artificial weathering of mica: In *Experimental Pedology* (Edited by Hallsworth, E. G. and Crawford, D. V.), pp. 40–72. Butterworths, London.

Reed, M. G. and Scott, A. D. (1962) Kinetics of potassium release from biotite and muscovite in sodium tetraphenylboron solutions: *Soil Sci. Soc. Am. Proc.* **26**, 437–440.

Robie, R. A. and Waldbaum, D. R. (1968) Thermodynamic properties of minerals and related substances at 298.15°K (25.0°C) and one atmosphere (1.013 Bars) pressure and at higher temperatures: *Geol. Survey Bull.* 1259, 256 p.

Roth, C. B., Jackson, M. L., Lotse, E. G. and Syers, J. K. (1968) Ferrous–ferric ratio and CEC changes on

deferration of weathered micaceous vermiculite: *Israel J. Chem.* **6**, 261–273.
Routson, R. C. (1970) *Illite solubility*. Ph.D. Thesis, Washington State University, Pullman, Washington.
Sawhney, B. L. and Voigt, G. K. (1969) Chemical and biological weathering in vermiculite from Transvaal: *Soil Sci. Soc. Am. Proc.* **33**, 625–629.
Schofield, R. K. and Taylor A. W. (1954) The hydrolysis of aluminum salt solutions: *J. Chem Soc.* **18**, 4445–4448.

Sridhar, K., Jackson, M. L. and Syers, J. K. (1972) Cation and layer charge effects on blister-like osmotic swelling of micaceous vermiculite: *Amer. Min.* **57**, 1832–1848.
Wells, C. B. and Norrish, K. (1968) Accelerated rates of release of interlayer potassium from micas: *Trans. 9th Int. Congr. Soil Sci.* **2**, 683–694.
White, D. E., Hem, J. D. and Waring, G. A. (1963) Chemical composition of sub-surface waters. In *Data of Geochemistry* (Edited by Fleischer, M.). U.S. Geol. Survey Prof. Paper 440–F.

Résumé—Des déterminations de stabilité ont été effectuées par des méthodes de solubilité sur deux vermiculites dérivées de micas trioctaédriques. La vermiculite dérivée de la phlogopite est instable dans des conditions de solution acide pour lesquelles les stabilités de la montmorillonite, de la kaolinite et de la gibbsite ont été déjà déterminées. On a essayé ensuite de localiser l'éventuel point triple montmorillonite vermiculite–silice amorphe. Ce point triple implique des conditions de pH alcalin, pH$_4$SiO$_4$ élevé et Mg^{2+} élevé. Ce sont des conditions dans lesquelles les vermiculites dérivées de phlogopite et biotite doivent très probablement contrôler l'équilibre si ce sont des minéraux stables. Les échantillons montmorillonite vermiculite–silice amorphe ont évolué vers le point triple montmorillonite magnésite silice amorphe, ce qui ne laisse subsister aucune aire de stabilité pour les vermiculites. Les vermiculites à cristaux de grande taille, dérivées de micas trioctaédriques, apparaissent instables dans toutes les conditions de T et P ambiantes.

On présente des arguments qui indiquent que les micas sont instables dans la plupart des environnement d'altération. On fait l'hypothèse que les vermiculites dérivées de micas découlent de l'unique voie selon laquelle les micas instables se dégradent dans ces conditions. On considère que la vermiculite provient d'une série de réactions dont les vitesses relatives entraînent souvent l'abondance de vermiculite. Ces vitesses relatives de réaction sont lentes pour la dissolution du mica, rapides pour l'extraction de K et pour les autres réactions concourant à la formation de la vermiculite, et lentes pour la dissolution de la vermiculite. En termes de chimie, les vermiculites dérivées de mica peuvent être considérées comme des intermédiaires instables à formation rapide.

Kurzreferat—An zwei aus trioktaedrischen Glimmern entstandenen Vermiculiten wurden Stabilitätsbestimmungen mit Löslichkeitsmethoden durchgeführt. Der phlogopitbürtige Vermiculit erwies sich als instabil unter sauren Lösungsbedingungen, bei denen früher die Stabilität von Montmorillonit, Kaolinit und Gibbsit bestimmt wurde. Als nächstes wurde versucht, einen gemeinsamen Schnittpunkt für die drei Phasen Montmorillonit–Vermiculit–amorphe Kieselsäure ausfindig zu machen. Dieser Schnittpunkt war durch alkalisches pH, hohes pH$_4$SiO$_4$ und hohe Mg^{2+}-Gehalte gekennzeichnet. Dieses sind Bedingungen, unter denen am ehesten mit einer Beeinflussung der Gleichgewichte durch phlogopit- und biotitbürtige Vermiculite zu rechnen ist, soweit diese stabile Minerale darstellen. Die aus Montmorillonit Vermiculit–amorpher Kieselsäure bestehenden Proben gelangten zum Montmorillonit Magnesit–amorphe Kieselsäure–Schnittpunkt, ohne irgendein Stabilitätsfeld für die Vermiculite offenzulassen. Diese grobkörnigen, trioktaedrischen, glimmerbürtigen Vermiculite scheinen unter allen Bedingungen von Raumtemperatur und -druck instabil zu sein.

Es werden Argumente dafür beigebracht, daß Glimmer nahezu in jedem Verwitterungsmilieu instabil sind. Eine Hypothese wird vorgestellt, daß zufolge glimmerbürtige Vermiculite das Ergebnis einer spezifischen Umwandlung sind, der instabile Glimmer unter solchen Bedingungen beim Abbau unterliegen. Es wird angenommen, daß Vermiculite aus einer Folge von Reaktionen entstehen, deren relative Raten oft zu einem Vermiculitüberschuß führen. Diese relativen Reaktionsraten sind langsam für die Glimmerauflösung, schnell für die K-Freisetzung und andere mit der Vermiculitbildung verbundene Reaktionen, sowie langsam für die Vermiculitauflösung. In der chemischen Terminologie können die glimmerbürtigen Vermiculite als sich schnell bildende, instabile Zwischenstufen betrachtet werden.

Резюме — Методами растворения определялась устойчивость двух переотложенных из слюды вермикулитов. В условиях кислотных растворов ранее применявшихся для определения устойчивости монтмориллонита, каолинита и гидраргиллита, переотложенный из флогопита вермикулит оказался неустойчивым. Затем сделали попытку определить местонахождение тройной точки аморфного кремнезема-монтмориллонита-гермикулита. Эта тройная точка включала условия щелочного pH, высокого pH$_4$SiO$_4$ и высокого Mg^{2+}. В этих условиях устойчивые вермикулиты, переотложенные из флогопита и биотита, наверно, более всего регулируют *равновесное состояние*. Образцы аморфного кремнезема-вермикулита-монтмориллонита перешли на тройную точку аморфного кремнезема-магнезита-монтмориллонита совсем не оставив устойчивой области для вермикулита. Эти крупнозернистые триоктаэдральные вермикулиты, отложенные из слюды, очевидно, неустойчивы при всех условиях комнатной температуры и давления.

Приводят доводы указывающие, что слюды являются неустойчивыми почти что во всех условиях выветривания. Выдвигается гипотеза, что отложенные из слюды вермикулиты результируются вследствие необыкновенного способа распадения слюд во всех фациях. Предполагают, что отложение вермикулита происходит вследствие целого ряда интенсивных взаимных реакций, в результате которых часто образуется много вермикулита. Эти относительные степени реакции медленно растворяют слюду, быстро удаляют К и другие реакции следующие за формованием вермикулита, и медленно растворяют вермикулит. В химической терминологии образующиеся из слюды вермикулиты можно считать быстрообразующимися неустойчивыми переходными типами пород.

Denis M. Shaw

Encl.

DMS/lh

JOURNAL OF
THE GEOCHEMICAL SOCIETY THE METEORITICAL SOCIETY

GEOCHIMICA ET COSMOCHIMICA ACTA

DR. DENIS M. SHAW / *Executive Editor*

Department of Geology / McMaster University / 1280 Main Street West / Hamilton, Ontario, Canada L8S 4M1

Telephone: 416-525-9140 Ext. 4395
Telex: 061-8347.

July 4, 1986

Dr. Ward Chesworth,
Department of Land Resource Science,
University of Guelph,
Guelph, Ontario N1G 2W1.

Dear Ward,

The enclosed book was sent to me 'gratis' as a GCA article appears on p.228. I thought you might find it a useful addition to your library.

With best wishes.

Yours sincerely,

11

Copyright © 1983 by the Royal Netherlands Academy of Agricultural Science
Reprinted from *Plant and Soil* 75:283-308 (1983)

Acidification and alkalinization of soils

N. VAN BREEMEN, J. MULDER and C. T. DRISCOLL*
Department of Soil Science and Geology, Agricultural University, P.O. Box 37, 6700 AA Wageningen, The Netherlands

Received 13 May 1983. Revised August 1983

Key words Acid neutralizing capacity Assimilation of cations and anions H^+-budget Mineralization of organic matter Mineral weathering N-cycle Oxidation reduction cycles Soil acidification Soil acidity Soil alkalinization Soil pH

Abstract Acidification or alkalinization of soils occurs through H^+ transfer processes involving vegetation, soil solution and soil minerals. A permanent change in the acid neutralizing capacity of the inorganic soil fraction ($ANC_{(s)}$), *i.e.* soil acidification ($\Delta ANC < 0$) or soil alkalinization ($\Delta ANC > 0$), results from an irreversible H^+ flux. This irreversible H^+ flux can be caused either by direct proton addition or depletion, by different mobility of components of the $ANC_{(s)}$ or by a permanent change in redox conditions. The contributions of (a) acidic atmospheric deposition, (b) nitrogen transformations, (c) deprotonation of CO_2 and of organic acids and protonation of their conjugate bases, (d) assimilation of cations and anions by the vegetation, (e) weathering or reverse weathering of minerals and (f) stream output to changes in the $ANC_{(s)}$ are illustrated by means of H^+ budgets for actual soils and watersheds.

1. Introduction

Numerous papers have been published on soil acidity and strongly alkaline soils, subjects of great practical importance because of the problems associated with plant growth at extremes in soil pH. These publications generally address the nature of soil acidity/alkalinity and methods to characterize and to amend these conditions. Ironically, research pertaining to the processes responsible for acidification or alkalinization of soils is lacking. Perusal of textbooks on soil science and geochemistry shows that the subject of soil acidification is often neglected. While the effect of protons on weathering, cation exchange processes, pH buffering, and properties of acidic soils receive much attention, the ultimate cause of soil acidity is often loosely attributed to 'dissociation of CO_2 or organic acids, uptake of cations, or nitrification'.

In view of (a) the practical importance of soil acidification and alkalinization and (b) the fact that acidification, or, to a lesser extent alkalinization, occurs in all soils, a more systematic survey of such processes is highly relevant. Apart from the evaluation by Mattson and

---

* Present address: Department of Civil Engineering Syracuse University, Syracuse NY 13210, USA

Koutler-Andersson[18] of the role of the vegetation in soil acidification/alkalinization, there was little early research on this subject. Only recently have evaluations of the quantitative importance of atmospheric acidic deposition compared to natural processes of soil acidification prompted further research on this subject[8, 10, 20, 23, 24, 27, 28]. Inspired by these publications, we will review concepts and processes that are important in H^+-production and -consumption in soils and discuss several examples of specific H^+-transfer processes that have lead to the formation of acidic or alkaline soils.

2. Concepts related to acidification and alkalinization of soil and water

According to Brønsted, an acid is a substance that can donate a proton (a proton donor) and a base is a substance that can accept a proton (a proton acceptor). So, proton transfer occurs when an acid (HA) reacts with a base (B^-).

$$\begin{aligned} HA &= A^- + H^+ \\ H^+ + B^- &= HB \\ \hline HA + B^- &= HB + A^- \end{aligned} \quad (1)$$

HA and A^-, or HB and B^- are called conjugate acid-base pairs. Examples of conjugate acid-base pairs in soil-water systems include $H_3O^+–H_2O$, $Al(OH)_2^+–Al(OH)_3$, $H_2CO_3–HCO_3^-$ and $NH_4^+–NH_3$. In order to evaluate the acidification of soil and water it is necessary to distinguish between intensity and capacity factors. Intensity factors are determined by chemical properties and are independent of the quantity or size of the system considered while, in contrast, capacity factors are a function of quantity or size of the system. For instance, the strength of an acid or a base (the tendency to donate or accept a proton) is an intensity factor, while the amount of acid or base present in a given system is a capacity factor.

The strength of acid/base systems (conjugate acid/base pairs) is quantified through proton dissociation constants (K_{HA}).

$$HA + H_2O = H_3O^+ + A^-; \quad K_{HA} = \frac{(H_3O^+)(A^-)}{(HA)};$$

$$pH = pK_{HA} + \log\frac{(A^-)}{(HA)}$$

The stronger the acid system, the greater the extent of proton dissociation and the greater the value of K_{HA}.

Soil is a mixture of acid/base systems. These acid/base systems

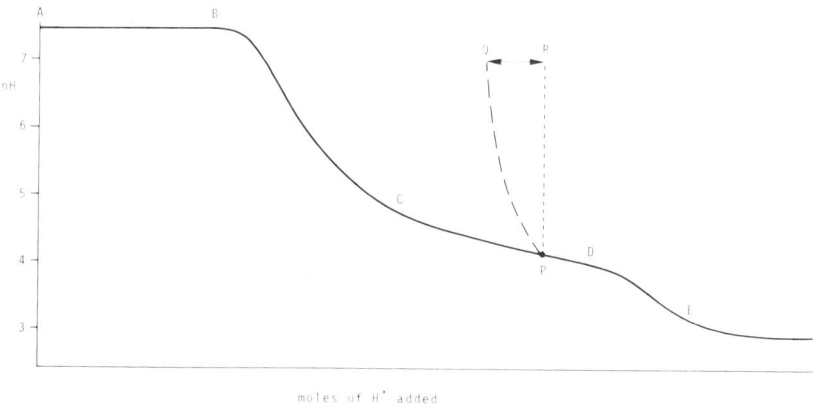

Fig. 1. Hypothetical titration curve of a well drained calcareous clay soil at constant pCO_2, titrated with a strong acid of pH 3.

can be thought of as proton energy levels. The proton energy level of an acid/base system is consistent with the K_{HA} of that system. An acid/base system with a low K_{HA} has a low proton energy level and therefore a high affinity for association with protons. Upon addition of protons to a mixture of acid/base systems, the added protons will occupy the lowest proton energy level available. In other words, added protons will associate with the conjugate base of the weakest acid in the mixture. With subsequent addition of protons, successive proton levels are filled (conjugate bases become protonated).

To illustrate this concept, consider the addition of strong acid (HA) to a calcareous clayey soil (Fig. 1). We will assume that the system is freely draining so that the concentration of A^- and the partial pressure of CO_2 remain constant. Protons initially added to the system will associate with $CaCO_3$, resulting in mineral dissolution.

$$CaCO_{3(s)} + H^+ + A^- \rightarrow Ca^{2+} + HCO_3^- + A^- \qquad (2)$$

As long as $CaCO_3$ is present added protons will be consumed by the dissolution reaction, and the pH of the soil solution will remain constant (A → B in Fig. 1). In the presence of $CaCO_3$ the soil is well buffered (resistant to changes in pH). When the $CaCO_3$ is depleted by addition of protons, the solution pH will decrease (B → C) until the added protons associate with the next available proton energy level. In our example this might be exchange sites on clay and ultimately the dissolution of clay minerals. These processes also result in pH buffering (C → D). When the reservoir of clay minerals is exhausted, further addition of protons result in an increase in solution hydrogen ion concentration or decrease in solution pH (D → E).

This titration curve summarizes many important aspects of soil acidification. pH is an intensity factor while the moles of strong acid added is a capacity factor. The titration curve illustrates that due to the presence of acid/base systems ($CaCO_3$, clay) the addition of strong acid does not directly result in a change in pH ($A \to B, C \to D$ in Fig. 1). The amount of strong acid required to reduce the pH of a system to a reference pH value is termed acid neutralizing capacity (ANC).

In aqueous solutions $ANC_{(aq)}$ is defined as the base equivalence less the strong acid equivalence of the system and is determined by strong acid titration to a reference pH. Generally the methyl orange equivalence point around a pH value of 4–5 is used. In most natural environments, $ANC_{(aq)}$ is largely attributed to the bicarbonate concentration. However at extremes in pH or with the presence of additional weak bases (A^- if pK_{HA} exceeds the reference pH) such as natural organic anions, other species may contribute to $ANC_{(aq)}$ (Eqn. 3).

$$ANC_{(aq)} = [HCO_3^-] + 2[CO_3^{2-}] + [OH^-] + [A^-] - [H^+] \quad (3)$$

In this discussion we will be concerned with the $ANC_{(aq)}$ of water draining from soils.

The ANC of most soils is associated with silicate minerals which have very slow dissolution kinetics, so titration by strong acid is not a practical way to determine $ANC_{(s)}$. We define 'soil' as all inorganic matter including soil solution, solid particles and adsorbed ions. To simplify calculations we have excluded all organic matter (both living and dead) from the soil system. In practice the best way to estimate the $ANC_{(s)}$ is by component composition,

$$ANC_{(s)} = 6(Al_2O_3) + 2(CaO) + 2(MgO) + 2(K_2O) + 2(Na_2O)$$
$$+ 4(MnO_2) + 2(MnO) + 6(Fe_2O_3) + 2(FeO) - 2(SO_3)$$
$$- 2(P_2O_5) - (HCl) \quad (4)$$

where brackets denote molar concentration.

The concentration of these components is usually determined by a total elemental analysis of the mineral soil. Solutes in the soil solution are considered as part of $ANC_{(s)}$. The contribution of individual basic and acidic components can be illustrated through the following reactions.

$$CaO_{(s)} + 2H^+ \to Ca^{2+} + H_2O \quad (5)$$

$$P_2O_{5(s)} + 3H_2O \to 2H_2PO_4^- + 2H^+ \quad (6)$$

Note that the number of moles of protons accepted or donated per mole of component is consistent with the stoichiometry of $ANC_{(s)}$ (Eqn. 4).

$ANC_{(s)}$ can be thought of as the component base equivalent less the component strong acid equivalence (pK_{HA} < reference pH) of the system. The weak acid components (pK_{HA} > reference pH) like CO_2 and SiO_2 do not contribute to the ANC of a soil.

$$CO_{2(s)} + H_2O \rightarrow H_2CO_{3(aq)} \qquad (7)$$

$$SiO_{2(s)} + H_2O \rightarrow H_4SiO_{4(aq)} \qquad (8)$$

These components are in a fully protonated form at the reference pH. Therefore they do not contribute to the base equivalence of the system nor can they contribute protons to serve as strong acid equivalence in the system. Because inorganic nitrogen is normally a minor component of soil, we have not considered it in our definition of $ANC_{(s)}$.

The components that contribute to $ANC_{(s)}$ depend on the reference pH chosen. Any reference pH may be used to define $ANC_{(s)}$, but the value selected should be consistent with the nature of the soil of interest. For example because of the general sensitivity of crops to acidic conditions, a reference pH of 5 might be appropriate for agricultural soils. The dissolution of Fe_2O_3, Al_2O_3 and MnO_2 is negligible above pH values of 5 so $ANC_{(s)}$ may be simplified (Eqn. 9)

$$ANC_{(s, \text{ref. pH} = 5)} = 2(CaO) + 2(MgO) + 2(K_2O) + 2(Na_2O)$$
$$+ 2(MnO) + 2(FeO) - 2(SO_3) - 2(P_2O_5) - (HCl) \qquad (9)$$

For forest soils a reference pH of 3 might be more reasonable. Under these conditions appreciable quantities of aluminium may dissolve. Therefore $ANC_{(s)}$ should be modified to include aluminium (Eqn. 10).

$$ANC_{(s; \text{ref. pH} = 3)} = 6(Al_2O_3) + 2(CaO) + 2(MgO) + 2(K_2O)$$
$$+ 2(Na_2O) + 2(MnO) + 2(FeO) - 2(SO_3) - 2(P_2O_5) - (HCl) \qquad (10)$$

Under extremely low pH conditions, *e.g.* in certain acid sulfate soils, significant dissolution of ferric iron and manganic manganese will occur. As a result, the complete expression of $ANC_{(s)}$ (Eqn. 4) is appropriate for very acidic soils.

The dissolution of relatively soluble ferrous and manganeous minerals contributes to the depletion of soil ANC and therefore these reduced components have been included in our definition. However transport of these substances will only occur, to any extent, under reducing (anoxic) or acidic conditions. In the presence of molecular oxygen these reduced components will generally be oxidized to less soluble components

$$2FeO_{(s)} + \tfrac{1}{2}O_{2\,(aq)} \rightarrow Fe_2O_{3\,(s)} \tag{11}$$

$$MnO_{2\,(s)} + O_{2\,(aq)} \rightarrow MnO_{4\,(s)} \tag{12}$$

Therefore in aerobic, non-acid soils FeO and MnO_2 do not significantly contribute to $ANC_{(s)}$.

When oxidized sulfur (SO_3) is leached from the soil, sulfuric acid, a strong acid is produced.

$$SO_{3\,(s)} + H_2O \rightarrow SO_4^{2-} + 2H^+ \tag{13}$$

Unlike oxidized sulfur, reduced sulfur (H_2S) is a weak acid ($pK_1 = 7.0$) and therefore does not directly contribute to the ANC of a soil. However in the presence of molecular oxygen reduced sulfur will be oxidized to sulfuric acid. H_2S should be viewed as potential acidity which is realized under aerobic conditions. To illustrate the role of reduced sulfur in soil ANC consider a hypothetical calcium-reduced sulfur system. Prior to any transformation the $ANC_{(s)}$ of the system is equivalent to the calcium content of the soil. If the sulfur volatilizes from the soil as H_2S gas there is no change in the soil ANC.

$$CaO \cdot H_2S_{(s)} \rightarrow CaO_{(s)} + H_2S_{(g)} \tag{14}$$

In the presence of molecular oxygen the reduced sulfur may oxidize and deplete the ANC of the soil.

$$CaO \cdot H_2S_{(s)} + 2O_{2\,(aq)} \rightarrow CaO \cdot SO_{3\,(s)} + H_2O \tag{15}$$

If the oxidized sulfur dissolves and is transported from the soil, the $ANC_{(s)}$ will return to its initial level and the drain water will acidify.

$$CaO \cdot SO_{3\,(s)} + H_2O \rightarrow CaO_{(s)} + 2H^+ + SO_4^{2-} \tag{16}$$

However, it is more likely that the sulfuric acid will transfer protons to the calcium oxide resulting in congruent dissolution.

$$CaO \cdot SO_{3\,(s)} \rightarrow Ca^{2+} + SO_4^{2-} \tag{17}$$

Through two of these transformations (Eqn. 15 and 17) the $ANC_{(s)}$ is reduced from its initial level.

As in aqueous systems[26] $ANC_{(s)}$, unlike pH, is a conservative parameter and fundamental to soil acidification. We define soil acidification as a decrease in the ANC of the soil, and soil alkalinization as an increase in soil ANC. Although the addition of acid often results in a decrease in both pH and ANC of a soil, this is by no means always so. Some examples of these factors may serve to illustrate some of the basic differences between these terms.

In a hydrologically closed system, wollastonite, $CaSiO_3$, weathers to $CaCO_3$:

$$CaSiO_{3(s)} + H_2CO_{3(aq)} + H_2O \rightarrow CaCO_{3(s)} + H_4SiO_{4(aq)} \quad (18)$$

The weak acid H_2CO_3 transfers two moles of H^+ per mole of $CaSiO_3$, causing the solution pH to decrease from about 10 to about 8[34]. However $ANC_{(s)}$ does not change because CaO is conserved in the transformation, and H_4SiO_4, being a weak acid, does not contribute to $ANC_{(s)}$.

In a hydrologically open (freely draining) system, weathering of $CaCO_3$ by H_2CO_3 and subsequent leaching results in a soil system depleted with respect to CaO and $ANC_{(s)}$.

$$CaCO_{3(s)} + H_2CO_3 \rightarrow Ca^{2+} + 2HCO_3^- \quad (19)$$

Buffering by $CaCO_3$ will maintain the solution pH near 8 even though the system is depleted with respect to $ANC_{(s)}$.

The depletion of ANC from a soil should not be equated with formation of soil acidity as exchangeable H^+ and Al^{3+}. Soil acidity is measured by titrating a soil with a strong base (*e.g.* $Ca(OH)_2$) to a reference pH (*e.g.* [7]). This is qualitatively illustrated in our titration curve (P → Q in Fig. 1). While application of an equivalent amount of base to a soil may increase the pH, it represents only a small fraction of the $ANC_{(s)}$ lost by soil acidification.

In the remainder of this paper we will discuss processes responsible for additions and depletions of protons from a soil and the corresponding changes in $ANC_{(s)}$ (*i.e.* the abscissa of Fig. 1). To what extent changes in $ANC_{(s)}$ affect soil pH depend on many factors that are beyond the scope of this paper (see *e.g.* [2, 5, 33, 34]).

3. H^+-transfer process

All proton transfer processes that are quantitatively important in soil-water-vegetation systems as well as the H^+-indifferent processes involving gaseous exchange between organisms (or dead organic matter) and the atmosphere are summarized in Table 1.

The H^+-transfer reactions (reactions 25–43 in Table 1) have been categorized into those proceeding between biota and a surrounding aqueous phase (*e.g.* the soil solution), those within the aqueous phase or between the aqueous phase and the atmosphere, and those between solids and an aqueous solution.

When processing from left to right (A) or from right to left (B),

Table 1. Reaction equations of H^+ transfer processes and related processes involving biota

Process from left to right	Reaction equation		Process from right to left	
	H^+-indifferent processes			
	biota/atmosphere			
20A photosynthesis	$CO_2 + H_2O$	$= CH_2O + O_2$	respiration	20B
21A N_2-fixation	$N_2 + H_2O + 2R \cdot OH$	$= 2R \cdot NH_2 + \frac{3}{2}O_2$	—	21B
22A NH_3-uptake	$NH_3 + R \cdot OH$	$= R \cdot NH_2 + H_2O$	volatilization of NH_3	22B
23A H_2S-uptake	$H_2S + R \cdot OH$	$= R \cdot SH + H_2O$	volatilization of H_2S	23B
24A SO_2-uptake	$SO_2 + R \cdot OH$	$= R \cdot SH + \frac{3}{2}O_2$	—	24B
H^+-source	**H^+-transfer process**		**H^+-sink**	
	biota/solution			
25A uptake of cations	$M^+ + R\overset{O}{O}H$	$= R\overset{O}{O}M + H^+$	mineralization of M^+	25B
26A uptake of NH_4^+	$NH_4^+ + R \cdot OH$	$= R \cdot NH_2 + H_2O + H^+$	mineralization of org. N	26B
27A mineralization + nitrification of organic N	$R \cdot NH_2 + 2O_2$	$= 2 \cdot OH + NO_3^- + H^+$	uptake of NO_3^-	27B
28A mineralization + oxidation of organic S	$R \cdot SH + \frac{3}{2}H_2O + \frac{7}{4}O_2$	$= R \cdot OH + SO_4^{2-} + 2H^+$	uptake of SO_4^{2-}	28B
29A mineralization of P	$R \cdot H_2PO_4 + H_2O$	$= R \cdot OH + H_2PO_4^- + H^+$	uptake of P	29B

Table 1 (continued)

	Solution or solution/atmosphere		
30A dissociation of H_2O	$2H_2O$	$= OH^- + H^+$	protonation of OH^-	30B
31A dissociation of CO_2	$CO_2 + H_2O$	$= HCO_3^- + H^+$	protonation of HCO_3^-	31B
32A dissociation of org. acids	R_{OH}^O	$= R_O^O + H^+$	protonation of org. anions	32B
33A complexation of metal ions L = org. ligand or OH^-	$HL + M^+$	$= ML + H^+$	decomplexation of metal ions	33B
34A oxidation of H_2S	$H_2S + 2O_2$	$= SO_4^{2-} + 2H^+$	sulfate reduction	34B
35A oxidation of SO_2	$SO_2 + \frac{1}{2}O_2 + H_2O$	$= SO_4^{2-} + 2H^+$		35B
36A nitrification of NH_4^+	$NH_4^+ + 2O_2$	$= NO_3^- + H_2O + 2H^+$		36B
37A nitrification of NO_x	$NO_x + \frac{1}{4}(5-2x)O_2 + \frac{1}{2}H_2O$	$= NO_3^- + H^+$	denitrification	37B
38A nitrification of N_2 (see text)	$N_2 + \frac{5}{2}O_2 + H_2O$	$= 2NO_3^- + 2H^+$	denitrification	38B

	solids/solution		
39A reverse weathering	$M^{n+}\frac{n}{2}H_2O$	$= \frac{n}{2}M_2O + nH^+$	weathering	39B
40A M^{n+}/H^+ exchange	$M^{n+} + nH \cdot exch$	$= M \cdot exch + nH^+$	H^+/M^{n+} exchange	40B
41A oxidation of Fe^{2+}	$Fe^{2+} + \frac{1}{4}O_2 + \frac{5}{2}H_2O$	$= Fe(OH)_3 + 2H^+$	reduction of $Fe(OH)_3$	41B
42A oxidation of FeS	$FeS + \frac{9}{4}O_2 + \frac{5}{2}H_2O$	$= Fe(OH)_3 + SO_4^{2-} + 2H^+$	reduction of $Fe(OH)_3$ and SO_4^{2-}	42B
43A desorption of SO_4^{2-}	$exch\ SO_4 + 2H_2O$	$= exch\ (OH)_2 + SO_4^{2-} + 2H^+$	adsorption of SO_4^{2-}	43B

most reactions represent naturally occurring opposing processes (*e.g.* photosynthesis versus respiration, assimilation versus mineralization of cations etc.). Reactions 25 to 43 have been written in terms of H^+ as a product (25A to 43A) or a reactant (25B to 43B), so that the role of these processes in H^+-transfer is apparent. In nature, however, proton donating and proton accepting processes are always coupled so that the pool of free H^+ (or rather H_3O^+) is extremely small relative to amount of H^+-transfer. For example, even during the extreme proton production (resulting in pH values < 3) associated with the oxidation of sulfides in acid sulfate soils (reaction 42A), more than 99% of the H^+ produced is consumed (buffered!) in weathering and cation exchange processes[30] (reactions 39B and 40B).

We have written all oxidation-reduction reactions with O_2 as the electron acceptor. In nature, reduction reactions (21A, 24A, 27B, 28B, 34B, 37B, 41B and 42B) normally proceed with organic matter as the electron donor. Organic matter is implicitly present in our tabulation of reduction reactions through the production of O_2 that occurs during formation of organic matter by photosynthesis (1A). For example, if the reduction of sulfate (34B) occurs with photosynthetically produced organic matter (20A) as the electron donor, the overall process is illustrated by combining the two reactions as follows:

$$SO_4^{2-} + 2H^+ \rightarrow H_2S + 2O_2 \qquad (34B)$$

$$\underline{2CH_2O + 2O_2 \rightarrow 2CO_2 + 2H_2O} \quad + \qquad (20B)$$

$$SO_4^{2-} + 2CH_2O + 2H^+ \rightarrow H_2S + 2CO_2 + 2H_2O \qquad (44)$$

Because atmospheric gases are uncharged, their uptake by biota does not involve H^+ transfer (reactions 20–24). However the resulting organic compounds may participate in proton transfer processes, when organic matter is mineralized (reactions 26B, 27A, 28A and 29A). There is no individual process that counters N_2-fixation (21B), but it is equivalent to mineralization of organic nitrogen followed by nitrification and denitrification to N_2.

The assimilation of cations by plants (or microorganisms) is accompanied by transfer of H^+ from the plant to the surrounding soil solution while H^+ is transferred into the plant during assimilation of anions (Reactions 25 to 29). Plants regulate the pH of the cell solution during ionic uptake by increasing or decreasing the carboxylate content (*e.g.*[6,11]). When plants assimilate more anions than cations (*e.g.* when all nitrogen is supplied as NO_3^-), the removal of H^+ from near-neutral soils is generally realized by deprotonation of CO_2 (or more appropriately H_2CO_3) leading to an increase of HCO_3^- in the soil solution (combination of reactions 27B and 31A). When cation assimilation exceeds anion assimilation (*e.g.* when most nitrogen is supplied as NH_4^+), the reaction by-product H^+ is generally neutralized by weathering or ion exchange reactions (39B and 40B). So the N-nutrition of plants may play a key role in the proton budget of a soil-vegetation ecosystem. If the extent of organic nitrogen mineralization to NO_3^- and/or NH_4^+ is equivalent to the original assimilation however, then these H^+ transfer processes balance and the overall process is of no importance for the H^+ budget of ecosystems. Moreover, because nitrogen fixation is not a H^+ transfer reaction and nitrogen cycling within soils with vegetation cover is usually very efficient, transformations of nitrogen are generally unimportant in the overall H^+ cycling of an ecosystem. However, when nitrogen is supplied by fertilizers (*e.g.* as $(NH_4)_2SO_4$, $NaNO_3$ or urea) or atmospheric pollution, or when the natural nitrogen cycle is radically disturbed (*e.g.* by clearcutting[15] or climatic change) nitrogen reactions may become extremely significant in the H^+ budget of ecosystems.

When processes involving nitrogen are not quantitatively important to the overall proton flux between plants and soil, ionic uptake by plants results in a net proton flux to the soil because uptake of cations far exceeds that of $H_2PO_4^-$ and SO_4^{2-}.

There are two types of H^+-transfer processes within the solution phase or between the solution and the atmosphere: protonation-deprotonation reactions (30–33) and oxidation-reduction processes (34–38). Weak acids such as CO_2 and most organic acids are important proton sources only when pH values approach and exceed their respective pK_{HA} values (> 5). CO_2 is quantitatively the most important weak acid for aqueous proton donor reactions. Through above-ground photosynthesis and oxidation of photosynthetically formed organic matter below the ground surface, plants transport CO_2 from the dilute reservoir in the atmosphere to high concentrations in the gas phase of the soil. Dissolved organic acids are intermediate (thermodynamically unstable) decomposition by-products that will ultimately be oxidized to CO_2. Some of these organic acids are quite strong (*e.g.* oxalic acid) and may act as a temporary proton source before they are completely oxidized[17]:

$$2CH_2O + \tfrac{3}{2}O_2 \rightarrow HC_2O_4^- + H^+ + H_2O \tag{45}$$

and

$$HC_2O_4^- + H^+ + \tfrac{1}{2}O_2 \rightarrow 2CO_2 + H_2O \tag{46}$$

Formation of metal-organic complexes (reaction 33A) is often a more important source of protons than simple dissociation of H^+ from organic acids (32A). Central metal cations, such as iron and aluminium that are common components in soils, have a high affinity for organic acids. Driscoll[7] observed that only 30% of the functional sites on natural dissolved organic acids are capable of direct proton transfer, the balance of the sites were complexed with iron and aluminium. Our list of protonation and deprotonation reactions could be expanded by including weak acids such as boric acid and NH_4^+. However, these are quantitatively unimportant and processes involving NH_3 and NH_4^+ are implicit in reactions 26 and 27.

In the presence of ambient levels of molecular oxygen, sulfate and nitrate are the thermodynamically stable forms of sulfur and nitrogen. So, compounds containing N and S in lower oxidation states are ultimately oxidized to NO_3^- and SO_4^{2-}, thereby releasing equivalent amounts of H^+ to the environment (reactions 34A–37A). Most of these oxidation reactions are microbially mediated.

Although oxidation of molecular nitrogen (N_2) by molecular oxygen is thermodynamically favoured at ambient atmospheric concentrations (which at equilibrium would result in a nitric acid ocean with a pH 1.1!), this process only removes about $10^{-3}\%$ of N_2 in the atmosphere each year[12]. The net effect of N_2-fixation followed by mineralization and nitrification of fixed organic N (reactions 21A plus 27A) is identical to direct oxidation of N_2 (38A). Over a geological time scale these processes could indeed lead to excessive acidification of soils and waters. But through efficient use of HNO_3 formed in the atmosphere and in the soil, and by denitrification of nitrate trapped under anoxic conditions (reaction 38B), the biosphere apparently maintains the atmospheric N_2–O_2 disequilibrium and the more comfortable low concentrations of HNO_3 in soils and waters[16].

In the absence of molecular oxygen, sulfate and nitrate can be used as electron acceptors in microbial respiration, leading to the production of reduced forms of S and N and the consumption of protons (reactions 34B, 37B and 38B). As discussed earlier, O_2 generated in these reduction reactions represents a by-product of the photosynthetic production of the actual electron donor, *i.e.* organic matter. Favourable conditions for microbial growth in terms of pH and availability of moisture and nutrients are necessary for microbial reduction and oxidation processes.

The last category of H^+-transfer processes involves reactions between solid and aqueous phases. Weathering of oxygen-coordinated metals bound in silicate or carbonate minerals is quantitatively the single most important sink for H^+-ions in the earth crust (reaction 39B). A typical example is the weathering of feldspar to kaolinite:

$$2KAlSi_3O_{8(s)} + 2H^+ + 9H_2O \rightarrow 2K^+ + Al_2Si_2O_5(OH)_{4(s)} + 4H_4SiO_4 \qquad (47)$$

This reaction is equivalent to a depletion of $ANC_{(s)}$ through the dissolution of K_2O (reaction 39B, if M represents K):

$$K_2O + 2H^+ \rightarrow 2K^+ + H_2O \qquad (48)$$

The protons required in weathering reactions can be provided by a host of processes (Table 1). Dissolution of Ca, Mg, K and Na by H^+ predominates at relatively high pH (5–9), whereas Al_2O_3 (either bound in silicates or as free or hydrous oxide) is an important proton sink at low pH values (pH 3–5).

New formation of silicate or carbonate minerals can be considered as reverse weathering[9] and results in an increase in $ANC_{(s)}$ (reaction 39A). Because formation of silicates or carbonates near the earth's surface occurs only at relatively high pH values (> 7), the H^+ produced in these reactions is generally neutralized by the protonation of HCO_3^- (reaction 31B). This is illustrated by the precipitation of calcium carbonate

$$Ca^{2+} + 2HCO_3^- \rightarrow CaCO_{3(s)} + CO_{2(g)} + H_2O \qquad (49)$$

Cation exchange reactions (reactions 40A and 40B) resemble weathering and reverse weathering in their effect on the solution composition. With increasing pH values the negative charge of exchangers increases due to increased deprotonation of functional groups. Humic substances represent an important source of protons in these types of transformations. Conversely, at very low pH values (< 5) anion adsorption, especially on hydrous oxides of Fe and Al, becomes an important mechanism to remove dissolved H^+ (Reaction 43B). Sulfate adsorption is stoichiometrically equivalent to the formation of basic sulfates such as jurbanite ($AlOHSO_4 \cdot 5H_2O$) and jarosite ($KFe_3(SO_4)_2(OH)_6$). These processes store acidity as SO_3 within the solid phase that can be released to solution (Reaction 43A) when the pH is increased[32].

Reduction of $Fe(OH)_3$ and of SO_4^{2-} to Fe^{2+} and Fe-sulfides (Reactions 41B and 42B) are confined to anaerobic conditions and require organic matter as an energy source (electron donor). The uptake of protons through these processes is an important mechanism for the increase in pH values often observed upon submergence and reduction of mineral soils. During the oxidation of Fe^{2+} and of FeS (or FeS_2) (Reactions 41A and 42A) these protons are released again.

4. Quantifying soil acidification and alkalinization: H^+-budget calculations and measurements

The addition of strong acid to a soil system results in an immediate depletion of $ANC_{(s)}$. Weak acid (pK_{HA} > reference pH) transformations do not directly contribute to changes in $ANC_{(s)}$, unless the conjugate base of the weak acids and its associated dissolved cation is transported from the soil. These transformations, in effect, are the result of an irreversible H^+ flux, or as termed by Ulrich[28] a spatial and/or temporal decoupling of H^+ producing and consuming processes. Irreversible H^+ fluxes occur by a permanent separation of the soil with another phase (gaseous, aqueous or biomass) participating in the H^+ transfer process. It is noteworthy, that the portion of the biomass not effectively involved in nutrient cycling can be considered as permanently separated from the soil system.

A few examples may serve to illustrate these principles. SO_2 deposited from the atmosphere to an aerobic soil, will oxidize to H_2SO_4 and directly deplete the ANC of a soil. This deposition is an irreversible input of H^+.

The dissolution of calcium carbonate by CO_2,

$$CaCO_{3(s)} + CO_{2(aq)} + H_2O \rightarrow Ca^{2+} + 2HCO_3^- \quad (50)$$

will result in a decrease in $ANC_{(s)}$ only when the calcium bicarbonate solution has been transported from the soil. Transport of the conjugate base (HCO_3^-) from the soil is required to accomplish an irreversible H^+ flux in weak acid proton transfer reactions.

The formation of H_2S by reduction of $SO_{3(s)}$ will increase $ANC_{(s)}$. If the H_2S is retained in the soil and reoxidized to $SO_{3(s)}$, $ANC_{(s)}$ will decrease to the initial value. However if H_2S volatilizes from the soil, the increase in $ANC_{(s)}$ is permanent.

After an irreversible H^+ flux has occurred, counter processes may proceed in adjacent ecosystems. For example the dissolved calcium bicarbonate may enter a lake and precipitate as $CaCO_3$, thereby increasing the $ANC_{(s)}$ of the lake sediments. The H_2S emanating from the soil may deposit in an adjacent aerobic soil, oxidize and deplete the ANC of that soil.

Because irreversible H^+ fluxes may vary over time, the 'time window' through which the soil system is observed is important. For example, during short periods of high water flux (snowmelt) solute transport is elevated and therefore rates of soil acidification are high. Conversely during dry periods limited water flow restricts changes in $ANC_{(s)}$.

Due to effects of assimilation by the vegetation and mineralization of organic matter the $ANC_{(s)}$ may also change seasonally. The net H^+-production and -consumption may vary from year to year, depending on variation in climate and biological activity. Monitoring over long periods, in the order of a decade or so, is often necessary to obtain a representative picture of either soil acidification or alkalinization with time[28].

It is noteworthy that alkalinization (or acidification) of water percolating through the soil is a counter process to soil acidification (or alkalinization); when a proton is transferred from solution to the soil $ANC_{(s)}$ is depleted while $ANC_{(aq)}$ is increased. To illustrate the concept consider the transport of a neutral salt, NaCl, through a hypothetical proton-saturated soil. The exchange of Na^+ from the solution for H^+ on the soil results in an alkalinization of the soil (due to the addition of Na_2O) while the drain water is acidified (due to the addition of H^+).

Soil acidification or alkalinization can be quantified in two ways: one approach is historic, the other is actualistic. The historic approach is to analyse the ANC of old and young soil material, by sampling either a chronosequence of soils or soil and its – presumed – parent material, or by sampling the same soil at different times. The actualistic approach involves the measurement of inputs and outputs of all substances involved in H^+-transfer processes, so that a so-called H^+-budget can be calculated[8]. The advantages of the H^+-budget approach are obvious: it is very sensitive compared to analyses of the $ANC_{(s)}$, and there are no problems associated with dating or determinating the authenticity of parent material. In our discussion of soil acidification we will only use the H^+-budget approach.

The principal pathways and reactions that should be considered in computing a H^+-budget are illustrated in Fig. 2: inputs of acid and potentially acidic substances from the atmosphere, export of potentially acidic gases to the atmosphere and solutes in drainage water. The sources and sinks of H^+ within the soil and H^+-transfer processes between the soil and living and dead biomass must also be evaluated.

The measurements required to compile a H^+-budget should include a) the flux of solutes entering and leaving the system, b) net assimilation (or mineralization) of solutes by living and dead biomass, c) net quantities of potentially acidic gases entering or leaving the system and d) net adsorption (*c.q.* precipitation) or desorption (*c.q.* dissolution) of solutes in the soil.

There are many ecosystem studies in which quantitatively important cations (Ca^{2+}, Mg^{2+}, K^+, Na^+, NH_4^+, H^+ and sometimes also Fe^{2+}, Al^{3+}

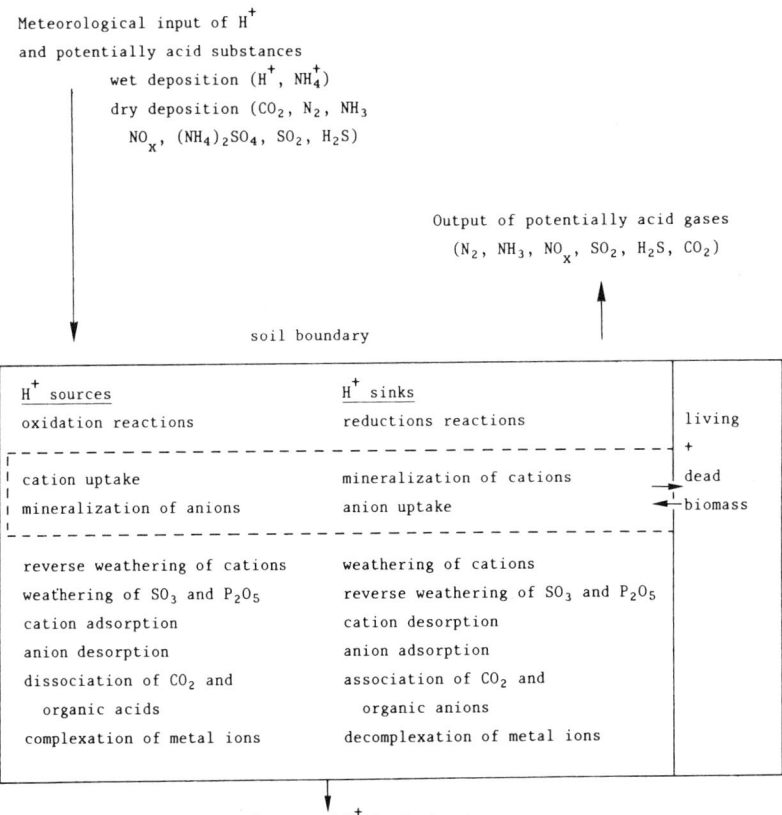

Fig. 2. Conceptual model of the H$^+$-budget of a soil (modified after Driscoll and Likens[8])

and Mn^{2+}) and anions (Cl$^-$, SO$_4^{2-}$, NO$_3^-$, HCO$_3^-$, H$_2$PO$_4^-$ and sometimes organic anions) have been monitored in incoming precipitation and outflowing drainage water. Only few of such studies have considered in detail the role the soil and vegetation in nutrient cycling (using soil solution collectors and soil-hydrologic models) rather than a whole watershed as a black-box. In most mass balance calculations for watersheds nutrient assimilation by vegetation is either ignored, or assumed to be in a steady state. The few studies that have considered ionic uptake in the biomass have shown (*e.g.*[3,14,19,25,28]) that such an assumption may lead to serious errors.

The influence of cation cycling (specifically calcium) on the H$^+$ budget of an ecosystem is illustrated in Figure 3A. The net output of cations (M^{n+} output + M^{n+} net biomass assimilation − M^{n+} input) is equivalent to the H$^+$ neutralized through weathering and ion exchange reactions, and to the depletion of cationic ANC components from the soil.

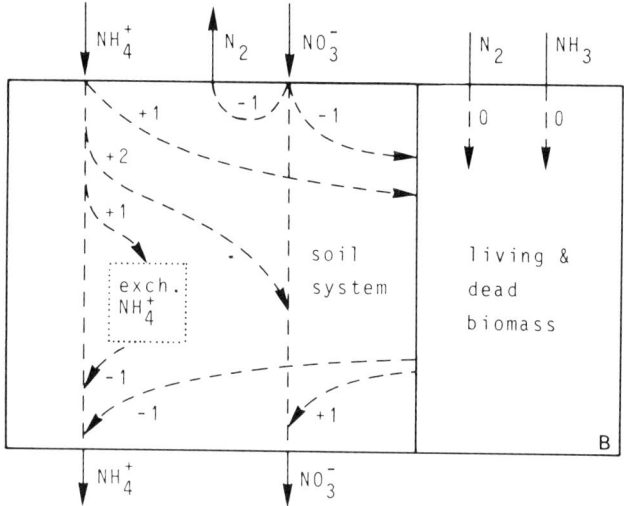

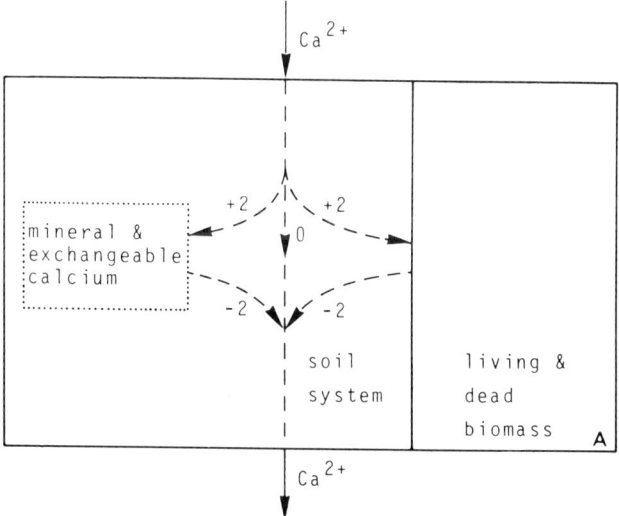

Fig. 3. The role of the Ca-cycle and the N-cycle in the H^+-budget of a soil. Numbers refer to protons formed (+) or consumed (−) per ion Ca^{2+} or NO_3^-/NH_4^+.

The contributions of volatile elements (C, N, S) to a H^+-budget are difficult to determine. Various procedures have been used to measure or estimate the magnitude of dry deposition of gases and aerosols[21,28], but a complete inventory of net gaseous exchange (Fig. 2) for a soil or watershed ecosystem is not currently possible. Fortunately the effect of these fluxes on the H^+-budget can be determined by indirect measurements. For example the amount of

HCO_3^- (+ $2CO_3^{2-}$) transported from a system may be equated to proton dissociation from CO_2 resulting in CO_2 weathering and the associated depletion of soil $ANC_{(s)}$.

Similarly the net proton flux attributed to nitrogen transformations can be determined (Fig. 3B) by net output of ammonium less net output of nitrate to the system [(NH_4^+ output − NH_4^+ input) − (NO_3^- output − NO_3^- input)]. A negative value for this quantity suggests that nitrogen transformations are generating protons to the ecosystem, while a positive value suggest that protons are being consumed from the ecosystem. This expression is valid regardless of the amount of N transported to and from the ecosystem as neutral gases (NO_x, N_2, NH_3). As discussed previously, neutral gases (Reactions 21, 22) do not directly participate in proton transfer reactions. Rather only through formation or consumption of NH_4^+ or NO_3^-, will these forms of nitrogen ultimately participate in the H^+ cycle.

Transformations of sulfur are among the most difficult to evaluate for budget calculations. Gaseous inputs of SO_2 and H_2S will oxidize within an ecosystem and generate protons. Uptake of sulfate by soil or biomass will serve as a proton sink. As a result sulfur inputs, outputs, assimilation by vegetation and retention by soils should be determined for budget calculations.

5. Examples of acidification or alkalinization of soils under different ecological regimes

To illustrate the concept of soil acidification and alkalinization we will discuss three groups of hydrologically different soils: 1. well-drained soils with excess rainfall over evapotranspiration, 2. periodically and permanently reduced soils, and 3. soils of semi-arid or arid areas. To facilitate this discussion we have compiled H^+ budgets for ecosystems of each of the above soil groups (Table 2). Sources of H^+ are tabulated as a) atmospheric inputs of H^+ and SO_2, b) nitrogen transformations, c) anion weathering (plus any reverse cation weathering), d) the deprotonation of CO_2 or organic acids and e) net assimilation of cations by biomass (or net anion mineralization). Sinks of H^+ include a) nitrogen transformations (net nitrate input less net ammonium output) b) weathering of cations (plus any reverse weathering of anions), c) protonation of bicarbonate or organic anions, d) net anion assimilation by biomass (or net cation mineralization) and e) export of H^+ by water draining from the ecosystem. The weathering rate of each system is expressed as the change in $ANC_{(s)}$ (anion

Table 2. Examples of H⁺ budgets for watersheds and soils (units are $kmol.ha^{-1}.yr^{-1}$). Data used in the calculations are from the references cited

Site, ref.	Soil, vegetation	H⁺ sources						H⁺ sinks						$\Delta ANC_{(s)}$
		$H^+ + SO_2$ inputs	N-trans-formations	Weather-ing	CO_2 + Org. deproton.	Bio-mass	Σ^+	Σ^-	N-trans-formations	Weather-ing	HCO_3^- proton.	Bio-mass	Drain-age	
HBEF[8], USA	Fragiorthod, pH 3.6–5.0 Acer, Fagus Betula	1.3	0.1	0.1	0.1	1.0	2.6	2.5	0.0	2.2	0.0	0.2	0.1	− 2.0
Hackfort B[36] Netherlands	Dystochrept pH 3.7–3.9 Quercus, Betula	0.4	3.9*	0.8	0.0	1.1	6.2	5.8	0.0	5.6	0.0	0.1	0.1	− 4.8
Castricum[22] Netherlands	Psamment pH 7–8 Quercus	1.6	0.6	0.2	12.4	0.1	15.0	15.1	0.6	14.5	0.0	0.0	0.0	− 14.3
Thoreau's Bog[10], USA	Sphagnum bog	1.0	0.0	0.0	0.4	0.2	1.5	1.5	0.3	0.3**	0.0	0.3	0.5	0.0
Lake Tchad[4] Tchad	Alkaline lake	0.0	0.0	3.4	0.0	0.0	3.4	3.3	0.0	0.0	3.3	0.0	0.0	+ 3.4

* 2.7 $kmol.ha^{-1}.yr^{-1}$ of H⁺ were due to net NH_4^+ input and reflect H⁺ accepted by atmospheric NH_3 from acidic atmospheric substances (mainly H_2SO_4 derived from SO_2); 1.2 $kmol.ha^{-1}.yr^{-1}$ of H⁺ were due to nitrification of atmospherically derived NH_3 and of organic N (net nitrate output).
** sulfate reduction

weathering less cation weathering), which is equivalent to the amount of H^+ neutralized by the soil system. Also the sum of all H^+ sources and sinks are tabulated. Total sources and sinks should theoretically be equal. Therefore the magnitude of the budget discrepancy provides an estimate of the accuracy of the budget. Budget discrepancies may be due to errors in measurements or to H^+ transfer processes that have been overlooked.

5.1 Well-drained soils with excess precipitation over evapotranspiration

Soils in this category cover the largest portion of the earth's surface. Ecosystems with these soils vary widely with respect to climate, vegetation type and stage of growth and geology. As a result the extent of soil acidification and the mechanisms causing this acidification are highly variable. To illustrate acidification of well drained soils with excess precipitation we will use data from three sites. The Hubbard Brook Experimental Forest (HBEF) in the United States is a forested ecosystem (*Fagus grandifolia, Acer saccharum, Betula alleghaniensis*) which is underlain by pelitic schist. The soils are well drained spodosols (Fragiorthods) which are acidic. Hackfort B is located in a woodland area, in the Netherlands, with vegetation dominated by oak (*Quercus robur*) and birch (*Betula pendula*). Most of the surrounding land is used for intensive grass production for stall fed dairy cattle. The parent material of the soil is sandy to loamy Pleistocene Rhine sediment. The river sediment was probably calcareous but the soil has been decalcified to 130 cm depth and is acidic. Castricum is located in coastal dunes in the Netherlands. It is an oak (*Quercus robur*) stand with calcareous (2.5% $CaCO_3$) sands.

The magnitude and nature of H^+ sources vary markedly for ecosystems with well drained soils and excess precipitation over evapotranspiration. At the HBEF, H^+ sources are largely due to atmospheric inputs of mineral acids and biomass assimilation of cations. Because the HBEF soils are acidic, the dissociation of CO_2 and organic acids contribute little to the total flux of H^+. At Hackfort B proton sources are also largely due to atmospheric inputs, but the nature of these inputs differ considerably from those at the HBEF. Agricultural activity in the area is responsible for considerable volatilization of ammonia to the atmosphere. Ammonia combines with acidic substances to produce ammonium (mainly ammonium sulfate). Inputs of ammonium to Hackfort are readily assimilated (Reaction 26A) or nitrified (Reaction 36A) which result in considerable acid production [35]. Most of this acidity in fact originated from atmospheric SO_2 that after oxidation protonated atmospheric NH_3 to NH_4^+ (see footnote of Table 2). Proton sources associated with biomass assimilation of cations are comparable to those observed at HBEF, however these inputs are small compared to the proton flux associated with atmospheric inputs of acidic and potentially acidic substances at Hackfort B. At Castricum total proton sources are considerably greater than either Hackfort B or the HBEF (15 > 5.8 and 2.6 kmol.ha^{-1}.yr^{-1} respectively), due to the presence of easily weatherable calcium carbonate. While proton contributions of atmospheric mineral acid inputs are comparable to those of the HBEF, the H^+ flux attributed to the dissociation of CO_2 is very high and dominates the H^+ sources at Castricum.

In well drained soils with excess drainage, H^+ neutralization primarily occurs through weathering reactions (Reaction 39B). As evidenced by HBEF, Hackfort B and Castricum data, the magnitude of those fluxes can vary greatly with soil type and geology. At the HBEF there is much concern over atmospheric acidic deposition and stream acidification. However the cause for this concern is not so much the loading of acidic deposition, indeed atmospheric H^+ inputs at Castricum exceed those at the HBEF (1.6 > 1.3 kmol.ha^{-1}.yr^{-1}), but rather the limited ability of HBEF soils to neutralize anthropogenic proton inputs.

Although less than Castricum, the soils at Hackfort B have considerable ability to neutralize H^+ fluxes (5.6 kmol.ha^{-1}.yr^{-1}). Even so, the proton flux associated with weathering of cationic components is smaller than the atmospheric proton inputs, resulting in the export of free protons with soil drainage water.

In summary well drained soils with precipitation exceeding evapotranspiration undergo acidification ($\Delta ANC_{(s)}$ is negative). In these systems a considerable flux of water through the soil and good contact of this water with the mineral soil facilitate mineral weathering. Most of the total H^+ inputs are neutralized through soil acidification associated with weathering. However the origin of this acidity varies greatly from ecosystem to ecosystem and depends on the soil pH, geology, type and growth rate of vegetation, and atmospheric chemistry of the region.

5.2 Periodically and permanently reduced soils

Strong acidification and alkalinization of soils may occur in reduced or periodically reduced soils. Well known cases are acid sulfate soils[31,32], ferrolysed soils[1] and alkaline soils formed as a result of sulfate reduction. We will briefly discuss the formation of these soils according to the principles outlined in this paper.

Soil solutions tend to be near-neutral to slightly alkaline when water saturated, because protons have been neutralized by reduction processes (Reactions 34B, 37B, 38B, 41B and 42B). Upon drainage and aeration reduced inorganic compounds are oxidized by O_2, resulting in a decrease in pH (Reaction 34A, 36A, 41A and 42A). Generally the reduced components FeO, MnO and H_2S are present under anaerobic conditions, whereas their oxidized counterparts Fe_2O_3, MnO_2 and SO_3 are dominant in aerobic conditions. As long as the soil remains closed for all components making up the $ANC_{(s)}$ the system may cycle from a reduced to an oxidized state and back without any net change in $ANC_{(s)}$. However, an irreversible H^+ flux caused by differential mobility of components during reduced and oxidized conditions may result in a permanent change in $ANC_{(s)}$.

Acid sulfate soils, which have pH values below 4, represent a dramatic example of soil acidification. They form in two stages (Fig. 4A). The first is characterized by the reduction of sulfates (generally conveyed by seawater) in tidal flats or seabottom sediments. Most of the sulfide formed is fixed in the soil or sediment as FeS_2, while $ANC_{(aq)}$ (HCO_3^-) formed during sulfate reduction (Reactions 34B + 31A) is removed by tidal turbulence or by diffusion into the overlying water. As a result mobile $ANC_{(aq)}$ (HCO_3^-) and immobile potential acidity (FeS_2) are separated. This leads to a permanent decrease in the $ANC_{(s)}$ after aeration and oxidation of FeS_2. Pyrite accumulates over several centuries in potential acid sulfate soils to concentrations typically in the order of 3–6%. If such pyritic sediments are drained artificially the amount of H^+ generated may be a staggering 10^3 kmol. $ha^{-1}.yr^{-1}$. This H^+ flux is two orders of magnitude greater than the total proton sources of any of our example ecosystems (Table 2), and results in extreme soil and water acidification.

Ferrolysis involves the formation of exchangeable ferrous iron during reduction and removal of the displaced cations with bicarbonate. This leads to a decrease in $ANC_{(s)}$ and an increase in $ANC_{(aq)}$. Oxidation of exchangeable iron to ferric oxide during the following aerobic stage leaves a partially H^+-saturated adsorption complex (Fig. 4B). Over many seasonal oxidation-reduction cycles strongly weathered and acid surface soils can be formed by this process[1].

If the alkalinity produced during sulfate reduction is not transported from the soil as $ANC_{(aq)}$, *e.g.* because the soils in question occur in a hydrologically closed basin, $ANC_{(s)}$ will increase (Fig. 4C). A classical example is Janitzky and Whittig's[13] description of formation

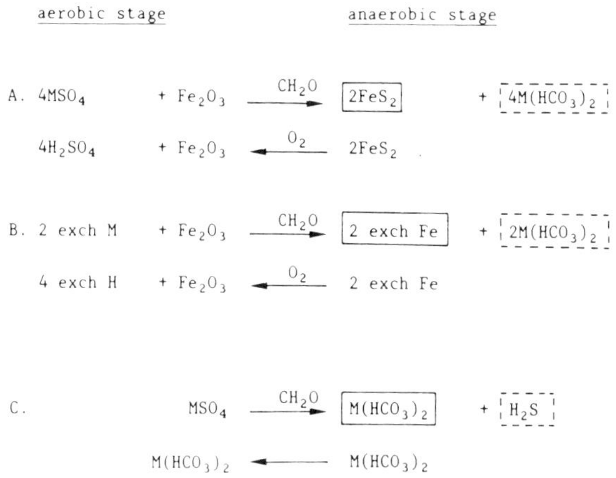

ACIDIFICATION AND ALKALINIZATION OF SOILS   305

```
        aerobic stage                    anaerobic stage

              CH₂O
A. 4MSO₄    + Fe₂O₃  ────►   2FeS₂    +   4M(HCO₃)₂

              O₂
   4H₂SO₄   + Fe₂O₃  ◄────   2FeS₂


              CH₂O
B. 2 exch M + Fe₂O₃  ────►   2 exch Fe +   2M(HCO₃)₂

              O₂
   4 exch H + Fe₂O₃  ◄────   2 exch Fe


                     CH₂O
C.         MSO₄     ────►   M(HCO₃)₂   +   H₂S

           M(HCO₃)₂ ◄────   M(HCO₃)₂
```

Fig. 4. Examples of permanent acidification or alkalinization in alternating aerobic and anaerobic soil systems, due to differential mobility of reduction products and resulting irreversible H^+-fluxes. Components that remain in the soil system have a solid outline, dissolved or gaseous components leaving the system have a broken outline. A. Acid sulfate soil formation. B. Ferrolysis. C. Alkalinization due to volatilization of H_2S from hydrologically closed areas.

of $NaHCO_3$ and Na_2CO_3 due to reduction of Na_2SO_4 in salt-affected soils. If the system is also closed to transport of reduced sulfur (*i.e.* all sulfide is precipitated in the soil) the $ANC_{(s)}$ would decrease to the original level after drainage and oxidation of the soils. However, any removal of sulfide by H_2S volatilization would lead to permanent soil alkalinization.

Hemonds study on Thoreau's Bog[10] illustrates the conditions of a permanently reduced soil that is not hydrologically closed. Thoreau's Bog is a Sphagnum bog located in Massachusetts, USA. Proton fluxes in this system were much lower than in the ecosystems discussed previously (Table 2). Proton inputs to the bog were mainly through atmospheric deposition and to a lesser extent by dissociation of organic acids. These inputs were largely neutralized by biomass assimilation of anions (S, N and P), denitrification, reduction of atmospherically derived sulfate and export of free protons. Unlike for well drained soils, the dissolution of oxide coordinated metals is not a significant mechanism for the neutralization of H^+ in Thoreau's Bog. Proton neutralization booked under 'weathering' is entirely due to sulfate reduction (Reaction 34B). Because reduced sulfur does not contribute to $ANC_{(s)}$ and because the $ANC_{(aq)}$ formed during sulfate reduction was not retained in the bog, there was no soil acidification or alkalinization. If sulfide was produced as a solid phase, *e.g.* as FeS or organic S, it would represent potential acidity. This acidity could be realized upon drainage and oxidation of the bog.

5.3 Soils of arid and semi-arid areas

Arid and semi-arid areas are characterized by excess evaporation over rainfall, so inputs of dissolved substances, both from rain and from stream- or irrigation water, tend to exceed outputs. Moreover, if water entering the system contains $ANC_{(aq)}$ (such as practically all ground and surface waters), arid or semi-arid soils will increase in $ANC_{(s)}$. By reverse weathering process (Reaction 39A) $ANC_{(s)}$ can be generated in the soil system, either as carbonate or as silicate minerals (*e.g.* smectites, sepiolite, zeolites or authigenic feldspars). Whether or not arid soils become highly alkaline (pH > 9) due to formation of sodium carbonates depends on the relative amounts of divalent cations (Ca^{2+} and Mg^{2+}) and $ANC_{(aq)}$ (HCO_3^-) in the incoming water. If divalent cations exceed $ANC_{(aq)}$, sodium carbonate will not form and the pH remains below 8.5, if the reverse is true, sodium carbonate will ultimately be produced[29]. Of course, the increase in $ANC_{(s)}$ depends only on the quantity of cations associated with $ANC_{(aq)}$ that have undergone reverse weathering, rather than the type of the basic cations.

Although our H^+-balance of the Lake Tchad basin[4], was developed for a lake with surrounding intermittently flooded land rather than to a soil-system, it may serve to illustrate soil-alkalinization in semi-arid areas (Table 2). In this system the supply of $ANC_{(aq)}$ and basic cations by river water exceeded the removal by seepage. Evaporation caused solute concentrations to increase and the solubility product of certain carbonate and silicate minerals was exceeded. As calcite and smectite were formed within the basin, the H^+ generated in these reverse weathering reactions was neutralized by bicarbonate (Reaction 31B). The result of this reverse weathering was an increase in the $ANC_{(s)}$.

5.4 Conclusions

Soil acidification results from a depletion of $ANC_{(s)}$, while soil alkalinization is an increase in $ANC_{(s)}$. These processes are the result of irreversible reactions which consume or produce H^+, respectively. The extent and causes of soil acidification/alkalinization vary considerably from ecosystem to ecosystem, but may be quantified for a given ecosystem by developing a H^+ budget.

Acknowledgements We thank Dr. R. Brinkman for his comments on the first draft of the manuscript, and greatfully acknowledge the enthusiastic help of a group of 24 students, who followed a course on H^+ budgets in soil-vegetation systems, in calculating H^+ budgets for case studies. This study was funded in part by the EEC environmental programme, EUR contract ENV-650-NL, by the Directorate General for the Environment, The Hague, project nr. 351042 and the National Science Foundation, USA (DEB 82-06980).

References

1. Brinkman R 1970 Ferrolysis; a hydromorphic soil forming process. Geoderma 3, 199–206.
2. Bruggenwert M G M 1972 Adsorptie van Al-ionen aan het kleimineraal montmorilloniet. Versl. Landb. Onderz. 768. Pudoc, Wageningen, 120 p.
3. Bruynzeel L A 1983 Hydrological and biogeochemical aspects of man-made forests in south-central Java, Indonesia. Ph. D. -thesis Free Univ. Amsterdam.
4. Carmoure J P 1976 La régulation hydrogéochimique du lac Tchad. Trav. et Docum. de L.ORSTROM No. 58. ORSTOM, Paris.
5. Coleman N T and Thomas G W 1967 The basic chemistry of soil acidity. *In* Soil acidity and liming. The American Society of Agronomy. Madison. Wisconsin: 1–41.
6. De Wit C T, Dijkshoorn W and Noggle J C 1963 Ionic balance and growth of plants. Versl. Landbouwk. Onderz. 69. 15. Wageningen.
7. Driscoll C T 1980 Chemical characterization of some dilute acidified lakes and streams in the Adirondack region of New York State. Ph. D. Thesis, Dept. of Environmental Engineering, Cornell Univ.
8. Driscoll C T and Likens G E 1982 Hydrogen ion budget of an aggrading forested ecosystem. Tellus 34, 283–292.
9. Garrels R M and Mackenzie F T 1971 Evolution of Sedimentary Rocks. Norton, New York.
10. Hemond H F 1980. Biogeochemistry of Thoreau's Bog, Concord, Massachusetts. Ecological Monographs 50, 507–526.
11. Hoagland D R and Broyer T C 1936 General nature of the process of salt accumulation by roots, description of experimental methods. Plant Physiol. 11, 472–507.
12. Holland H D 1978 The Chemistry of the Atmosphere and Oceans. John Wiley. New York
13. Janitzky P and Whittig L D 1964 Mechanisms of formation of Na_2CO_3 in soils. II. Laboratory study of biogenesis. Soil Sci. 15, 145–157.
14. Likens G E, Bormann F H, Pierce R S, Eaton J S and Johnson N M 1977 Biogeochemistry of a forested ecosystem. Springer Verlag, New York, 146 p.
15. Likens G E, Bormann F H, Johnson N M, Fisher D W and Pierce R S 1970 Effects of forest cutting and herbicide treatment on nutrient budgets in the Hubbard Brook watershed-ecosystem. Ecol. Monographs 40, 23–47.
16. Lovelock J E 1979 Gaia. A new Look at Life on Earth. Oxford Univ. Press. Oxford.
17. Mattson S 1938 The constitution of the pedosphere. Landbrukshögskolans Ann. 5, 261–276.
18. Mattson S and Koutler-Andersson E 1941 The acid-base condition in vegetation, litter and humus: I. Acids, acidoids and bases in relation to decomposition. Landbrukshögskolans Ann. 9, 1–26.
19. Matzner E, Khanna P K, Meiwes K J, Lindheim M, Prenzel J und Ulrich B 1982 Elementflüsse in Waldökosystemen im Solling-Datendokumentation-Göttinger Bodenk. Ber. 71, 267 p.
20. Matzner E and Ulrich B 1980 The transfer of chemical elements within a heath-ecosystem (*Calluna vulgaris*) in Northwest Germany. Z. Pflanzenernaehr. Bodenkd. 143, 666–678.
21. Miller H G and Miller J D 1980 Collection and retention of atmospheric pollutants by vegetation. *In* Ecological impact on acid precipitation, Eds. D Drablos and A Tollan. SNSF-project, Oslo, 33–40.
22. Minderman G and Leeflang K W F 1968 The amounts of drainage water and solutes from lysimeters. Plant and Soil 28, 61–80.
23. Reuss J O 1976 Chemical and biological relationships relevant to the ecological effects of acid rainfall. *In* Proc. first int. symposium on acid precipitation and the forest ecosystem. Eds. L S Dochinger and T A Seliga U.S.D.A. For. Serv. Gen. Tech. Rep. NE-23, 791–813.
24. Rosenqvist I T 1977 Sur jord-surt vann. Ingenφrforlaget, Oslo, 123 p.
25. Sollins P, Grier C C, McCorison F M, Cromack K, Fogel R and Frederiksen R L 1980

The internal element cycles of and old-growth douglas-fir ecosystem in western Oregon. Ecol. Monographs 50, 261–285.
26 Stumm W and Morgan J J 1970 Aquatic Chemistry. Wiley-Interscience, New York, 583 p.
27 Tollan A (Ed.) 1977 Acid precipitation and some alternative sources as the cause of the acidifying of water sources. S.N.S.F. project. Oslo, Norway.
28 Ulrich B, Mayer R und Khanna P K 1979 Deposition von Luftverunreinigungen und ihre Auswirkungen in Waldökosystemen im Solling. J D Sauerländer's Verlag, Frankfurt am Main.
29 Van Beek C G E M and Van Breemen N 1973 The alkalinity of alkaline soils. J. Soil Sci. 24, 129–136.
30 Van Breemen N 1973 Soil forming processes in acid sulfate soils. *In* Dost H (Ed.) Acid Sulphate Soils, ILRI Publ. 18, 1973, Vol. I, 66–130, Wageningen, the Netherlands.
31 Van Breemen N 1975 Acidification and deacidification of coastal plant soils as a result of periodic flooding. Soil Sci. Soc. Am. Proc. 39, 1153–1157.
32 Van Breemen N 1976 Genesis and solution chemistry of acid sulfate soils in Thailand. Agric. Res. Rep. 848, Pudoc, Wageningen. Netherlands. 263 p.
33 Van Breemen N and Wielemaker W G 1974 Buffer intensities and equilibrium pH of minerals and soils. I. The contribution of minerals and aqueous carbonate to pH-buffering. Soil Sci. Soc. Am. Proc. 38, 55–60.
34 Van Breemen N and Wielemaker W G 1974 Buffer intensities and equilibrium pH of minerals and soils. Soil Sci. Soc. Am. Proc. 38, 61–66.
35 Van Breemen N, Burrough P A, Velthorst E J, Van Dobben H F, Toke de Wit, Ridder T B and Reynders H F R 1982 Soil acidification from atmospheric ammonium sulphate in forest canopy throughfall. Nature London 299, 548–550.
36 Van Breemen N, Van Grinsven J J M and Jordens E R 1983 H^+ budgets and nitrogen transformations in woodland soils in the Netherlands influenced by high inputs of atmospheric ammonium sulfate. Proc. Int. Conf. Acid Precipitation – Origin and Effects VDI Düsseldorf, (*In press*).

ERRATA

Page 288, equation (12) should read:

$$MnO_{(S)} + \tfrac{1}{2}O_{2(aq)} - MnO_{2(S)}$$

Page 291, equation (39A) should read:

$$M^{n+} + \tfrac{n}{2}H_2O - \tfrac{2}{n}M_{\tfrac{2}{n}}O + nH^+$$

Page 300, line 6 from bottom should read: " . . . (net nitrate input plus net ammonium output) . . ."

HIGH RESOLUTION ELECTRON MICROSCOPY OF FELDSPAR WEATHERING

RICHARD A. EGGLETON

Geology Department, Australian National University
P.O. Box 4, Canberra, ACT 2600, Australia

PETER R. BUSECK

Departments of Chemistry and Geology, Arizona State University
Tempe, Arizona 85281

Abstract—High resolution imaging by transmission electron microscopy has revealed a mechanism for the weathering of intermediate microcline in a humid, temperate climate. Dissolution of the feldspar begins at the boundary of twinned and untwinned domains and produces circular holes which enlarge to form negative crystals. Amorphous, ring-shaped structures develop, about 250 Å in diameter, within the larger holes. These rings, in turn, crystallize to an arcuate phase having a 10-Å basal spacing and then to crinkled sheets of illite or dehydrated montmorillonite. The 10-Å layer silicate shows an irregular stacking sequence, including 10-, 20-, and 30-Å sequences. Included plagioclase crystals show a similar mechanism of weathering and, moreover, are more intensely weathered.

Key Words—Feldspar, Illite/montmorillonite, Ion thinning, Microcline, Transmission electron microscopy, Weathering.

INTRODUCTION

The weathering products of feldspar are the major constituents of soil. Whereas there have been extensive studies of feldspar weathering in bulk samples and a great deal is known about the chemistry of the process and its end products, relatively little is known about the earliest stages of feldspar breakdown. The present study uses the methods of high-resolution transmission electron microscopy (HRTEM) to investigate feldspar weathering at its onset, before the reaction products have reached the degree of crystallinity and crystal size necessary for identification by X-ray diffraction or other bulk methods.

Petrović (1976), in his discussion of rate control in feldspar dissolution, summarized results obtained in aqueous solution over a range of pH and temperature. The alteration phases found in such experiments are aluminosilicate gel particles, <200 Å in diameter, and the crystalline products gibbsite, fibrous boehmite, kaolinite, and halloysite.

Weathered feldspar develops a brownish "turbid" appearance, very familiar to optical microscopists. Folk (1955) suggested that turbidity resulted from the development of sub-micrometer size vacuoles. This form of alteration develops at some distance from visible cleavages or cracks within the body of the feldspar, where flowing water cannot penetrate; thus it is evident that weathering does not only proceed from exposed surfaces. The clay minerals that ultimately form during feldspar weathering must owe their origin to the processes occurring at this early stage.

Scanning electron microscopy of weathered feldspar grains (Parham, 1969; Berner and Holdren, 1977; Keller 1976, 1978; Nixon, 1979) shows that crystallographically controlled solution pits form on the surface of feldspar as it weathers. Keller (1976) stated: "Micropitting of feldspar, presumably by incongruent dissolution, is a common pre- or early-stage, of kaolinization of feldspar." Berner and Holdren (1977) expressed ". . . no doubt . . . that the dissolution of feldspar proceeds by selective etching of the surface, probably along dislocations."

Suttner et al. (1976) concluded that kaolin forms from feldspar under humid conditions, whereas smectite forms in semi-arid climates, the difference resulting from more complete flushing of K^+ ions under humid weathering. Vacuoles occur in all of their optically examined material, but are most prominent in the western arid samples. They also found that plagioclase altered more rapidly, or more extensively, than microcline.

EXPERIMENTAL

Sample

The feldspar selected for this investigation was from a suite of perthitic K-feldspar megacrysts from the Kameruka granodiorite in southern New South Wales, Australia. These crystals have been the subject of an extensive study (Eggleton, 1979). By examining such well known material, it is easier to separate weathering effects from phenomena associated with the feldspar's earlier igneous history. The weathering took place in a humid climate, with an annual rainfall of about 1000

Table 1. Chemical analyses of feldspars.

	ABF28	ABF67	Ksp/Ill
SiO_2	64.9	60.5	60.6
Al_2O_3	18.6	23.5	23.3
K_2O	15.8	14.2	14.0
Na_2O	0.7	0.9	0.5
H_2O	—	—	1.5
Total	100.0	99.1	99.9

ABF 28 = Microcline.
ABF 67 = Weathered microcline.
Ksp/Ill = Calculated composition for a mixture of 72% ABF 28 and 28% illite.

mm and a temperature range of 0°–40°C. Sample ABF28 was collected from a cut at the north end of the Brogo Dam wall, where the granodiorite is heavily weathered and decomposed. Here, megacrysts 5 to 6 cm long may be picked out by hand. Optical examination showed that all of the megacrysts have unaltered regions, and that some have highly altered turbid regions. Plagioclase crystals within the K-feldspar are everywhere more altered than the host. Analysis by energy dispersive electron microprobe of optically clear areas gave a megacryst composition of Or_{93}, in good agreement with the X-ray diffraction composition of Or_{94} determined using the relation of Stewart and Wright (1974). X-ray powder diffraction films established that the K-feldspar is an intermediate microcline ($\gamma = 89.66°$). The perthite albite lamellae have a composition of $Ab_{96}Or_4$; the included plagioclase crystals are An_{31} at the core and are zoned to An_{15} at the margin.

Analyses of weathered feldspar are limited by the need to select well polished areas, which automatically excludes severely altered material. The most highly altered microcline analyzed (ABF67 in Table 1) is from a maximum microcline megacryst where Jingo Creek cuts the Wyndham Burragate Road. This analysis shows the expected loss of K_2O and SiO_2 and gain in Al_2O_3 with weathering. The cation proportions of ABF67 correspond closely to an illite:microcline mix in the proportions 28:72, assuming an illite composition of $K_{0.67}Al_{2.67}Si_{3.33}O_{10}(OH)_2$. Helgeson (1971) showed that at 25°C and high K^+ activity ($\log(aK^+/aH^+) > 6$), K-mica is in equilibrium with microcline at one atmosphere and unit activity of water.

High resolution transmission electron microscropy

Material selected for examination was taken from 20-μm thick optical thin sections, and also scratched from areas on megacryst surfaces. Samples were chosen to show a range of weathering, from optically clear through moderately and densely turbid microcline to nearly opaque plagioclase. Sections cut near (001) and (100) were thinned by argon ion bombardment at 5 kV

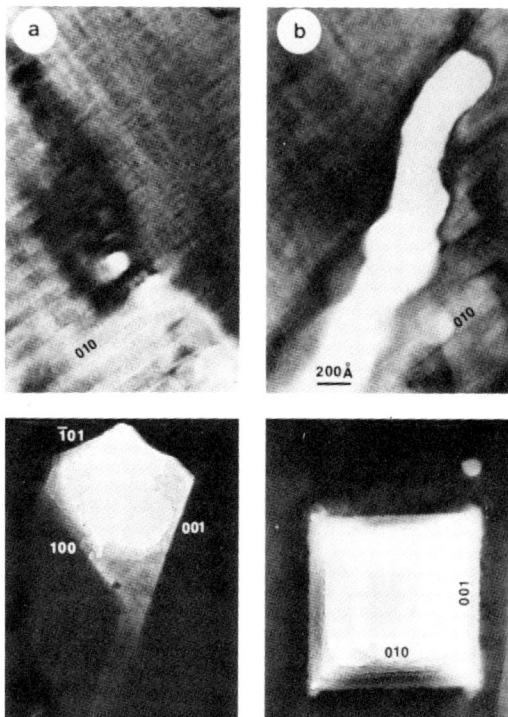

Figure 1. Development of vacuoles. (a) [102] section showing small vacuoles at the boundary between twinned (lower left) and untwinned regions of K-feldspar. (b) Coalescence of vacuoles along a similar boundary region; [102] section. (c) [010] section showing a negative crystal fragment with contained amorphous material. (d) [100] section showing a negative crystal. The scale bar of (b) applies to all four figures.

and 4 μA, with a final 'cleaning' bombardment at 1 kV for 6 hr. Fragments were also dug from thin sections and from the megacryst surface, crushed, and examined after dispersal on holey carbon films. Observations were made on a JEOL JEM100B electron microscope, operating at 100 kV. The instrumental and experimental conditions and the procedures are those described by Buseck and Iijima (1974) and Veblen and Buseck (1979). The crystals were oriented using a top-entry tilt-rotate stage, and the diffraction patterns indexed using a gold-coated crystal as standard, and confirmed by tilting to other orientations and then relating all observations by stereographic projection. Such standardization is particularly important to insure the distinction between dimensionally similar albite and K-feldspar.

The slightly weathered feldspar examined here is subject to damage by an electron beam to the extent that the time required to tilt a crystal to a desired orientation is considerably greater than the crystal's life expectancy. For ion-milled samples, this poses no particular problem, as the operator can translate to an undamaged area after orientation is achieved. Small fragments are more difficult to orient without damage, and chance orientations were used mostly. The crystalline weathering products are similarly short lived and have generally only been seen in grain mounts. No orientation studies were made of these fragments.

RESULTS

In all but the freshest feldspar, the electron microscope revealed holes, or vacuoles, most commonly in ion-thinned specimens, but also in crushed grains. In an ion-milled specimen, initially 20 μm thick, but viewed after thinning to about 100 Å, an observed hole is unlikely to be the "actual" hole present before milling, but a hole that was "translated" through the crystal by the milling process itself. Because the ion beam strikes the sample surface at a glancing angle of only 25° or 30°, many surface irregularities will be smoothed out during milling. Even so, it is to be expected that the observed incidence of holes in a particular section will be greater than was originally present in the 100-Å thick layer exposed to examination. "Translated" holes, if not originally round, will be smoothed and rounded by the milling process, and only holes that are exposed just as milling concludes can be expected to show their true morphology. Similarly, occluded nonfeldspar phases within the holes will only be preserved if they reach the surface as milling concludes.

Many of the vacuoles seen in ion-thinned sections lie at the ends of albite twin lamellae. Interplanar angles measured from Figure 1a and similar photographs showed that the twinned region is close to maximum microcline, whereas the untwinned area is, within experimental error, dimensionally monoclinic. Between these lattices of different geometry must be a region of high strain, or high dislocation density. In fresh feldspar both the orientation changes expected in a strained crystal, and edge dislocations, were seen in such boundary regions. It was therefore concluded that weathering of feldspar commences by the solution of material from these high strain volumes within the crystal. Larger holes are commonly bounded by planes whose traces are parallel to crystallographic directions; thus these holes are negative crystals.

Material showing no diffraction contrast is present within some negative crystals and is accordingly regarded as amorphous material. By electron imaging it is impossible to determine the origin of the lack of crystallinity, whether by ion-beam damage or as the result of weathering. Amorphous material, yielding a broad diffraction halo between 3 and 4 Å, is common in grain mounts of more highly altered feldspar, and these two amorphous phases can be correlated on a tentative basis.

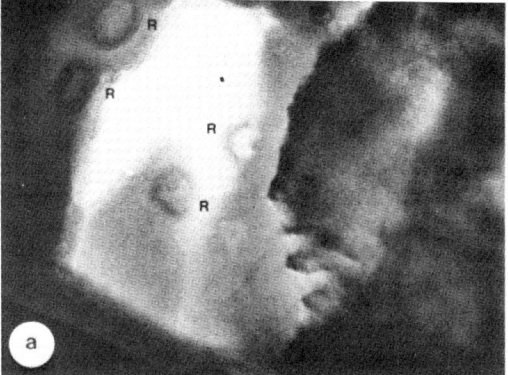

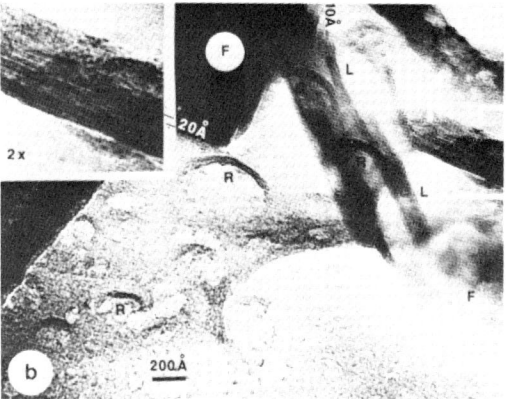

Figure 2. Ring structures (R) and their development to 10-Å layer silicate. (a) [100] feldspar section, showing a negative crystal with enclosed amorphous material and ring, (R). (b) Fragment suspended on holey carbon film. F = relict feldspar, L = layer silicate with 10-Å fringes. The framed crystal on the right is enlarged 2× on the upper left to reveal the 20-Å, 2-layer repeat.

In some of the amorphous material, particularly in the holes, ring texture is visible. The rings are about 250 Å in diameter with a dark (more strongly diffracting) margin (Figure 2a). Under the electron beam, they move slightly, pulsating or "fluttering." By combining observations of sections and crushed grains, a progression can be traced from very unstable, almost structureless rings, to more stable arcuate laths with 10-Å lattice planes, to a well-crystallized 10-Å layer silicate (Figure 2b). In other areas there is no evidence for an intermediate ring structure; a poorly to well-crystallized 10-Å layer silicate occurs directly against, or very close to feldspar (Figure 3).

Figure 4a shows a 3-fringe "bridge" across a vacuole. O'Keefe and Buseck (1979) showed that intuitive identification of high resolution images of silicates is only possible close to optimum defocus (~900 Å for very thin crystals), and even then caution is needed.

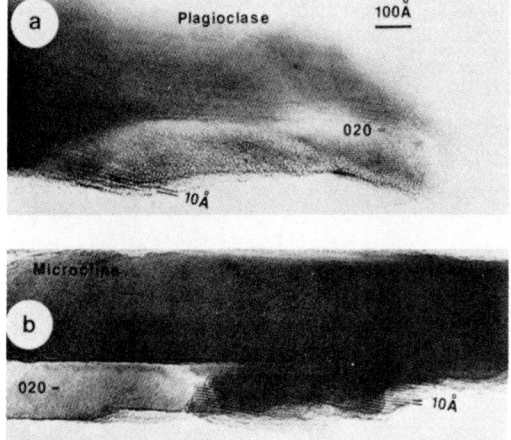

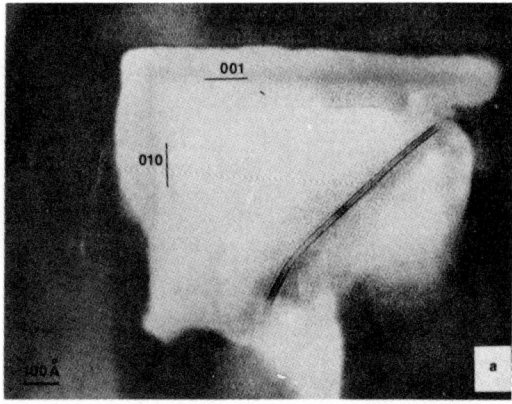

Figure 3. 10-Å layer silicate formed adjacent to feldspar. (a) Plagioclase viewed parallel to (010) showing 6.4-Å (020) fringes and curved 10-Å layer silicate along the lower edge. (b) Comparable microcline crystal with well developed 10-Å phase adjacent.

The wide fringes in Figure 4a are about 16 Å apart, a spacing not found in any feldspar alteration product, except perhaps in fully hydrated montmorillonite, an unlikely survivor in the high vacuum of the microscope. Computation of images to be expected from thin mica crystals (Eggleton and O'Keefe, in preparation) using the methods of O'Keefe and Buseck (1979) shows that at an underfocus of 350 Å, a 2-layer illite crystal, a few unit cells wide and oriented with (001) parallel to the electron beam, can give the image shown in Figure 4b. Such an image could be produced from a curled 2-layer crystal in which the plane of the layer silicate sheet is turned up parallel to the electron beam, but only for a depth of a few unit cells. Many such curled illite sheets have been seen in grain mounts, and, although they are uncommon in ion-thinned crystals, examples such as that shown in Figure 4 reveal that crystallization of alteration products can occur directly in the solution holes.

According to Helgeson (1971), microcline should alter initially to illite or montmorillonite, and plagioclase should alter to montmorillonite. Page and Wenk (1979) found that montmorillonite is an early product in the hydrothermal alteration of plagioclase. Meunier and Velde (1979) reported that "initial weathering in the lowest-massive levels of the (granite) profile produces first illitic mica . . . ." Both feldspars studied here show abundant 10-Å phases in their more altered regions (Figures 5, 6). In grain mounts, this phase appears as extremely thin flakes curled at the edges and crinkled or corrugated in several directions (Figure 5a). Projections of these flakes close to [001] shows a regular 4.5-

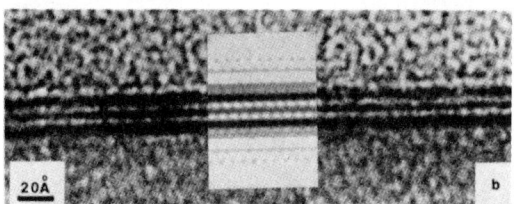

Figure 4. Negative crystal in [100] section of K-feldspar showing 6.4-Å (020) and (001) feldspar-lattice fringes bridged by a crystallite of two sheets of 10-Å layer silicate. (b) The 10-Å layer silicate, at 5 times the magnification of (a), is compared to a computed image calculated at 350-Å under focus from two dioctahedral 2:1 layers with K in ⅔ of the interlayer sites. At this defocus, the image bears little similarity to the projected charge density of the crystal.

Å hexagonal lattice in a few examples (Figure 5c), but more commonly the xz plane is modulated into linear domains about 10 unit cells across (Figure 6b); lattice rows across domain boundaries change direction by less than 5°. In other flakes subgrains exist only 3 or 4 unit cells wide in any direction. Many examples have been seen of extremely thin (2 or 3 layer) sheets peeling from a substrate having the typical diffraction pattern of a layer silicate a^*b^* section (Figure 6a). Where such sheets curl to present an edge to the electron beam, the layers have been imaged and the 10-Å layer spacing identified in flakes so small that no electron diffraction pattern was detectable. Iijima and Buseck (1978) showed that mica layer sequences can be determined from images obtained from slightly misoriented crystals. Many of the flakes imaged here have 20-Å, 2-layer repeats; many are 1-layer; and others are mixed (Figure 5b).

It is difficult to decide from morphology alone whether a particular crystal is illite, montmorillonite, or muscovite. It seems possible that a complete range through these 10-Å layer silicates is present, the thin crinkled crystal of Figure 6a being montmorillonite, the modu-

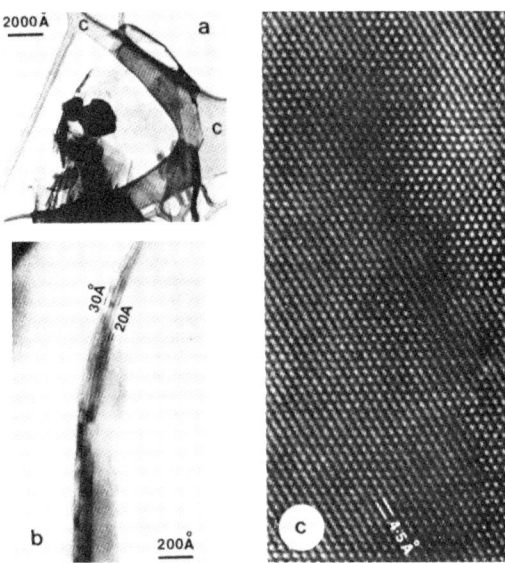
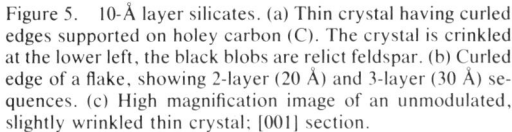
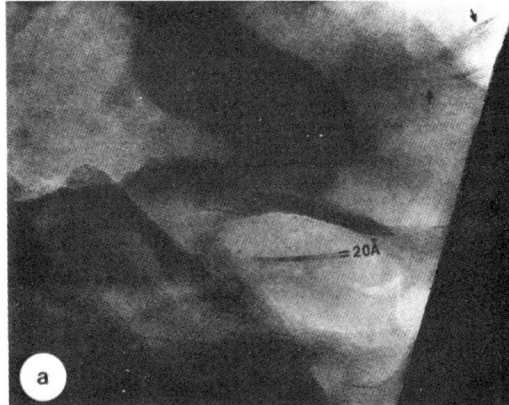
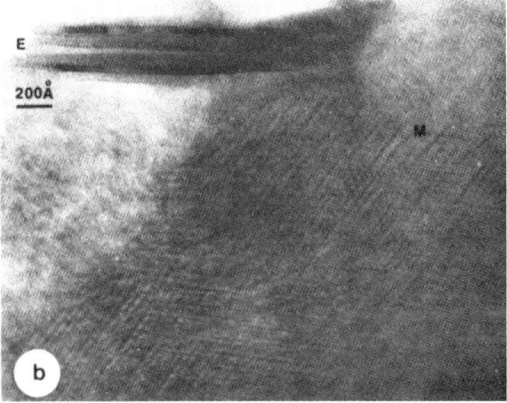

Figure 5. 10-Å layer silicates. (a) Thin crystal having curled edges supported on holey carbon (C). The crystal is crinkled at the lower left, the black blobs are relict feldspar. (b) Curled edge of a flake, showing 2-layer (20 Å) and 3-layer (30 Å) sequences. (c) High magnification image of an unmodulated, slightly wrinkled thin crystal; [001] section.

Figure 6. (a) Pairs of 2:1 "talc" layers (arrowed) and a triple (20 Å) sequence peeled up from an unresolved layer-silicate substrate. The dark triangular area on the right is a large feldspar fragment. (b) Layer-silicate crystal (M) with modulations on ab, and a curled edge (E) showing 10-Å fringes. The modulations are spaced at about 40 Å; the diameter of the curled edge is about 250 Å.

lated crystal of Figure 6b illite, and the flatter, uniform crystal of Figure 5c 1Md muscovite. At this level of crystallization it is also possible that these distinctions have little meaning, the modulations of one crystal may reflect variations in chemistry across it, giving rise to a 10-Å layer that is "illite" in one area and "montmorillonite" in another. Such modulated structures may be the precursors to interstratified illite/montmorillonite of soils, as the K and Al proportions in a given layer dictate whether it will ultimately become expandable. Alcover et al. (1977) described short-range order of the large cations in vermiculitized muscovite, a phenomenon that may be similar to the modulations observed in this study.

Despite extensive searching and examination of more than a thousand micrographs, no evidence has been found for the existence of any other alteration product. In particular, kaolinite or halloysite, gibbsite, or boehmite were sought, but without success. Apparently the movement of water into the regions of feldspar studied was slow enough that the concentrations of K, Na, and Si remained high, and equilibrium was maintained between feldspar and the 10-Å layer silicate. It is unlikely, on the basis of these observations, that the formation of a weathering crust would have any significant effect on the subsequent rate of feldspar weathering to illite or montmorillonite unless such crust were more impervious to water than feldspar itself.

CONCLUSIONS

Under the weathering conditions studied here, it is not appropriate to consider reactions as occurring simply at the crystal surface, with feldspar inside and reaction product + water outside. Rather, there is an initial reaction between feldspar and diffusing water, beginning at structural defects, and gradually expanding within the crystal. At this early stage, the only crystalline phases are feldspar and 10-Å layer silicate, and it must be assumed from Helgeson's work (1971), that these are in equilibrium. Thus, the beginning of feldspar weathering is, in this instance, an equilibrium process, leading to the formation of an illite/montmorillonite mixed-layer mineral.

ACKNOWLEDGMENTS

We are grateful to Drs. D. Veblen and I. Mackinnon for helpful discussion during the course of this work, and to M. A. O'Keefe for assistance with electron image computations. RAE particularly thanks J. Wheatley for instruction and help in the use of the microscope. The research was supported in part by grant EAR 77-00128 from the Earth Sciences Division of the National Science Foundation.

REFERENCES

Alcover, S. F., Gatineau, L., Mering, J., and Kodama, H. (1977) The distribution of Ba cations in vermiculite and vermiculitized micas: unpubl. manuscript.

Berner, R. A. and Holdren, C. R. (1977) Mechanism of feldspar weathering: some observational evidence: *Geology* **5,** 369–372.

Buseck, P. R. and Iijima, S. (1974) High resolution transmission electron microscopy of silicates: *Amer. Mineral.* **59,** 1–21.

Eggleton, R. A. (1979) The ordering path for igneous K-feldspar megacrysts: *Amer. Mineral.* **64,** 906–911.

Folk, R. L. (1955) Note on the significance of turbid feldspars: *Amer. Mineral.* **40,** 356–357.

Helgeson, H. C. (1971) Kinetics of mass transfer among silicates and aqueous solutions: *Geochim. Cosmochim. Acta* **35,** 421–469.

Iijima, S. and Buseck, P. R. (1978) Experimental study of disordered mica structures by high resolution electron microscopy: *Acta Cryst.* **A34,** 709–719.

Keller, W. E. (1976) Scan electron micrographs of kaolins collected from diverse environments or origin—I: *Clays & Clay Minerals* **24,** 107–113.

Keller, W. E. (1978) Kaolinization of feldspars as displayed in scanning electron micrographs: *Geology* **6,** 184–188.

Meunier, A. F. and Velde, B. (1979) Weathering mineral facies in altered granites. The importance of small-scale local equilibria: *Mineral. Mag.* **43,** 261–268.

Nixon, R. A. (1979) Differences in incongruent weathering of plagioclase and microcline—Cation leaching versus precipitates: *Geology* **7,** 221–224.

O'Keefe, M. A. and Buseck, P. R. (1979) Calculation of the high resolution TEM images of minerals: *Trans. Amer. Cryst. Assoc.* **15** (in press).

Page, R. and Wenk, H. R. (1979) Phyllosilicate alteration of plagioclase studied by transmission electron microscopy: *Geology* **7,** 393–397.

Parham, W. E. (1969) Formation of halloysite from feldspar: low temperature artificial weathering versus natural weathering: *Clays & Clay Minerals* **17,** 13–22.

Petrović, R. (1976) Rate control in feldspar dissolution—II. The protective effect of precipitates: *Geochim. Cosmochim. Acta* **40,** 1509–1521.

Suttner, L. J., Mack, G., James, W. C., and Young, S. W. (1976) Relative alteration of microcline and sodic plagioclase in semi-arid and humid climates: *Geol. Soc. Amer. Abstracts Programs* **8,** 512.

Veblen, D. R. and Buseck, P. R. (1979) Chain width order and disorder in biopyriboles: *Amer. Mineral.* **64,** 687–700.

(Received 27 August 1979; accepted 19 December 1979)

Резюме—Высоко разрешающие изображения, полученные трансмиссионной электронной микроскопией, позволили выявить механизм выветривания промежуточного мироклина во влажном умеренном климате. Растворение полевого шпата начинается на границе сдвоенных и несдвоенных доменов, образуя круглые выемки, которые, увеличиваясь, формируют отрицательные кристаллы. В пределах больших выемок образуются аморфные, кольцеобразные структуры около 25 Å в Диаметре. Эти кольца, в свою очередь, кристаллизуются в дугообразную фазу, имеющую базальные промежутки в 10 Å и затем в морщинистые листы иллита или обезвоженного монтмориллонита. 10-Å слоистый силикат проявляет ненормальную последовательность слоев, включая последовательности 10, 20, и 30 Å. Окклюдированные плагиоклазовые кристаллы проявляют сходный механизм выветривания и, к тому же, выветриваются более интенсивно. [N.R.]

Resümee—Die hohe Auflösung, die durch die Transmissionselektronenmikroskopie erreicht wird, zeigt einen Verwitterungsmechanismus von intermediärem Mikroklin in einem humiden, gemäßigten Klima. Die Auflösung von Feldspat beginnt an der Grenze zwischen verzwillingten und nichtverzwillingten Domänen und erzeugt kreisförmige Löcher, die größer werden und negative Kristalle bilden. Es entstehen amorphe, ringförmige Strukturen mit einem Durchmesser von etwa 25 Å in den größeren Löchern. Diese Ringe wiederum kristallisieren zu einer bogenförmigen Phase, die einen Basisabstand von 10 Å hat und anchließend zu runzligen Blättchen aus Illit oder dehydratisiertem Montmorillonit. Das 10 Å-Schichtsilikat zeigt eine unregelmäßige Stapelungsfolge, die 10 Å-, 20 Å- und 30 Å-Folgen beinhaltet. Eingeschlossene Plagioklaskristalle zeigen einen ähnlichen Verwitterungsmechanismus und sind darüberhinaus intensiver verwittert. [U.W.]

Résumé—Des images à haute résolution obtenues par microscopie à transmission d'électrons ont révélé un mécanisme pour l'altération de la microcline intermédiaire dans un climat humide et tempéré. La dissolution du feldspath commence à la séparation des domaines jumelés et non-jumelés et produit des trous circulaires qui s'aggrandissent pour former des cristaux négatifs. Des structures amorphes d'environ 25 Å de diamètre, en forme d'anneau, se développent dans les trous les plus grands. Ces anneaux, à leur tour, se cristallisent en une phase arguée ayant un espacement de base de 10 Å et ensuite en des lames froncées d'illite ou de montmorillonite deshydratée. Le silicate à couches-10 Å montre une séquence irrégulière d'empilement comprenant des séquences de 10, 20, et 30 Å. Des cristaux de plagioclase occlus montrent un mécanisme d'altération semblable, et, de plus, sont plus intensément altérés. [D.J.]

Part III

CHEMICAL PRINCIPLES

Editor's Comments
on Papers 13 Through 18

13 GARRELS
Some Free Energy Values from Geologic Relations

14 KITTRICK
Soil Minerals in the Al_2O_3 - SiO_2 - H_2O System and a Theory of Their Formation

15 KITTRICK
Solubility of Two High-Mg and Two High-Fe Chlorites Using Multiple Equilibria

16 TARDY and GARRELS
A Method of Estimating the Gibbs Energies of Formation of Layer Silicates

17 HELGESON, GARRELS, and MACKENZIE
Evaluation of Irreversible Reactions in Geochemical Processes Involving Minerals and Aqueous Solutions - II: Applications

18 TSUZUKI
Solubility Diagrams for Explaining Zone Sequences in Bauxite, Kaolin, and Pyrophyllite-Diaspore Deposits

For many years students of soil mineral weathering were caught up in the details of minerals and their analyses. While weathering sequences and other generalizations systemized information on mineral occurrence, what was not clear at the time is that soil mineral weathering is an attempt to reach equilibrium with the soil solution. The equilibrium solution composition for minerals can be described if the necessary standard free energies of formation (ΔG_f^0) of minerals and ions are known. In 1957, ΔG_f^0 values for common soil minerals were scarce and unreliable. Garrels (Paper 13, published in 1957) shows how some of these ΔG_f^0 values can be estimated from known relations in earth surface environments. It is irrelevant that those particular values are obsolete now. The idea, not the numbers, was the important thing. Garrels also showed that ΔG_f^0 values can often be checked for reasonableness and that when several ΔG_f^0 values are known, other ΔG_f^0 values or environmental conditions can be estimated. Taken as a whole, the work in Paper 13 demonstrates the power and the scientific beauty of applying elementary thermodynamics to soil mineral weathering.

By 1969, preliminary ΔG_f^0 values for smectites were combined by Kittrick (Paper 14), with reasonably good values for gibbsite, kaolinite, and amorphous silica. Paper 14 includes a simple stability diagram that is in remarkably good agreement with natural relations. It made possible the prediction of which mineral pairs could form and which were thermodynamically "forbidden." It showed how gibbsite, kaolinite, and smectite, in turn, control solution Al-ion activity at low levels, depending upon H_4SiO_4 levels known to exist in the soil solution. Kittrick also suggests that soil solution H_4SiO_4 control is primarily kinetic. It is evident now from Papers 7 and 12 that detailed mineral equilibria are likely to be successfully related only to microenvironments described in chemical detail. However, Paper 14 indicates that there is also an approximate correspondence between secondary soil minerals and the general chemical environment. This paper demonstrated to many soil scientists for the first time that elementary thermodynamics can describe mineral weathering.

The ΔG_f^0 values necessary for describing soil mineral weathering have been slow in coming. There are two experimental methods, calorimetric and solubility, and neither is easy. The main difficulty with the calorimetric method is estimating the composition and amount of impurities, both of which can significantly contribute to heat measurements. The main difficulty with the solubility method is that the sample *must* be brought to equilibrium. This method requires solution conditions in which the mineral is stable and where the mineral controls at least one common ion. Sometimes there is a bonus in that the equilibration process of the solubility method is similar to aspects of weathering, thereby contributing additional information. Relatively few ΔG_f^0 values are available for the more complex soil minerals, but at least the perceived need has progressed from "get me a ΔG_f^0 value" to "get me a *good* ΔG_f^0 value."

In Paper 15 Kittrick gives the first complete experimental ΔG_f^0 for chlorite and tackles the most difficult situation yet for the solubility method. Because chlorite is stable only at alkaline pH where Fe and Al in solution could not be analyzed, gibbsite, kaolinite, and hematite were added to control these ions at calculable levels. Eh measurements were needed to apportion Fe^{2+} and Fe^{3+}. In some cases five different indicators of equilibrium were used. This attention to the necessary equilibrium requirements contrasts with many investigations where the samples were equilibrated until the investigator became restless, or the study obtained similar results for successive solubility experiments performed in exactly the same way (confusing precision with accuracy).

Prediction of mineral equilibria is dependent upon reliable ΔG_f^0 values for the ions and minerals involved. Unfortunately, some of the most important minerals in soil mineral weathering are solid solutions (for example chlorites, smectites, illites, micas). Values of ΔG_f^0 for all members of these solid solution series will never be available. Some way is needed to estimate ΔG_f^0 values for layer and fibrous silicates, using relatively few experimental values as check points. Tardy and Garrels (Paper 16), recognized that these minerals contain essentially the same elements in the same basic structural arrangements. In Paper 16, they assumed that the silicates could be represented by multiples of oxide and hydroxide components possessing constant ΔG_f^0 values. This assumption permitted them to obtain good agreement with experimentally determined values. Other schemes have since been developed, but this is still a good one.

While it is true that thermodynamics cannot predict a detailed weathering path, Helgeson, Garrels, and Mackenzie (Paper 17) show that a series of partial equilibrium steps can chart the major minerals in an irreversible course of events such as weathering. They calculate solution composition changes and the appearance and disappearance of mineral phases as protons react with primary minerals in closed system (protons from hydrolysis) or in open systems (protons from CO_2). Paper 17 is the first to successfully predict the extent of weathering reaction by using thermodynamics. The authors evaluated their model for reasonableness by comparing their results with water analysis from the Sierra Nevada by Feth, Roberson and Polzer (1964). As a bonus, Helgeson et al. derived some valuable weathering rate estimates. Paper 17 was also one of the first in which a computer model was used to handle the large amount of data and calculations.

When the necessary ΔG_f^0 values have been determined, as in Paper 15, and the general calculation methods demonstrated, as in Paper 17, it remains for researchers to innovatively apply them in order to understand a particular situation. Tsuzuki (Paper 18) simplified the reaction path approach of Helgeson, Garrels, and Mackenzie (Paper 17) for a particular application. Tsuzuki followed only the critical constituents, Al^{3+} and H_4SiO_4, which are the coordinates of his stability diagrams. This method required a series of diagrams for different K^+ and pH levels, but simplified the calculations enormously. A qualitative interpretation of these diagrams was sufficient to Tsuzuki's purpose of explaining zoned mineral sequences in certain deposits. This is the kind of direct application of ΔG_f^0 values and calculation methods to natural systems that is required to advance our understanding of soil mineral weathering.

REFERENCE

Feth, J. H., C. E. Roberson, and W. L. Polzer, 1964, Sources of Mineral Constituents in Water from Granitic Rocks, Sierra Nevada, California, and Nevada, *U. S. Geol. Surv. Water-Supply Pap. 1535-I,* 170 pp.

13

Copyright © 1957 by the Mineralogical Society of America
Reprinted from *Amer. Miner.* **42**:780-791 (1957)

SOME FREE ENERGY VALUES FROM GEOLOGIC RELATIONS

R. M. GARRELS, *Harvard University, Cambridge, Massachusetts.*

ABSTRACT

Mineral associations in earth surface environments have been used to obtain basic thermochemical data for reactions and compounds of geologic interest. From a study of weathering processes, standard free energies of formation from the elements at 25° C. and one atmosphere total pressure have been estimated for kaolinite (−883 kcal.), muscovite (−1298 kcal.), and K-feldspar (−856 kcal.). From relations in the zone of oxidation of ore deposits a correction has been made to the published value for hydrocerussite (from −409.1 kcal. to −406 kcal.) and a new value for malachite (−217 kcal.). It is suggested that the methods used can be applied to obtain useful free energy values for many other compounds.

INTRODUCTION

Since 1952, when Latimer published the second edition of his classic "Oxidation Potentials," sufficient free energy data on compounds and ions have been easily available to calculate equilibrium relations for many low temperature-low pressure systems of geologic interest. Although these data are strictly applicable only at 25° C. and 1 atmosphere total pressure, the error in applying them to natural environments at or near the earth's surface is usually small.

Eh-pH diagrams utilizing such data were first used by Pourbaix (1949) as a convenient and provocative method of showing interrelation between solids and dissolved ionic species with special reference to problems of metallic corrosion. Since then, similar diagrams have been used to depict approximate equilibrium relations between minerals and the ions in equilibrium with them in aqueous solution, and the results have been compared to natural relations, especially in problems of atmospheric oxidation of ores, and those of primary chemical precipitation (cf. Garrels and Huber, 1953).

When such diagrams were first constructed, it was with hope that they would bear a faint resemblance to nature, and conceivably might be used like an "ideal gas" or "ideal solution"—as hypothetical models useful in showing how far complex natural relations departed from simple systems involving chemical compounds instead of minerals, and containing only those ions or other dissolved species for which thermochemical information happened to be available.

It has been a surprise to find that these naive chemical analogs of nature are directly useful in many instances. That is to say that numerous calculated solubilities fit those deduced from geologic relations, predicted assemblages of chemical compounds are reflected by identical assemblages of their nearest mineral analogs, and the environment of their occurrence, expressed on a pH-potential grid, corresponds to that found in the field. In fact, natural systems reflect the diagrams much better than most experimental ones!

The reasons for this agreement seem to be: (1) There is sufficient time available for achievement of near-equilibrium in many natural low temperature aqueous environments. Even though the low temperature activation energies for transformations of silicate structures, for example, are high, and experimental work is well-nigh impossible, the months and years during which natural systems fluctuate through a small range of conditions permit a close approach to equilibrium. (2) The amount of interaction in natural systems is less than might be anticipated from their complexity. For example, the error in calculating the free energy of formation of calcite from solubility data on sea water is small if all the currently known interactions are considered. (3) The effect of biological activity is to add complexity on the one hand, and to catalyze reaction on the other. The net effect seems to be one of increasing reaction rates of well-established reactions, and hence of helping rather than hindering the investigator. (4) In a considerable number of instances, the difference in free energy between a pure synthetic compound and its mineral analog is not large (although the difference may be extremely important for some processes!)

Because of the close correspondence between natural relations and those calculated from free energy data—that is, from experiments carefully designed to approach equilibrium, the interesting possibility arose of obtaining free energy data directly from observations on natural systems. Numerous checks were made by calculating free energies of reaction or of formation for reactions or compounds for which complete free energy data already were available. The results were so encouraging that some values for free energy of formation of minerals hitherto unknown were attempted.

Check Calculations

To illustrate the methods used, and to show the degree to which there is quantitative agreement between free energy values calculated from natural relations and those determined by various experimental methods, a few examples are given here of calculation of free energy of formation of substances for which experimental values already have been published.

ΔF_{CaCO_3}

The argument has raged for years as to whether the low latitude oceans are saturated or supersaturated with calcium carbonate. Thus the system seems to be one that approaches equilibrium. According to

Sverdrup et al. (1942, p. 205) the average calcium content of the oceans is about 0.0102 mols per liter, the average HCO_3^- is about 0.0018 equivalents per liter, the average pH about 8.2, the ionic strength is 0.7, and the temperature in the vicinity of 25° C.

Assuming equilibrium, we can write:

$$\frac{a_{Ca^{++}_{aq}} a_{CO_3^=_{aq}}}{a_{CaCO_3 \text{ solid}}} = k_{CaCO_3} \tag{1}$$

$$\frac{a_{H^+} + a_{CO_3^=}}{a_{HCO_3^-}} = k_{HCO_3^-} \tag{2}$$

$$\Delta F°_{\text{Reaction}} = RT \ln k = -1.364 \log k_{(25°)}. \tag{3}$$

$$\Delta F°_{\text{Reaction}} = \Delta F°_{Ca^{++}_{aq}} + \Delta F°_{CO_3^=_{aq}} - \Delta F°_{CaCO_3 \text{ solid}} \tag{4}$$

But a_{CaCO_3} solid at 25° C. and 1 atmosphere pressure is unity by convention, and $a_{Ca^{++}} = \gamma_{Ca^{++}} \cdot m_{Ca^{++}}$, and $\gamma_{HCO_3^-} = a_{HCO_3^-} m_{HCO_3^-}$, where m represents molality and γ the activity **coefficient**. From Garrels and Dreyer (1952, p. 234), $\gamma_{Ca^{++}}$ in seawater is about 0.26, and $\gamma_{HCO_3^-}$ about 0.36. At pH = 8.2, $a_{H^+} = 10^{-8.2}$.

Then, from No. 2:

$$a_{CO_3^=} = \frac{k_{HCO_3^-} \gamma_{HCO_3^-} m_{HCO_3^-}}{a_{H^+}}$$

Substituting numerical values:

$$a_{CO_3^=} = \frac{10^{-10.34} 10^{-0.44} 10^{-2.75}}{10^{-8.2}} = 10^{-5.33} \tag{5}$$

Also:

$$a_{Ca^{++}} = m_{Ca^{++}} \cdot \gamma_{Ca^{++}}$$

Using the analyzed value of calcium for $m_{Ca^{++}}$ (this assumes no important interactions of Ca^{++}) and substituting a numerical value for $\gamma_{Ca^{++}}$:

$$a_{Ca^{++}} = 10^{-2.01} 10^{-0.59} = 10^{-2.60} \tag{6}$$

Substituting the values from No. 5 and No. 6 in No. 1:

$$\frac{10^{-2.60} 10^{-5.33}}{10^0} = k_{CaCO_3} = 10^{-7.93}$$

Then this value of k can be substituted in No. 3:

$$\Delta F°_{\text{reaction}} = -1.364 \log 10^{-7.93}$$

$$\Delta F°_{\text{reaction}} = 10.8 \text{ kcal}.$$

Finally, using this value, and values for $\Delta F°_{Ca^{++}}$ and $\Delta F°_{CO_3^-}$ from Latimer (1952) and substituting in No. 4:

$$10.8 = -132.18 - 126.2 - \Delta F°_{CaCO_3 \text{ solid}}$$

$$\Delta F°_{CaCO_3 \text{ solid}} = -269.2$$

This compares with the published value for calcite of -269.78 and of aragonite of -269.53 (Latimer, op. cit.).

The calcium carbonate-sea water system is an unusually complex one, owing to the high concentration of salts which makes it difficult to obtain reliable γ values for the ions involved. Yet the check is within one kilocalorie on the basis of a fairly crude calculation. Obviously no information has been obtained on calcite-aragonite relations, for which (Latimer, 1952):

$$\Delta F°_{calcite} - \Delta F°_{aragonite} = -0.27 \text{ kcal.}$$

$\Delta F°_{UO_2CO_3}$

Another example is the calculation of $\Delta F°$ for UO_2CO_3 (rutherfordine). The mineral is sometimes intimately associated with various uranyl oxide hydrates, or uranyl hydroxyhydrates. The stable hydrate in water is $UO_2(OH)_2 \cdot H_2O$ (Bullwinkel, 1954, p. 7). From the reaction:

$$UO_2(OH)_2 \cdot H_2O + CO_2 = UO_2CO_3 + 2H_2O$$

it is apparent that the existence of rutherfordine at equilibrium with $UO_2(OH)_2 \cdot H_2O$ in nearly pure water at a fixed temperature can occur at a single value for the partial pressure of CO_2. Thus in the weathering environment, which is close to 25° C., the coexistence of these two species suggests that the equilibrum partial pressure of CO_2 is close to that of CO_2 in the atmosphere ($\cong 10^{-3.5}$ atmospheres). For those conditions the equilibrium constant for the reaction is:

$$k = \frac{1}{P_{CO_2}} = \frac{1}{10^{-3.5}}$$

The free energy is:

$$\Delta F°_R = -1.364 \log 10^{-3.5} = 4.8 \text{ kcal.}$$

Using Bullwinkel's value of $\Delta F°$ for $UO_2(OH)_2 \cdot H_2O$ (op. cit., p. 30) and Latimer's values for the others:

$$\Delta F°_R = \Delta F°_{UO_2CO_3} + 2\Delta F°_{H_2O} - \Delta F°_{UO_2(OH)_2 \cdot H_2O} - \Delta F°_{CO_2(gas)}$$
$$4.8 = \Delta F°_{UO_2CO_3} - 113.4 + 391 + 94.3$$
$$\Delta F°_{UO_2CO_3} = -377.1$$

Bullwinkel's listed value is -377. Actually, his value and the one calculated here have to agree, because he found experimentally that the equilibrium partial pressure of CO_2 was almost exactly that of the atmosphere, but this experimental determination does not destroy the validity of the preceding geologic reasoning.

These examples illustrate the *general* relation that geologic coexistence of two chemically precipitated phases usually indicates that the free energy of the reaction to form one from the other in their environment is about 2 kilocalories or less. The phases in question may be compatible or incompatible pairs.

Relations Deduced From Weathering Processes

Students of weathering have worked out, from various lines of evidence, a general sequence of mineral stability as follows:

$$\text{K-feldspar} \rightarrow \text{K-mica} \rightarrow \text{kaolinite} \rightarrow \genfrac{}{}{0pt}{}{\text{diaspore}}{\text{gibbsite}}$$

$$\text{Na-feldspar} \rightarrow \text{montmorillonite} \rightarrow \text{kaolinite} \rightarrow \genfrac{}{}{0pt}{}{\text{diaspore}}{\text{gibbsite}}$$

(c.f. Keller, 1957, Reiche, 1945, Goldich, 1938, Mohr and Van Baren, 1954).

In other words, in a system open to rain water, the rock-forming silicates tend to alter to last residue of aluminum oxide hydrates (and ferric oxide where femic minerals originally are involved). In an idealized system, we can visualize a vertical soil profile under conditions of high rainfall and continuous downward drainage as a steady state condition with zones of stability of the minerals from unaltered silicates at the bottom to aluminous residue near the top.

We have fairly good information on the free energy of formation of gibbsite and boehmite (Deltombe and Pourbaix, 1956, p. 3), at one end of the reaction series, and at the other end only a value for $\Delta H°$ for orthoclase (Yoder and Eugster, 1955, p. 262). The plan here is to start with gibbsite, and attempt to calculate the free energies of formation of the intermediate products, using $\Delta H°$ for orthoclase as an approximate check-point at the other end of the series.

Relations Among the Aluminum Oxide Hydrates

Lateritic soils contain a variety of aluminum oxide hydrates. Among the chief minerals are diaspore ($Al_2O_3 \cdot H_2O$), its dimorph boehmite; and gibbsite ($Al_2O_3 \cdot 3H_2O$). All three of these are found in important quantities, and are intimately associated. Boehmite is probably unstable relative to diaspore under geologic conditions (G. MacDonald, personal communication), but the free energy difference between these minerals is probably small, so that boehmite can form and persist. Recent work by Deltombe and Pourbaix (1956) indicates that gibbsite is stable relative to boehmite in aqueous solution at room temperature.

In summary, the geologic occurrence and thermochemical information are in harmony if it is assumed that gibbsite forms if equilibrium is attained in the presence of nearly pure liquid water, and diaspore and boehmite are the products of disequilibrium, or of heating, and drying. Gibbsite can be looked upon ideally as the product of wet leaching of silicates; boehmite and diaspore as its dehydration products. Undoubtedly there is further important control by grain size and other complicating factors, but where soil leaching takes place in well drained soils in the presence of nearly pure water, gibbsite is the expected stable phase.

Relation of Kaolinite to Aluminum Oxide Hydrates

In the weathering of feldspar, kaolinite may or may not be an intermediate product in the formation of aluminum oxide hydrates. Goldman (1955) shows convincingly that feldspar altered directly to gibbsite in the Arkansas bauxite deposits. He further interprets that gibbsite altered to cliachite (fine-grained aluminum oxide hydrates, probably chiefly monohydrate). This cliachite is, in turn, selectively altered to kaolinite. Elsewhere, as in the alteration of shales to bauxite, kaolinite is a definite precursor of the aluminum oxide hydrates. Gordon and Tracey (1951, p. 32) tie in some of the resilication of bauxite in Arkansas with waters from overlying swamps.

Thus the reaction:

$$\text{"Bauxite"} + \text{silica} = \text{kaolinite}$$

is clearly reversible under geologic conditions, going to the right when silica content of water is typical of that of swamp water, and to the left at some lower value. Furthermore, the monohydrates seem to be more easily kaolinized than the tri-hydrate, which follows from their relative stabilities in water, but both apparently can be resilicated. A reasonable estimate of the equilibrium condition would be kaolinization of gibbsite at a dissolved silica content of about 10 parts per million at 25° C. Tropical streams carry about three times this amount, average streams a little less (Clarke, 1927, Chap. III). From this the relation can be written:

$$\underset{\text{Gibbsite}}{Al_2O_3 \cdot 3H_2O} + \underset{\substack{\text{10 ppm. (0.00017 }m\text{)} \\ \text{silica in solution}}}{2SiO_{2\,aq}} = \underset{\text{kaolinite}}{H_4Al_2Si_2O_9} + \underset{\substack{\text{nearly pure} \\ \text{liquid water}}}{H_2O}; \quad \Delta F = 0 \quad (7)$$

According to Krauskopf (1956, p. 23) dissolved silica is saturated with respect to its amorphous polymers (silica glass) at a concentration of about 140 ppm. (0.0024 m) at 25° C. Thus:

$SiO_{2\,glass} = SiO_{2\,aq\,(140\,ppm)}; \quad \Delta F = 0$

$k = a_{SiO_2} = 0.0023$ (assuming activity = molality for a molecular species in dilute solutions)

$\Delta F°_{reaction} = -1.364 \log 0.0023 = 3.5$ kilocalories

Therefore:

$\Delta F°_{SiO_2\,aq} - \Delta F°_{SiO_2\,glass} = 3.5$ kilocalories

$\Delta F°_{SiO_2\,aq} = 3.5 + \Delta F°_{SiO_2\,glass} = 3.5 - 190.9 = -187.4$ kcal.

For reaction No. 7, the equilibrium constant is:

$$k = \frac{1}{a^2_{SiO_2}} \frac{1}{(0.00017)^2} = 10^{7.54}$$

$$\Delta F°_R = -1.364 \log 10^{7.54} = -10.3 \text{ kcal.}$$

From the standard free energy of the reaction, and the standard free energy of formation of the reactants and products, it is possible to solve for $\Delta F°$ of kaolinite.

$$\Delta F°_R = \Delta F°_{kaol.} + \Delta F°_{H_2O} - \Delta F°_{gibbsite} - 2\Delta F°_{SiO_2\,aq}$$

$$\Delta F°_{kaol.} = \Delta F°_R + \Delta F°_{gibbsite} + 2\Delta F°_{SiO_2\,aq} - \Delta F°_{H_2O}$$

$$\Delta F°_{kaol.} = -10.3 - 554.6 - 374.8 + 56.7 = -883.0 \text{ kcal.}$$

In passing it should be noted that $\Delta F°_{kaolinite}$ is not particularly sensitive to the value of soluble silica chosen—a tenfold larger or smaller value would change $\Delta F°_{kaolinite}$ by ± 2.6 kcal.

INTERRELATIONS OF KAOLINITE, MICA, AND K-FELDSPAR

From the field evidence, kaolinization of feldspar seems to take place under almost any soil condition, even at pH values as high as 8 or 9, suggesting that such conditions are sufficient to convert feldspar into mica, and mica into kaolinite. If the feldspar-mica alteration is slow, and the mica-kaolin reaction fast, mica might not even be observed as an intermediate product; in fact it might not form at all as a crystalline material if conditions under which kaolinite is stable are superimposed on a feldspar grain. However, very careful work shows mica as an intermediary between feldspar and kaolin (Sand, 1956). The total evidence indicates that K-feldspar, quartz and K-mica are in equilibrium at a pH and K^+ content of the system close to but higher than those at which K-mica and kaolin are stable.

Hydrolysis of mica yields maximum pH values of the order of 9.3 (Garrels and Howard) in solutions containing K^+ ions at activity of about $10^{-3.0}$ (about 40 ppm.). The reaction involved is a surface reaction transforming K-mica to H-mica, but should be a guide to the boundary between K-mica and kaolin. At any rate the K-mica-kaolin boundary will not be at lower pH at the same K^+. Tentatively we can place the K-mica-kaolin boundary at pH 9.5 at a K^+ activity of 10^{-3}, and the K-feldspar-K-mica boundary at pH 10.5 at the same activity of K^+. These values fit geologic relations fairly well; waters at the lower limit of the zone of soil formation must represent conditions very close to equilibrium with feldspar, or at least conditions under which reaction does not take place at a finite rate over long periods of time. Not many pH measurements have been made of waters percolating through at this boundary, but values of 9 to 10 are not uncommon.

Tentatively we can write:

$$3H_4Al_2Si_2O_9 + 2K^+ = 2KAl_3Si_3O_{10}(OH)_2 + 3H_2O + 2H^+$$

$$K_{Ka-mica} = \frac{(H^+)^2}{(K^+)^2} = \frac{10^{-19}}{10^{-6}} = 10^{-13} \tag{8}$$

$$\Delta F°_R = -1.364 \log 10^{-13} = 17.7 \text{ kcal.}$$

and:

$$KAl_3Si_3O_{10}(OH)_2 + 2K^+ + 6SiO_2 = 3KAlSi_3O_8 + 2H^+$$

$$K_{mica-feldspar} = \frac{(H^+)^2}{(K^+)^2} = \frac{10^{-21}}{10^{-6}} = 10^{-15} \tag{9}$$

For (8): $\Delta F_R° = -1.364 \log 10^{-15} = 20.2$ kcal.

$$\Delta F°_R = 2\Delta F°_{mica} + 3\Delta F°_{H_2O} + 2\Delta F°_{H^+} - 3\Delta F°_{kaolin} + 2\Delta F°_{K^+}$$
$$2\Delta F°_{mica} = -3\Delta F°_{H_2O} - 2\Delta F°_{H^+} + 3\Delta F°_{kaolin} + 2\Delta F°_{H^+} + \Delta F°_R$$
$$2\Delta F°_{mica} = +170.1 + 0 - 2649 - 134.9 + 17.7$$
$$\Delta F°_{mica} = -1298 \text{ kcal.}$$

For reaction No. 9, assuming first equilibrium with silica glass:

$$\Delta F°_R = 3\Delta F°_{OR} + 2\Delta F°_{H^+} - \Delta F°_{mica} - 2\Delta F°_{K^+} - 6\Delta F°_{SiO_2(glass)}$$
$$3\Delta F°_{OR} = \Delta F°_R - 2\Delta F°_{H^+} + \Delta F°_{mica} + 2\Delta F°_{K^+} + 6\Delta F°_{SiO_2(glass)}$$
$$3\Delta F°_{OR} = 20.2 - 0 - 1298 - 134.9 - 1145.4$$
$$\Delta F°_{OR} = -853 \text{ kcal.}$$

If it is assumed that equilibrium is with quartz instead of with silica glass:

$$\Delta F°_{orthoclase} = -856 \text{ kcal.}$$

As a check, a value of $\Delta H°$ for the formation of orthoclase from the oxides can be calculated, and compared with that cited by Yoder and Eugster (1955, p. 263). From the relation:

$$\Delta H° = \Delta F° + T\Delta S°$$

the standard heat of formation can be obtained.

For the reaction to form orthoclase from the oxides:

$$\tfrac{1}{2}Al_2O_3 + \tfrac{1}{2}K_2O + 3SiO_2 = KAlSi_3O_8$$

the standard entropy of reaction is:

$$\Delta S°_R = S°_{OR} - \tfrac{1}{2}S°_{Al_2O_3} - \tfrac{1}{2}S°_{K_2O} - 3S°_{SiO_2}$$

Using values of $S°$ for the oxides from Latimer, and that for orthoclase (adularia) from Kelley, et al (1955, p. 11)

$$\Delta S°_R = 52.5 - 6.1 - 10.4 - 30 = 6 \text{ cal/mol.}$$

For the same reaction:

$$\Delta F°_R = \Delta F°_{OR} - \tfrac{1}{2}\Delta F°_{Al_2O_3} - \tfrac{1}{2}\Delta F°_{K_2O} - 3\Delta F°_{SiO_2}$$
$$\Delta F°_R = -856 + 188.3 + 38.1 + 577.2 = -52.4 \text{ kcal.}$$

Then:

$$\Delta H°_R = \Delta F°_R + T\Delta S°_R$$
$$\Delta H°_R = -52.4 + (298 \times 0.006) = -50.6 \text{ kcal.}$$

The value cited by Yoder and Eugster (*op. cit.*) is -56.4 kcal. The check, considering the length of the "bridge" used, the uncertainty in the value for gibbsite at one end of the bridge, and the uncertainty in the value for the heat of formation of orthoclase as cited by Yoder, as well as other possible sources of error, is good.

Summary

In summary, the following values have been obtained for standard free energies of reactions and of formation of compounds:

TABLE 1. STANDARD FREE ENERGIES

	$\Delta F°_{298.1}$ (kilocalories)	
Reaction		
$Al_2O_3 \cdot 3H_2O_c + 2SiO_{2aq} = H_4Al_2Si_2O_{9c} + H_2O_{liq}$	-10.3	All reaction values
$Al_2O_3 \cdot 3H_2O_c + 2SiO_{(glass)} = H_4Al_2Si_2O_{9c} + H_2O_{liq}$	-17.3	estimated plus or
$Al_2O_3 \cdot 3H_2O_c + 2SiO_{2quartz} = H_4Al_2Si_2O_{9c} + H_2O_{liq}$	-20.5	minus 2 kilocalories,
$3H_4Al_2Si_2O_{9c} + 2K_{aq}^+ = 2KAl_3Si_3O_{10}(OH)_{2c} +$		with strong possi-
$3H_2O_{liq} + 3H_{aq}^+$	$+18$	bility of larger er-
$KAl_3Si_3O_{10}(OH)_{2c} + 2K_{aq}^+ + 6SiO_{2quartz} = 3KAlSi_3O_{8c} +$		rors in values for
$2H_{aq}^+$	$+20$	minerals
Mineral	$\Delta F°_f$	
Kaolinite	-883	
K-mica	-1298	
K-feldspar	-856	

The error on individual determinations is estimated at one or two kilocalories, but the possibility of cumulative errors across the "bridge" are considerable. There still is serious doubt concerning the value for gibbsite, which has been used as a base, because it is based in turn on the somewhat questionable value of αAl_2O_3. But when a firmer value for $\Delta F°$ orthoclase becomes available as a check, it should be possible to fix mica and kaolinite within narrow absolute limits.

Standard Free Energy of Formation of $Pb_3(OH)_2(CO_3)_2$ and $Cu_2(OH)_2CO_3$

In addition to using natural occurrence to obtain $\Delta F°$ values for compounds for which values have not been obtained experimentally, it is possible to use it to correct values already determined or to choose between data from conflicting sources.

$Pb_3(OH)_2(CO_3)_2$

Latimer estimates a $\Delta F°$ value for $Pb_3(OH)_2(CO_3)_2$ of -409.1 kilocalories. This compound corresponds to the rare mineral hydrocerussite. In nature cerussite ($PbCO_3$) is the major lead carbonate. If the reaction is written:

$$3PbCO_{3c} + H_2O_{liq} = Pb_3(OH)_2(CO_3)_{2c} + CO_{2g}$$

then the equilibrium constant, assuming that the groundwater is nearly pure water, is:

$$k = P_{CO_2}$$
$$\Delta F°_R = \Delta F°_{Pb_3(OH)_2(CO_3)_2} + \Delta F°_{CO_2} - 3\Delta F°_{PbCO_3} - F°_{H_2O}$$
$$\Delta F°_R = -409.1 - 94.26 + 449.1 + 56.7 = 2.4 \text{ kcal.}$$

Then:
$$2.4 = -1.364 \log k$$
$$k = 10^{-1.8}$$

Thus the equilibrium partial pressure, according to these relations, is $10^{-1.8}$ atmospheres of CO_2. This clearly cannot be the case, for $PbCO_3$ would be unknown at equilibrium at the earth's surface. Instead, because of its few occurrences, it is much more likely that it forms at a P_{CO_2} somewhat less than that commonly observed—perhaps at 10^{-4} atmospheres. In this case:

$$k = 10^{-4}$$
$$\Delta F°_R = -1.364 \log 10^{-4} = 5.5 \text{ kcal.}$$

Then, assuming that the correction should be made to $Pb_3(OH)_2(CO_3)_2$:

$$5.5 = \Delta F°_{Pb_3(OH)_2(CO_3)_2} - 94.26 + 449.1 + 56.7$$
$$\Delta F°_{Pb_3(OH)_2(CO_3)_2} = -406 \text{ kcal.}$$

This value, 3.1 kilocalories larger than that of Latimer, has been shown to be a minimum by a prettily reasoned application of the phase rule to similar equilibria by W. L. McIntire (unpublished manuscript), but the straightforward application of known mineral occurrence probably suffices.

$Cu_2(OH)_2CO_3$

Basic copper carbonate, corresponding to the mineral malachite, occurs abundantly in the zone of weathering of copper deposits. In many instances it has been observed intimately intergrown with CuO (tenorite). From the relation:

$$Cu_2(OH)_2CO_{3c} = 2CuO_c + H_2O_{liq} + CO_{2g}$$

it is apparent that in dilute aqueous solution the equilibrium constant contains only P_{CO_2}. From the prevalence of malachite occurring alone, but in apparent equilibrium with tenorite in some deposits, it can be concluded that the equilibrium P_{CO_2} is perhaps slightly less than that of the normal atmosphere. If so:

$$k = 10^{-3.7}$$
$$\Delta F°_R = 1.364 \log 10^{-3.7} = 5.0 \text{ kcal}.$$

The value of $\Delta F°_{Cu_2(OH)_2CO_3}$ is not available, but accepting those for the other species as given in Latimer:

$$\Delta F°_R = 2\Delta F°_{CuO} + \Delta F°_{H_2O} + \Delta F°_{CO_2} - \Delta F°_{Cu_2(OH)_2CO_3}$$
$$5.0 = -60.8 - 56.7 - 94.26 - \Delta F°_{Cu_2(OH)_2CO_3}$$
$$\Delta F°_{Cu_2(CO)_2CO_3} = -217 \text{ kcal}.$$

The method could obviously be used to determine $\Delta F°$ values for other basic copper salts, and work is in progress by P. Hostetler at Harvard University that will compare experimentally determined values with those deduced in this manner.

Conclusion

The examples given here of the use of natural mineral occurrence to obtain basic thermochemical data suggest that many useful estimates could be made to supplement current values derived almost entirely from laboratory experiment, and further that experimental values can be checked by knowledge of natural occurrence. Many geologists have gone to the laboratory to obtain data to use in explaining geologic phenomena; the possible contributions that might be made by reversing the process seem to be somewhat neglected.

Acknowledgments

I want to express my gratitude to the members of the Geochemistry Seminar at Harvard University, who contributed largely to the development of the ideas presented here. I am especially indebted to P. B. Hostetler and to W. L. McIntire of that Seminar, to J. B. Thompson, and R. Siever, of the Division of Earth Sciences at Harvard, and to C. L. Christ, of the United States Geological Survey, who gave freely of their time in discussing many of the problems involved, and made many valuable suggestions.

References

BULLWINKEL, E. P. (1954), The chemistry of uranium in carbonate solutions: *U. S. Atomic Energy Commission*, R.M.O. **2614**, 59 pp.

CLARKE, FRANK WIGGLESWORTH (1924), The data of geochemistry: *U. S. Geol. Surv. Bull.* **770**, 841 pp.

DELTOMBE, E., AND POURBAIX, M. (1956), Comportement electrochimique de l'aluminum. Diagramme d'equilibre tension-pH du system Al-H$_2$O, a 25° C.: Rapport Technique, **42**, Centre Belge d'Etude de la Corrosion, Brussels, 14 pp.

GARRELS, R. M. AND HUBER, N. K. (1953), Relation of pH and oxidation potential to sedimentary iron mineral formation: *Econ. Geol.*, **48**, 337–358.

GARRELS, R. M. AND DREYER, R. M. (1952), Mechanisms of limestone replacement at low temperatures and pressures: *Bull. Geol. Soc. Amer.*, **63**, 325–379.

GARRELS, R. M. AND HOWARD, PETER, Reactions of feldspar and mica with water at low temperature and pressure: in press.

GOLDICH, S. S. (1938), A study in rock weathering: *Jour. Geol.*, **46**, 17–58.

GOLDMAN, MARCUS I. (1955), Petrography of bauxite surrounding a core of kaolinized nepheline syenite in Arkansas: *Econ. Geol.*, **50**, 586–610.

GORDON, MACKENZIE, JR. AND TRACEY, JOSHUA I. (1951), Origin of the Arkansas bauxite deposits: In *Amer. Inst. Min. Met. Eng. Symposium* "Problems of clay and laterite genesis," pp. 12–34.

KELLER, W. D. (1957), The principles of chemical weathering: Lucas Bros., Columbia, Mo., Revised Edition, 110 pp.

KELLEY, K. K., TODD, S. S., ORR, R. C., KING, E. G. BONNICKSON, K. R. (1953), Thermodynamic properties of sodium-aluminum and potassium-aluminum silicates: *U. S. Bur. Mines R. I.*, **4955**, 14 pp.

KRAUSKOPF, K. K. (1956), Dissolution and precipitation of silica at low temperatures: *Geochim. et Cosmochim. Acta*, **10**, 1–26.

LATIMER, WENDELL M. (1952), Oxidation Potentials: Prentice Hall, Inc., New York, Second Edition, 392 pp.

MOHR, E. C. J. AND VAN BAREN, F. A. (1954), Tropical Soils: Interscience Publishers, New York.

POURBAIX, M. J. N. (1949), Thermodynamics of dilute aqueous solutions: Edward Arnold and Co., London, 136 pp.

REICHE, PARRY (1945), Survey of weathering processes and products: *Univ. New Mex. Pub. in Geology*, No. **1**.

SAND, L. B. (1956), On the genesis of residual kaolins: *Am. Mineral.*, **41**, 28–40.

SVERDRUP, H. U., JOHNSON, MARTIN W., FLEMING, RICHARD H. (1942), The Oceans: Prentice Hall, Inc., New York, 1086 pp.

YODER, H. S. AND EUGSTER, H. P. (1955), Synthetic and natural muscovites: *Geochim. et Cosmochim. Acta*, **8**, 225–280.

ERRATA

Page 782, equation (2) should read:

$$\frac{(a_{H^+})(a_{CO_3^=})}{a_{HCO_3^-}} = k_{HCO_3^-}$$

Page 782, line 14 should read: "vention, and

$$a_{Ca^{++}} = \gamma_{Ca^{++}} \cdot m_{Ca^{++}}, \text{ and } \gamma_{HCO_3^-} = a_{HCO_3}/m_{HCO_3^-}, \text{ where } m$$

SOIL MINERALS IN THE Al_2O_3-SiO_2-H_2O SYSTEM AND A THEORY OF THEIR FORMATION*

J. A. KITTRICK†

Department of Agronomy, Washington State University, Pullman, Wash. 99163

(Received 22 January 1969)

Abstract – An attempt has been made to assemble the best thermodynamic information currently available for soil minerals in the Al_2O_3-SiO_2-H_2O system at 25°C and 1 atm. Montmorillonite is included by considering its aluminum silicate phase. Diagrams are presented so that the stability of the minerals can be visualized in relation to the ionic environment. Although the Al_2O_3-SiO_2-H_2O system is a very simple one compared to soils and sediments, the stability diagrams depict a mineral stability sequence and mineral pair associations that are in good agreement with natural relations.

According to the stability diagram, mineral pairs that can form in intimate association are gibbsite-kaolinite, kaolinite-montmorillonite, and montmorillonite-amorphous silica. Forbidden pairs are amorphous silica-kaolinite, amorphous silica-gibbsite, and montmorillonite-gibbsite. The formation of intimate mixtures of three or more of these minerals is also forbidden. The stability diagrams predict ion activity relationships that are in reasonable agreement with those obtained from soils and sediments.

Amorphous silica probably limits high silica levels, with montmorillonite also forming at high silica levels. Kaolinite forms at intermediate and gibbsite at low silica levels. These minerals in turn probably control the activity of aluminum ions at a level appropriate to the pH. The formation of gibbsite, kaolinite, montmorillonite and amorphous silica appears to be controlled by a combination of kinetics and equilibria. That is, the kinetic dissolution of unstable silicates appears to control the H_4SiO_4 level. The new mineral(s) most stable at that H_4SiO_4 level appear to precipitate in response to solution equilibria.

INTRODUCTION

IT HAS been shown that the composition of the ocean depends upon the composition of river waters, which in turn depends upon the ground waters whose composition depends upon the minerals they contact (Mackenzie and Garrels, 1966; Bricker and Garrels, 1967; Stumm and Leckie, 1967). The objective of this paper is to contribute to the knowledge of mineral formation in soils and sediments. The general hypothesis of this paper is that the ground water-mineral system is a two-way street. That is, not only do minerals help determine the composition of ground waters, the composition of the ground waters also determines the course of mineral formation in the soil or sediment.

There are too many important minerals that form in soils and sediments to consider them all at once, yet is very difficult to understand the formation of a single mineral in isolation from others that compete for the same elements. The necessary compromise is to consider only those minerals that seem most important in competing for a limited group of elements. The Al_2O_3-SiO_2-H_2O system at 25°C and 1 atm has been chosen for two reasons. First, it is a relatively simple system, yet contains minerals that make up large percentages of many soils and sediments. Second and most important, stability information is available for the minerals that need to be considered.

The organization of this paper involves an initial discussion of individual mineral stabilities within several mineral groups. The stability relations between minerals in these groups are displayed in a series of diagrams. Next, a single mineral from each group is selected for inclusion in a stability diagram representing the whole Al_2O_3-SiO_2-H_2O system. Finally, the apparent applicability of the simple Al_2O_3-SiO_2-H_2O system to the much more complicated soil system is considered.

AMORPHOUS SILICA AND QUARTZ

Quartz is an important mineral because it is so abundant in soils and sediments. The determination of quartz solubility has been difficult, partly because of the influence of more soluble particles adhering to a quartz surface which may itself have more soluble regions. When these more soluble

---

*This investigation was supported in part, by contract WP-01016 with the U.S. Public Health Service, and is published as Scientific Paper No. 3215 Washington Agr. Exp. Sta. Proj. 1885.
†Professor of Soils.

substances are removed chemically, a fundamental problem in determining quartz solubility remains. The problem appears to derive from the high activation energy required to alter the Si—O—Si bond (Stöber, 1967). This results in a very slow dissolution rate and a negligible precipitation rate for quartz at room temperature. The negligible precipitation rate prevents the establishment of a true thermodynamic equilibrium at room temperature. Perhaps the best procedure is to determine quartz solubility at higher temperatures, where dissolution and precipitation rates are appreciable, and to extrapolate these values to room temperature. Siever (1962) has extrapolated high temperature data of his own, together with those of Kennedy (1950), Fournier (1960) and van Lier (1959), and obtained a solubility for quartz at 25°C of 10·8 ppm SiO_2 or $1·80 \times 10^{-4}$ M. An independent extrapolation of their own data by van Lier, DeBruyn and Overbeek (1960) confirms this value, which is indicated by the quartz solubility line in Fig. 1 (upper left) intersecting the abscissa at a pH_4SiO_4 of 3·74 (the ordinate of this diagram is explained later).

The very slow dissolution rate and the negligible precipitation rate of quartz at room temperature resulting from the high activation energy of the Si—O—Si bond have two very important consequences in the weathering of quartz. The very slow dissolution rate means that quartz will be relatively "resistant to weathering" and the negligible precipitation rate means that quartz is unlikely to control any silica equilibria. One result is that many silica equilibria are controlled at silica levels considerably in excess of quartz solubility, some by amorphous silica.

Essentially, amorphous silica is an unstable intermediate that accumulates because it precipitates more readily than quartz (much amorphous silica is produced biochemically from undersaturated

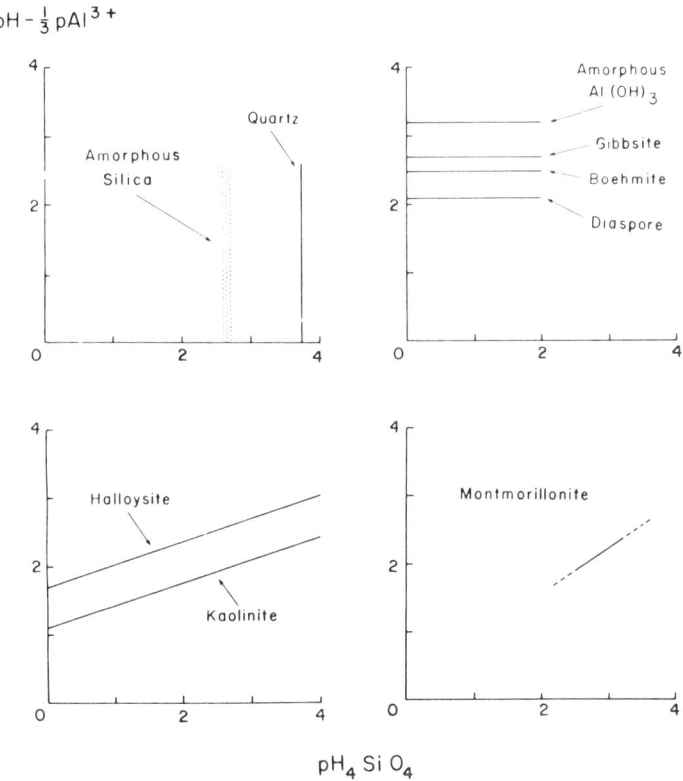

Fig. 1. Stability diagrams for some minerals in the Al_2O_3–SiO_2–H_2O system at 25°C and 1 atm. The area above or to the left of the solubility lines represents supersaturation, whereas the area below or to the right represents undersaturation.

solutions) and yet converts to more stable forms slowly (for example, opal in sea water that has not altered to quartz in over 60 million yr, Rex, 1967). The exact solubility of amorphous silica is difficult to determine. There is a tendency for samples to remain supersaturated indefinitely, probably because the precipitation reaction, although much faster than for quartz, is still slow on the time scale of laboratory experiments. As with quartz, the best approach appears to be high temperature equilibration with subsequent extrapolation to room temperature. Morey, Fournier and Rowe (1964) found the data of both Hitchen (1935) and Kitahara (1960) fit the same straight line, which extrapolates to 115 ppm at 25°C. The room temperature dissolution data of Morey et al. (1964) for amorphous silica indicate an initial supersaturation followed by a very slow approach to equilibrium at 115 ppm SiO_2. This same value is obtained from supersaturated hot spring waters (but only after months or years of equilibration and only in alkaline solutions, where precipitation seems to be more rapid).

Most solubility values for amorphous silica range between 115 and 140 ppm SiO_2 (Alexander, Heston and Ihler, 1954; Krauskopf, 1956). The writer has had a montmorillonite sample in contact with a solution of 151 ppm SiO_2 for over 3 yr at pH 2·74, with no apparent tendency for the concentration of SiO_2 to decrease. The range in values from 115 to 150 ppm SiO_2 probably represents the same range of concentrations that can be supported by amorphous silica in soils and sediments, hence its solubility is given as a range from pH_4SiO_4 of 2·60 to 2·72 in Fig. 1 (upper left).

ALUMINUM HYDROXIDE AND OXYHYDROXIDES

Consider the dissolution of aluminum hydroxide

$$Al(OH)_3 + 3H^+ = Al^{3+} + 3H_2O. \qquad (1)$$

The equilibrium constant, K, equals $Al^{3+}/(H^+)^3$, assuming the activity of aluminum hydroxide and water is unity. Taking negative logarithms, $pK = pAl^{3+} - 3pH$, Rearranging and dividing by 3 to permit direct use of experimental pH values, $pH - \frac{1}{3}pAl^{3+} = -\frac{1}{3}pK$. This is the equation of a straight line of slope zero and an intercept of $-\frac{1}{3}pK$ on the ordinate in Fig. 1 (upper right). The quantity $pH - \frac{1}{3}pAl^{3+}$, related to the chemical potential of aluminum hydroxide (Schofield and Taylor, 1955) is sometimes referred to as the aluminum hydroxide potential. It is a convenient way of representing two variables on one ordinate.

Amorphous aluminum hydroxide. From (1),

$$\Delta G_r = \Delta G_{Al^{3+}} + 3\Delta G_{H_2O} - 3\Delta G_{H^+} - \Delta G_{Al(OH)_3(amorph.)}$$

where ΔG_r is the standard free energy of reaction. The ΔG of amorphous aluminum hydroxide is approximately $-271·9$ kcal per mole*, so that $\Delta G_r = -115·0 + 3(-56·7) - 3(0·0) - (-271·9) = -13·2$ kcal. From the equation $\Delta G_r = 1·36$ pK, $pK = -13·2/1·36 = -9·7$ and $-\frac{1}{3}pK = 3·2$, which is the point of intersection of the amorphous aluminum hydroxide solubility line with the ordinate in Fig. 1 (upper right).

Because most mineralogical identification techniques require crystalline material, there is little direct evidence to indicate that amorphous aluminum hydroxide is an important constituent in soils and sediments. Amorphous aluminum hydroxide is commonly the first solid phase to form during the precipitation of aluminum hydroxides and oxyhydroxides in the laboratory and one might anticipate similar results in the field. However, previous work of Frink and Peech (1962) and computations by the author from the data of Pierre et al. (1932) have seldom indicated a value of $pH - \frac{1}{3}pAl^{3+}$ for soil solutions as high as that supported by amorphous aluminum hydroxide. This does not necessarily mean that amorphous aluminum hydroxide is not present in these soils, but it does indicate that, if present, it does not control the aluminum hydroxide potential.

Gibbsite, also $Al(OH)_3$, is a common constituent of soils and sediments.† A range of stabilities between amorphous aluminum hydroxide and well-crystallized gibbsite has been reported, apparently depending upon the crystallinity of the gibbsite. Well-crystallized gibbsite is much less soluble than amorphous aluminum hydroxide, with a ΔG of $-274·2$ kcal per mole (Kittrick, 1966a). From (1) we may compute ΔG_r for the dissolution of gibbsite to be $-10·9$ kcal and $-\frac{1}{3}pK$ to be 2·7 as indicated in Fig. 1 (upper right).

Boehmite. The oxyhydroxides of aluminum appear to be even less soluble than the hydroxides, although their variation in stability with crystallinity is not well established. Consider the equation

$$AlOOH + 3H^+ = Al^{3+} + 2H_2O. \qquad (2)$$

As before, $pH - \frac{1}{3}pAl^{3+} = -\frac{1}{3}pK$. The ΔG for boehmite is given as $-217·5$ kcal per mole by Latimer (1952) and by Fyfe and Hollander (1964).

---

*The source of thermodynamic values not otherwise indicated is Appendix 2 of Garrels and Christ (1965).

†Bayerite, also $Al(OH)_3$, has a slightly different structure from that of gibbsite. It can be synthesized in the laboratory and with a ΔG of about $-274·0$ kcal per mole (Hem and Roberson, 1967), it appears to be about as stable as gibbsite. Because it appears to be rare (Wayman, 1963) or nonexistent (Federickson, 1952) in nature, it need not be considered in this investigation.

A recent NBS Technical Note (Wagman et al. 1968) gives it as −218·2 kcal per mole. Using the most recent value, $\Delta G_r = -115\cdot0 + 2(-56\cdot7) -3(0\cdot0) - (-218\cdot2) = -10\cdot2$ kcal, pK = $-10\cdot2/1\cdot36 = -7\cdot50$, and $-\frac{1}{3}$pK = 2·5, as shown in Fig. 1 (upper right).

Diaspore. Wagman et al. (1968) indicate the ΔG of diaspore to be −220 kcal per mole of AlOOH (Fyfe and Hollander, 1964, give −219·5 kcal). From (2) ΔG_r is computed to be −6·2 kcal and $-\frac{1}{3}$pK to be 2·1 for diaspore, as shown in Fig. 1 (upper right). Thus, diaspore appears to be the most stable mineral of the group.*

Relative stabilities and the influence of time. All explanations for the distribution of the aluminum hydroxides and oxyhydroxides in soils and sediments ultimately depend upon what one believes their relative stability to be. A typical compilation of data on the hydroxides and oxyhydroxides published a few years ago indicated boehmite and diaspore to be of approximately equal stability, with gibbsite more stable than either. The recent work previously cited has provided more accurate data, especially for gibbsite, indicating the stability sequence, amorphous aluminum hydroxide < gibbsite < boehmite < diaspore.†

Considering that gibbsite is less stable than boehmite or diaspore and yet is much more common in such geologically-young features as soils, it appears that on a geologic timescale gibbsite is a metastable fast-former that alters to more stable forms relatively slowly. The product of the laboratory dehydration of gibbsite is normally boehmite rather than the more stable diaspore (Deer, Howie and Zussman, 1962). A similar sequence appears to take place with time in nature where Bridge (1952) observes that gibbsite predominates in Cenozoic bauxites, with some admixture of boehmite. Bauxites of Mesozoic age are principally boehmite, with some admixture of gibbsite, especially in the younger Mesozoic. Diaspore deposits appear to be generally confined to the Paleozoic.

Factors determining the alteration rate from amorphous aluminum hydroxide to gibbsite to boehmite to diaspore may be such things as local climate and associated substances with catalytic properties. These factors are unlikely to be uniform over the earth or over vast time periods. Thus, in spite of a few exceptions noted by Bridge and by Keller (1964) in the time sequence, the present correlation of thermodynamic stability and time seems remarkably good.

KAOLINITE

Consider the equation

$$Al_2Si_2O_5(OH)_{4(kaolinite)} + 6H^+ = 2Al^{3+} + 2H_4SiO_4 + H_2O. \quad (3)$$

Assuming the activity of water and kaolinite is unity, pK = $2pAl^{3+} + 2pH_4SiO_4 - 6pH$, and pH $-\frac{1}{3}pAl^{3+} = \frac{1}{3}pH_4SiO_4 - \frac{1}{6}$pK. This is the equation of a line of slope $\frac{1}{3}$ with an intercept of $-\frac{1}{6}$pK.

For the most crystalline kaolinite encountered so far, $\Delta G = -903\cdot8$ kcal per mole (Kittrick, 1966b). Taking $\Delta G_{H_4SiO_4}$ to be −313·0 kcal per mole (computed from the data of van Lier et al., 1960; Wise, et al., 1963), from (3), $\Delta G_r = 2(-115\cdot0) + 2(-313\cdot0) + (-56\cdot7) - 6(0\cdot0) - (-903\cdot8) = -8\cdot9$ kcal. Then pK = $-8\cdot9/1\cdot36 = -6\cdot5$ and $-\frac{1}{6}$pK = 1·1, which is the intersection of the kaolinite solubility line of slope $\frac{1}{3}$ with the ordinate in Fig. 1 (lower left).

As a first approximation, it is assumed that the lower limit of crystallinity for kaolinite may be represented by the halloysite of Barany and Kelley (1961), from a deposit near Bedford, Indiana. Adding −15·0 kcal per mole to their calorimetrically-determined value, to take into account a revised heat of solution for quartz, one obtains −898·6 kcal per mole for the ΔG of halloysite.*

---

*The other mineral that might be considered in this group is corundum. There are indications that it forms in small amounts in bauxite deposits (Keller, 1964). With a ΔG of −378·2 kcal per mole of Al_2O_3 (Wagman, et al, 1968), the intersection of its solubility curve with the pH $-\frac{1}{3}pAl^{3+}$ ordinate in Fig. 1 (upper right) is 2·7. Thus, the stability of corundum is essentially identical to that of highly crystalline gibbsite. It may form at the same time in nature as gibbsite, but at a much slower rate. A correspondingly slower rate of alteration may explain its persistence in nature, even after gibbsite has altered to more stable minerals.

†Reesman and Keller (1968) present extensive original ΔG data on the high-alumina and clay minerals. Their paper should be consulted, but for experimental or computational reasons their ΔG values for minerals may not be directly comparable to those used in this paper. For example, for a more direct comparison one would have to use the same values for $Al(OH)_3$ and H_4SiO_4 in the computations.

---

*Nacrite and dickite are two minerals with identical formulas, and structures basically similar to kaolinite. Both are relatively rare and are thought to be of hydrothermal origin. The ΔG of dickite is −902·4 kcal per mole (Barany and Kelley, 1961, corrected for the revised heat of solution of quartz). Thus dickite is somewhat less stable than crystalline kaolinite at 25°C, but much more stable than halloysite.

For the dissolution of halloysite according to equation (3), $\Delta G_r = -10.4$ and $-\frac{1}{6}pK = 1.7$, resulting in the solubility line in Fig. 1 (lower left). It can be seen that halloysite is unstable relative to crystalline kaolinite.

Values of ΔG down to -902.5 kcal per mole were found to be due to kaolinite of proportionately lesser crystallinity than the material that gave -903.8 kcal per mole (Kittrick, 1966b). However, many kaolinite samples contained small X-ray-amorphous particles whose stability ranged down to -899.2 kcal per mole, approximately that of the halloysite from Bedford, Indiana. Thus, there not only appears to be a gradation from halloysite to crystalline kaolinite for samples from different sources, but also, a similar gradation within a sample from a single source. Based upon structural and field evidence, it has often been suggested that kaolinite may form from halloysite. Similarly, the stability evidence strongly suggest that halloysite is a fast-forming metastable precursor to crystalline kaolinite.

MONTMORILLONITE

Montmorillonite cannot be completely defined in the Al_2O_3–SiO_2–H_2O system, because it contains at least one other element. However, because it often occurs associated with kaolinite and particularly with amorphous silica, it would be of great value to represent montmorillonite on the same ordinates chosen for the minerals previously discussed. This may be done as a first approximation by considering the other element(s) (usually Mg) in montmorillonite as an impurity substituting for aluminum in what is essentially an aluminum silicate with the pyrophyllite formula $Al_2(Si_2O_5)_2(OH)_2$.

Consider the following dissolution equation for montmorillonite (Mt.)

$$Al_2(Si_2O_5)_2(OH)_{2(Mt.)} + 4H_2O + 6H^+ = 2Al^{3+} + 4H_4SiO_4.$$

Assuming the activity of water to be unity, $pK = 2pAl^{3+} + 4pH_4SiO_4 - 6pH^+ - pMt.$, and $pH - \frac{1}{3}pAl^{3+} = \frac{2}{3}pH_4SiO_4 - \frac{1}{6}(pK + pMt.)$. This is the equation of a line of slope $\frac{2}{3}$ whose intercept is $-\frac{1}{6}(pK + pMt.)$.

The intercept of the montmorillonite solubility line in the Al_2O_3–SiO_2–H_2O system will depend upon the activity of the aluminum silicate phase in the particular montmorillonite in question, but the slope of the solubility line will be $\frac{2}{3}$ if an aluminum silicate phase of the indicated composition is controlling the equilibrium. A solubility line of approximately two-thirds slope has been found for three montmorillonites.* Thus, to a first approximation at least, the solubility line for the aluminum silicate phase of montmorillonite may be displayed as shown in Fig. 1 (lower right).

STABILITY DIAGRAM

Having established the stabilities of some common soil minerals in the Al_2O_3–SiO_2–H_2O system, the next step is to compare their stabilities in a single diagram. Such a diagram would be applicable to soils and sediments where the minerals considered control the activity of H^+, or Al^{3+}, or H_4SiO_4, or some combination of the three. In a strict thermodynamic sense, the only two minerals that need be displayed on the stability diagram are quartz and diaspore. Amorphous silica, montmorillonite, kaolinite and gibbsite are all thermodynamically unstable fast-forming intermediates relative to quartz and diaspore. However, amorphous silica, montmorillonite, kaolinite and gibbsite predominate on the time scale of weathering of most soils and sediments and thus, it is *their* relative stabilities that are displayed in Fig. 2.

Phase relationships

The composition of each phase, P, in the present system can be expressed by the three components, C, which are Al_2O_3, SiO_2 and H_2O. A given stability line in Fig. 2 represents two phases, the mineral and solution. According to phase rule, $F = C + 2 - P$, so there are three degrees of freedom, F in the system. By limiting ourselves to the weathering environment we fix two degrees of freedom (temperature and pressure), so that $F = C - P$. Thus each stability line in Fig. 2 has only one degree of freedom. That is, for a selected value of pH_4SiO_4 there is a single equilibrium value of $pH - \frac{1}{3}pAl^{3+}$, and *vice versa*.

Where two minerals are at equilibrium with the same solution in Fig. 2, we have a three-phase system with no degrees of freedom. Thus the point where two solubility lines cross represents an invariant point. The values of both pH_4SiO_4 and $pH - \frac{1}{3}Al^{3+}$ are fixed for that mineral pair. This has been demonstrated experimentally for the

---

*The montmorillonites used were from Belle Fourche, South Dakota, from Otay, California, and from Aberdeen, Mississippi. The solubility of the 0·2–5 μ fraction after Fe-removal treatment was determined from undersaturation and from supersaturation. (Kittrick, J. A. Montmorillonite stability from solubility measurements in the Al_2O_3–SiO_2–H_2O system at 25°C and 1 atm. Presented before Div. S-9, American Society of Agronomy Meetings, Nov. 12, 1968. New Orleans, La.).

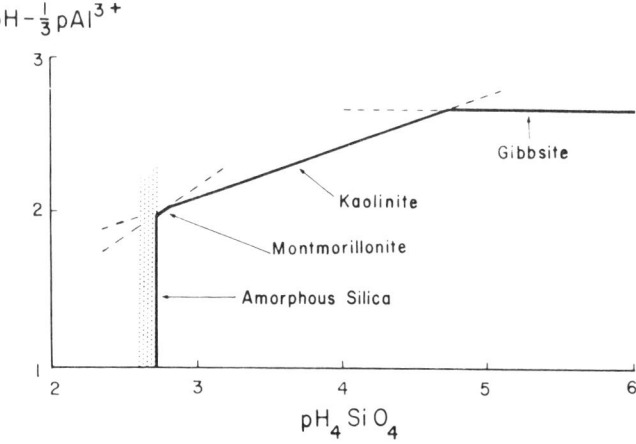

Fig. 2. Composite stability diagram for some minerals in the Al_2O_3–SiO_2–H_2O system at 25°C and 1 atm. The stability line of each mineral is solid where it is the most stable mineral of the group.

gibbsite–kaolinite pair (Kittrick, 1967) and for the montmorillonite–kaolinite pair.* In the present system, no more than two minerals can be in equilibrium at a time and they must be minerals whose solubility lines intersect.

Ion activity relationships

The interpretation of Fig. 2 is strongly influenced by the choice of ordinates. Subject to later justification, pH_4SiO_4 has been chosen as the independent variable, and $pH - \frac{1}{3}pAl^{3+}$ as the dependent variable. Thus, for a given activity of H_4SiO_4 in Fig. 2, the most stable mineral supports the lowest aluminum hydroxide potential. Where a given mineral is the most stable of the group, its solubility line is shown solid. The metastable extension is shown dotted. A solution composition above or to the left of the solid line is supersaturated with respect to at least one mineral, whereas a point below or to the right indicates undersaturation.

Notice the change in mineral stability with change in silica activity in Fig. 2. At the far left the system is supersaturated with respect to amorphous silica. As the H_4SiO_4 activity decreases ($pH_4SiO_4 > 2.7$), montmorillonite briefly supports the lowest aluminum hydroxide potential and is the most stable mineral. Kaolinite becomes more stable than montmorillonite at a pH_4SiO_4 of about 2.8 and then kaolinite supports the lowest aluminum hydroxide potential until the silica activity decreases to a pH_4SiO_4 of about 4.7. In the low silica environment above pH_4SiO_4 of 4.7, the most stable mineral of the group is gibbsite.

The logarithmic scale in Fig. 2 tends to obscure the real extent of the stability range of the minerals involved. Silica concentrations in natural waters are normally reported in terms of ppm SiO_2. In these terms, the stability range for amorphous silica in Fig. 2 is approximately 150 ppm SiO_2 and higher, the stability range for montmorillonite in Fig. 2 is from about 150 to 96 ppm SiO_2, the stability range for kaolinite is from about 96 to 1 ppm SiO_2, and the stability range for gibbsite is from 1 to 0 ppm SiO_2.

APPLICATION OF THE STABILITY DIAGRAM TO THE WEATHERING OF SOILS AND SEDIMENTS

The essential correctness of the ΔG values for amorphous silica, kaolinite and gibbsite is indicated by agreement between values derived from independent measurement methods (measurement of solubilities compared to measurement of enthalpy and entropy changes). Since ΔG for montmorillonite has been measured only by solubility methods so far, its validity must be evaluated by a comparison with natural relations.

Since the Al ion species is pH dependent, it would be reasonable to have the ordinate in Fig. 2 in terms of an Al ion species that is dominant in the pH range where most weathering occurs. That is approximately within a pH unit of 7. At the present time there is little definite that can be

---

*The montmorillonites used were from Belle Fourche, South Dakota, from Otay, California, and from Aberdeen, Mississippi. The solubility of the 0·2–5 μ fraction after Fe-removal treatment was determined from undersaturation and from supersaturation. (Kittrick, J. A. Montmorillonite stability from solubility measurements in the Al_2O_3–SiO_2–H_2O system at 25°C and 1 atm. Presented before Div. S-9, American Society of Agronomy Meetings, Nov. 12, 1968, New Orleans, La.).

said about the dominant Al ion species in the near-neutral pH range, except that it is polynuclear. This does not affect the *validity* of the equations or the graph derived from them. They hold at any pH, for however much of the total Al is Al^{3+}.

The *applicability* of the equations and the graph cannot be checked by direct analysis of soil solutions in the near-neutral pH range, because it is not possible to determine what portion of the total Al in solution is Al^{3+}. This drawback is more theoretical than actual, because most of the minerals support too little Al and certainly too little Al^{3+} in solution to measure in the near-neutral pH range. For example, the pAl^{3+} supported by amorphous aluminum hydroxide at pH 7 will be 11·4. Much lower levels of Al^{3+} will usually be supported by the other minerals. Fortunately, the applicability of Fig. 2 (and the applicability of mineral stabilities derived from equilibrium thermodynamics) to the weathering of soils and sediments can be tested by a comparison of predicted quantities with what is known of natural mineral relations. The considerable amount that is known about the mineral structures, their reactions, and especially their field relationships, will be a great advantage in such a comparison.

Phase relationships

General agreement between Fig. 2 and field relationships is indicated by the fact that as the H_4SiO_4 activity decreases, one proceeds through stages 9 (montmorillonite), 10 (kaolinite), and 11 (gibbsite) of the weathering sequence of Jackson *et al.* (1948). The minerals in the weathering sequence are adjacent and in the proper order.

A much more detailed comparison of theory with nature is permitted by mineral associations. The only minerals in Fig. 2 that can form in intimate association with one another are those adjacent to invariant points. Permitted associations are gibbsite–kaolinite, kaolinite–montmorillonite, and montmorillonite–amorphous silica. Non-permitted associations are gibbsite–montmorillonite, gibbsite–amorphous silica, kaolinite–amorphous silica, and three or more of these minerals.*

---

*Keep in mind the dependence of these associations upon the minerals selected for inclusion in the diagram. For example, if a situation exists where quartz (or a dehydrated silica phase of similar solubility) forms, then kaolinite–quartz is a permitted pair and montmorillonite has no stability area at all. The fact that montmorillonite is widespread indicates that this is not a common occurrence. If a situation exists where diaspore forms readily, then diaspore–montmorillonite is a permitted association and diaspore–kaolinite is forbidden, or vice versa, the stability line intersections are too close to state definitely. As mentioned previously, if quartz and diaspore both form readily, then none of the other minerals are stable.

By consulting a review article such as Jackson and Sherman (1953), it can be seen that some of the permitted associations are very common particularly montmorillonite–kaolinite and gibbsite–kaolinite. However, in order to prove the positive assertion, that the diagram in Fig. 2 holds for weathering in soils and sediments, one must show that the proper associations always occur. The negative assertion permits a simple, much more sensitive test, because only one exception is required. The author knows of no associations where non-permitted minerals have definitely been shown to precipitate in intimate mixture, but it is frequently a difficult thing to determine if minerals have formed in place or are merely a mechanical mixture transported from other locations. Where such a determination is possible, investigators can readily check the applicability of Fig. 2 to mineral weathering.

Ion activity relationships

It must be recognized at the outset that for several reasons the ion activity levels depicted in Fig. 2 can only be considered approximate. For example, the stability line for each mineral will vary somewhat depending upon the crystallinity of the mineral. Further, the lines represent the stability of mineral specimens as determined by dissolution and other methods. These methods determine the ion activities that the mineral specimen will support. Similar minerals may not necessarily *precipitate* in nature at exactly these ion activity levels. An important uncertainty still exists in some of the mineral stability determinations and also in some of the thermodynamic values required to compute ΔG values for these minerals. For example, the gibbsite and kaolinite ΔG values used in this paper are dependent upon the ΔG of Al^{3+}, which may have to be revised.

As more is learned about the stability of these minerals, their stability lines will be placed more exactly. It would be unwise at this time to insist upon detailed agreement between ion activity levels indicated in Fig. 2 and those found in natural situations. Fortunately, the theory of mineral formation developed from Fig. 2 will depend much more upon the general geometry of the graph than upon the exact placement of solubility lines.

Aluminum hydroxide potential as the dependent variable. The selection of pH $-\frac{1}{3} pAl^{3+}$ as the dependent variable was based upon the geochemistry of Al. Al is usually concentrated at the site of weathering. The activity of ionic Al (as opposed to Al complexed by organic substances) appears to be controlled at very low levels in natural waters. More Al appears in acid than in neutral waters, so the control is evidently pH-dependent. The precipitation of new minerals containing Al is the most likely reason for the concentration of Al at the site of weathering. These new minerals are probably responsible for the pH-dependent control of Al at low levels. It is likely that the aluminum hydroxide potential in soils and sediments often depends upon which of the minerals in Fig. 2 is controlling it. Thus, the aluminum hydroxide potential is the logical choice as the dependent variable in Fig. 2.

In order to compare aluminum hydroxide potential levels predicted in Fig. 2 with those found in nature, determinations of the latter must definitely be equilibrium measurements and must be made where mineralogical determinations indicate one or two of the minerals in Fig. 2 are forming. So far as the author is aware, none of the available data meet these requirements. It is encouraging to note, however, that much of the rather limited aluminum hydroxide potential data on soils and sediments do fall in the range indicated in Fig. 2 (see for example Lindsay, Peech, and Clark, 1959).

H_4SiO_4 as the independent variable. Whatever controls H_4SiO_4 in solution is extremely important, because as the independent variable, the activity of H_4SiO_4 determines which of the minerals in Fig. 2 will precipitate. The control of mineral precipitation by the H_4SiO_4 activity is in accord with field observations which suggest that H_4SiO_4 levels are an important factor in secondary mineral formation in soils and sediments. For example, high H_4SiO_4 levels appear to be required for montmorillonite formation (Jackson, 1964; Wildman, Jackson and Whittig, 1968).

The control of H_4SiO_4 activity seems likely to be mostly a matter of kinetics (see for example, Wollast, 1967). This should involve the rate of dissolution of unstable silicates, the rate of removal of H_4SiO_4 from the water (along with most of the Al from solution) by precipitation of new minerals, and the rate of movement of H_4SiO_4-bearing waters out of the system. Acquaye and Tinsley (1965) have shown that in some soils at least, the level of H_4SiO_4 is seasonal and may be strongly influenced by plant uptake.

Most soils and sediments contain unstable primary minerals that have a high ratio of Si to Al (such as olivines, pyroxenes, amphiboles and feldspars), whose dissolution rate could control the H_4SiO_4 level. Where appreciable primary minerals are not present (absent initially, or weathered out completely, or perhaps weathered out in the finer size fractions), the dissolution of a secondary silicate such as montmorillonite or kaolinite may control the H_4SiO_4 level that determines which mineral of a lesser Si content will form. This mechanism would require that normal weathering proceed from minerals of high silica content to minerals of low silica content. Considering the silica-containing minerals in the weathering sequence of Jackson *et al.* (1948), this appears to be exactly the case.

According to Fig. 2, the H_4SiO_4 levels over which gibbsite, kaolinite and montmorillonite may form range from zero to about 150 ppm SiO_2. Since the levels of H_4SiO_4 in solution in soils and sediments are postulated to be kinetically controlled, a comparison of H_4SiO_4 levels required in Fig. 2 with those found in nature is much easier than was a similar comparison for the aluminum hydroxide potential. The values for H_4SiO_4 in nature do not have to be at equilibrium and no evidence for the concomitant formation of the minerals in Fig. 2 is required. The measured silica activity in solutions in contact with acid soils generally ranges from less than 1 ppm up to about 40 ppm SiO_2 (McKeague and Cline, 1963; Miller, 1967). Because amorphous silica is precipitated in at least some soils and sediments, the silica activity must exceed approximately 150 ppm SiO_2 at times. This may happen when the soil solution is concentrated by drying, but can also happen if enough acid is available to dissolve unstable primary minerals sufficiently.* It is evident that the solution silica levels required in Fig. 2 have either been measured or are indicated indirectly in soils and sediments.

Kinetics and weathering intensity. Montmorillonite, kaolinite and gibbsite represent successive stages of increasing weathering intensity. If the stability of these minerals is dependent upon successively decreasing levels of H_4SiO_4 activity in solution, which in turn are a matter of kinetics, then kinetics and weathering intensity should be related. As mentioned previously, there are at least four factors that appear to be important in determining solution H_4SiO_4 levels, (1) the rate of dissolution of unstable silicates, (2) the rate of precipitation of stable silicates, (3) the rate of movement of H_4SiO_4-bearing solutions out of the system and (4), the rate of plant uptake. Rates (1), (2) and (4) and their variation from place to

---

*Two eastern Washington loess samples (Palouse C, Thatuna C) and an interbasaltic sediment containing abundant primary minerals were equilibrated to determine the silica activities they would support. 30·1g samples were treated for 2 months with water to which HCl was added to maintain approximately the same pH which had been used to determine the stabilities of montmorillonite, kaolinite and gibbsite (pH 3·2–3·5). The first solution analysis showed 245 to 275 ppm SiO_2. Five months later there were 209–218 ppm SiO_2, still supersaturated with respect to amorphous silica.

place are largely unknown, but rate (3) can be inferred from rainfall and drainage data.

As shown by Jackson *et al.* (1948), high rainfall is one of the characteristics of the more intense weathering situations where gibbsite forms. Thus, large amounts of water might pass through the system so rapidly as to permit only a low silica activity from the dissolution of unstable silicates. Further, relatively intense weathering may have already removed most of the more reactive silicates or their finer size fractions, which would also help to keep the silica activity low.

Jackson *et al.* also suggest impeded drainage to indicate a less intense weathering situation. The common association of montmorillonite with impeded drainage suggests a longer contact time between solution and unstable soil minerals, resulting in the higher H_4SiO_4 activities required for montmorillonite stability. High H_4SiO_4 activities could also be maintained by relatively static solutions in gas vesicles in minerals such as volcanic glass, with which montmorillonite is commonly associated. Montmorillonite may actually serve as a geologic indicator mineral of silica levels at near-saturation with respect to amorphous silica.

If the H_4SiO_4 activity is largely a matter of kinetics as suggested, areas where the minerals of Fig. 2 form do not necessarily have to be widely separated in time and space. If mineral stability depends in part upon rate of water movement, then different minerals may be stable during different seasons, or in closely adjacent parts of the landscape where different drainage conditions prevail (for example, Sherman and Uehara, 1956). In any event, it is evident from natural relations that kinetics and weathering intensity are related.

REFERENCES

Acquaye, D. K., and Tinsley, J. (1965) Soluble silica in soils: In *Experimental Pedology* (Edited by E. G. Hallsworth and D. V. Crawford). Butterworths, London, 126–148.

Alexander, G. B., Heston, W. M., and Ihler, R. K. (1954) The solubility of amorphous silica in water: *J. Phys. Chem.* **58**, 453–455.

Barany, R., and Kelley, K. K. (1961) Heats and free energies of formation of gibbsite, kaolinite, halloysite and dickite: *U.S. Bur. Mines R. I.* **5825**, 13 pp.

Bricker, O. P., and Garrels, R. M. (1967) Mineralogic factors in natural water equilibria: In *Principles and Applications of Water Chemistry* (Edited by S. D. Faust and J. V. Hunter). Wiley, New York, 449–469.

Bridge, J. (1953) Discussion: In *Problems of Clay and Laterite Genesis. AIME Symp.* 212–214.

Deer, W. A., Howie, R. A., and Zussman, J. (1962) *Rock Forming Minerals*, Vol. 5. *Non-Silicates*: Wiley, New York.

Fournier, R. O. (1960) Solubility of quartz in water in the temperature interval from 25°C to 300°C: *Bull. Geol. Soc. Am.* **71**, 1867–1868.

Frederickson, A. F. (1952) The genetic significance of mineralogy. In *Problems in Clay and Laterite Genesis: AIME Symp.* 1–11.

Frink, C. R., and Peech, M. (1962) The solubility of gibbsite in aqueous solutions and soil extracts: *Soil Sci. Soc. Am. Proc.* **26**, 346–347.

Fyfe, W. S., and Hollander, M. A. (1964) Equilibrium dehydration of diaspore at low temperatures: *Am. J. Sci.* **262**, 709–712.

Garrels, R. M., and Christ, C. L. (1965) Solutions, Minerals and Equilibria: Harper and Row, New York, 435 pp.

Hem, J. D., and Roberson, C. E. (1967) Form and stability of aluminum hydroxide complexes in dilute solutions: *USGS Water-Supply Paper* **1827-A**.

Hitchen, C. S. (1935) A method for experimental investigation of hydrothermal solutions, with notes on its application to the solubility of silica: *Trans. Inst. Mining Met.* **44**, 255–280.

Jackson, M. L. (1964) Chemical composition of soils: In *Chemistry of the Soil* (Edited by F. E. Bear). *Am. Chem. Soc. Monograph* **160**, 2nd Edn. Reinhold, New York, pp. 71–141.

Jackson, M. L., and Sherman, G. D. (1953) Chemical weathering of minerals in soils: *Advan. Agron.* **5**, 219–318.

Jackson, M. L., Tyler, S. A., Willis, A. L., Bourbeau, G. A., and Pennington, R. P. (1948) Weathering sequence of clay-size minerals in soils and sediments —I. Fundamental generalizations: *J. Phys. Coll. Chem.* **52**, 1237–1260.

Keller, W. D. (1964) The origin of high-alumina clay minerals—a review: *Clays and Clay Minerals* **12**, 129–151.

Kennedy, G. C. (1950) A portion of the system silica-water: *Econ Geol.* **45**, 629–653.

Kitahara, S. (1960) The polymerization of silicic acid obtained by the hydrothermal treatment of quartz and the solubility of amorphous silica: *Rev. Phys. Chem. Japan* **30(2)**, 131–137.

Kittrick, J. A. (1966a) The free energy of formation of gibbsite and $Al(OH)_4^-$ from solubility measurements: *Soil Sci. Soc. Am. Proc.* **30**, 595–598.

Kittrick, J. A. (1966b) Free energy of formation of kaolinite from solubility measurements: *Am. Mineralogist* **51**, 1457–1466.

Kittrick, J. A. (1967) Gibbsite-kaolinite equilibria: *Soil Sci. Soc. Am. Proc.* **31**, 314–316.

Krauskopf, K. B. (1956) Dissolution and precipitation of silica at low temperatures: *Geochim. Cosmochim. Acta* **10**, 1–26.

Latimer, W. M. (1952) *Oxidation potentials*: 2nd Edn. Prentice-Hall, Englewood Cliffs, New Jersey, 392 pp.

Lindsay, W. L., Peech, M. and Clark, J. L. (1959) Determination of aluminum ion activity in soil extracts: *Soil Sci. Soc. Am. Proc.* **23**, 266–269.

Mackenzie, F. T., and Garrels, R. M. (1966) Chemical mass balance between rivers and oceans: *Am. J. Sci.* **264**, 507–525.

McKeague, J. A., and Cline, M. G. (1963) Silica in soils: *Advan. Agron.* **15**, 339–396.

Miller, R. W. (1967) Soluble silica in soils: *Soil Sci. Soc. Am. Proc.* **31**, 46–50.

Morey, G. W., Fournier, R. O., and Rowe, J. J. (1964) The solubility of amorphous silica at 25°C: *J. Geophys. Res.* **69**, 1995–2002.

Pierre, W. H., Pohlman, G. G., and McIlvaine, T. C. (1932) Soluble aluminum studies—I. Concentration of aluminum in the soil solution of naturally acid soils: *Soil Sci.* **34**, 145–160.

Reesman, A. L., and Keller, W. D. (1968) Aqueous solubility studies of high-alumina and clay minerals: *Am. Mineralogist* **53**, 929–942.

Rex, R. W. (1967) Authigenic silicates formed from basaltic glass by more than 60 million yr contact with sea water, Sylvania Guyot, Marshall Islands: *Clays and Clay Minerals* **15**, 195–203.

Schofield, R. K., and Taylor, A. W. (1955) Measurements of the activities of bases in soils: *J. Soil Sci.* **6**, 137–146.

Sherman, G. D., and Uehara, G. (1956) The weathering of olivine basalt in Hawaii and its pedogenic significance: *Soil Sci. Soc. Am. Proc.* **20**, 337–340.

Siever, R. (1957) The silica budget in the sedimentary cycle: *Am. Mineralogist* **42**, 821–841.

Siever, R. (1962) Silica solubility, 0–200°C, and the diagenesis of siliceous sediments: *J. Geol.* **70**, 127–150.

Stöber, W. (1967) Formation of silicic acid in aqueous suspensions of different silica modifications: In *Equilibrium concepts in natural water systems* (Edited by R. F. Gould. *Advan. Chem. Ser.* **67**. American Chemical Society, Washington D.C. pp. 161–182.

Stumm, W., and Leckie, J. O. (1967) Chemistry of ground waters: Models for their composition: *Environ. Sci. Tech.* **1**, 298–302.

Van Lier, J. A. (1959) The solubility of quartz: Utrecht, Kermink en Zoon, 54 pp.

Van Lier, J. P., De Bruyn, P. L., and Overbeek, J. T. G. (1960) The solubility of quartz: *J. Phys. Chem.* **64**, 1675–1682.

Wagman, D. D., Evans, W. H., Halow, I., Parker, V. B., Bailey, S. M., and Schumm, R. H. (1968) Selected values of chemical thermodynamic properties. Tables for the first 34 elements in the standard order of arrangement: *NBS Technical Note* **720-3**, U.S. Government Printing Office, Washington, 264 pp.

Wayman, C. H. (1963) Solid-gas interface in weathering reactions: *Clays and Clay Minerals* **11**, 84–94.

Wise, S. S., Margrave, J. L., Feder, H. M., and Hubbard, W. N. (1963) Fluorine bomb calorimetry—V. The heats of formation of silicon tetrafluoride and silica: *J. Phys. Chem.* **67**, 815–821.

Wildman, W. E., Jackson, M. L., and Whitting, L. D. (1968) Iron-rich montmorillonite formation in soils derived from serpentinite: *Soil Sci. Soc. Am. Proc.* **32**, 787–794.

Résumé – On a tenté de rassembler toutes les informations thermodynamiques actuellement disponibles sur les minéraux terrestres dans le système $Al_2O_3-SiO_2-H_2O$ a 25°C et 1 atm. La montmorillonite y est incluse compte tenu de sa phase de silicate d'aluminum. Des diagrammes sont présentés de façon à ce que la stabilité des minéraux puisse être représentée en relation avec l'environnement ionique. Bien que le système $Al_2O_3-SiO_2-H_2O$ soit très simple par comparaison aux sols et sédiments, les diagrammes de stabilité décrivent une sequence de stabilité minérale et des associations minérales paires qui sont en accord avec les relations naturelles.

Selon le diagramme de stabilité, les paires minérales qui peuvent se former en association intime sont gibbsite-kaolinites, kaolinites-montmorillonites, et silice amorphe montmorillonite. Les paires interdites sont les silices amorphes kaolinites, silices amorphes gibbsites, kaolinites montmorillonites, et silices amorphes montmorillonites. La formation de mélanges intimes de 3 ou plus de ces minéraux est également interdite.

Le diagramme de stabilité prédit des rapports d'activité ionique en accord raisonable avec ceux obtenus à partir des sols et des sédiments. Les silices amorphes limitent probablement des niveaux de silice élevés, avec de la montmorillonite se formant également à ces memes niveaux de silice. Les kaolinites se forment à des niveaux intermédiaires et les gibbsites à des niveaux de silice très bas. A leur tour, ces minéraux contrôlent probablement les activités des ions d'aluminum à un niveau approprié au pH. La formation de gibbsite, de kaolinite, de montmorillonite et de silice amorphe paraît être contrôlée par une combinaison de cynétique et d'équilibre. C'est à dire que la dissolution cynétique de silices instables semble contrôler le niveau H_4SiO_4. Le nouveau minéral le plus stable à ce niveau H_4SiO_4, semble précipiter en réponse à la solution d'équilibre.

Kurzreferat – Es wurde versucht, die besten derzeit erhältlichen thermodynamischen Daten für Bodenminerale im $Al_2O_3-SiO_2-H_2O$ System bei 25°C und 1 Atmosphäre zu vereinigen. Montmorillonit wurde unter Berücksichtigung seiner Aluminiumsilikatphase miteingeschlossen. Aus den beigefügten Kurvenbildern lässt sich die Beständigkeit des Minerals in Bezug auf die Ionenumgebung beurteilen. Obwohl das $Al_2O_3-SiO_2-H_2O$ System im Vergleich mit Böden und Ablagerungen ein sehr einfaches ist, zeigen die Beständigkeitskurven eine Mineralbeständigkeitsfolge und Mineralpaarungen, die gut mit natürlichen Verhältnissen übereinstimmen.

Gemäss den Beständigkeitskurven können folgende Mineralpaare in enger Association geformt werden: Gibbsit-Kaolinit, Kaolinit-Montmorillonit und Montmorillonit-amorphe Kieselsäure. Verboten sind Paare von amorpher Kieselsäure-Kaolinit, amorpher Kieselsäure-Gibbsit und Montmorillonit-Gibbsit. Die Bildung enger Mischungen von drei oder mehr dieser Minerale ist ebenfalls ausgeschlossen. Die Beständigkeitskurven weisen auf Ionenaktivitätsbeziehungen hin, die recht gut mit den aus Boden- und Ablagerungsproben erhaltenen übereinstimmen.

Amorphe Kieselsäure schliesst wahrscheinlich hohe Kieselsäureniveaus aus während Montmorillonite auch bei hohen Kieselsäureniveaus geformt werden. Kaolinit bildet sich bei mittleren, und Gibbsit bei niedrigen Kieselsäureniveaus. Diese Minerale bestimmen wahrscheinlich die Aktivität der Aluminiumionen a einem dem pH entsprechenden Niveau. Die Bildung von Gibbsit, Kaolinit, Montmorillonit und amorpher Kieselsäure scheint durch eine Kombination von Kinetik und Gleichuf gewichten bestimmt zu werden, d.h. die kinetische Auflösung unbeständiger Silikate scheint das H_4SiO_4 Niveau zu bestimmen. Die neuen auf diesem H_4SiO_4 Niveau beständigsten Minerale scheinen durch die Lösungsgleichgewichte zur Ausfällung gebracht zu werden.

Резюме—Сделана попытка собрать наиболее достоверные термодинамические данные для минералов почв, образующихся в системе Al_2O_3-SiO_2-H_2O при 25°C и 1 атм. Монтмориллонит включен в связи с его алюмосиликатной составной частью. Диаграммы даны так, чтобы можно было составить представление о стабильности минералов в зависимости от их ионного окружения. Хотя система Al_2O_3-SiO_2-H_2O очень проста в сравнении с почвами и осадками, диаграммы эти изображают и последовательность стабильности и парные ассоциации минералов в хорошем согласии с природными соотношениями.

В соответствии с диаграммами стабильности, пары минералов, образующих тесные ассоциации, таковы: гиббсит—каолинит, каолинит—монтмориллонит, монтмориллонит—аморфный кремнезем. К запрещенным парам относятся: аморфный кремнезем—каолинит, аморфный кремнезем—гиббсит и монтмориллонит—гиббсит. Образование тонких смесей из трех (или более) минералов также запретно. Диаграммы стабильности позволяют предсказать соотношения активности ионов, удовлетворительно согласующиеся с найденными при изучении почв и осадков.

Аморфный кремнезем, вероятно, ограничивает верхний предел активности кремнезема, причем монтмориллонит образуется также при высокой активности кремнекислоты. Образование каолинита происходит при промежуточных, а образование гиббсита—при низких уровнях активности кремнезема. Эти минералы, вероятно, в свою очередь контролируют активность ионов алюминия в соответствии со значением pH. Образование гиббсита, каолинита, монтмориллонита и аморфного кремнезема, как кажется, контролируется совместным влиянием и кинетики и равновесия, т.е. кинетика растворения неустойчивых силикатов, по-видимому, контролирует уровень активности H_4SiO_4; новый минерал или минералы, которые наиболее устойчивы при этом уровне активности H_4SiO_4, по-видимому, осаждаются в соответствии с равновесными отношениями в растворе.

15

Copyright © 1982 by the Clay Mineral Society
Reprinted from Clays Clay Miner. 30:167–179 (1982)

SOLUBILITY OF TWO HIGH-Mg AND TWO HIGH-Fe CHLORITES USING MULTIPLE EQUILIBRIA[1]

J. A. KITTRICK

Department of Agronomy and Soils, Washington State University
Pullman, Washington 99164

Abstract—High-Mg chlorites from Vermont and Quebec and high-Fe chlorites from Michigan and New Mexico were equilibrated at room temperature in the near-neutral pH range. Gibbsite, kaolinite, and hematite of known stability were added to the samples to control unmeasurable variables at calculable levels. Equilibrium solution compositions were obtained from undersaturation and from supersaturation. Other indicators of equilibrium were good agreement between successive analyses over a long period of time, between duplicate samples, between independent systems, and between independent measures of equilibrium. All four chlorites were stable relative to brucite and, with a few exceptions, relative to talc under the conditions of study. When in equilibrium with gibbsite, the pH − ½pMg^{2+} value of the chlorites ranged from 6.3 to 6.5, at a pH$_4$SiO$_4$ value of 4.0. These values are in good agreement with prior estimates of chlorite stability. The calculated standard free energy of formation of the chlorites is dependent upon solution Fe^{2+} calculated from the sample Eh and assumed equilibrium with hematite, with the assumption that the Fe^{2+}-Fe^{3+} couple is at the same Eh as the sample.

Key Words—Chlorite, Free energy of formation, Iron, Magnesium, Solubility, Stability.

INTRODUCTION AND EXPERIMENTAL DESIGN

No actual calorimetric or solubility determinations of the standard free energy of formation (ΔG^0_f) of chlorite have been made, but several investigators have indirectly estimated ΔG^0_f of the magnesium end member (clinochlore), Mg$_5$Al$_2$Si$_3$O$_{10}$(OH)$_8$, as shown in Table 1. These ΔG^0_f values can be better understood in terms of solution compositions by considering the dissolution of clinochlore as follows (all equations herein involve crystalline mineral phases, aqueous ions, and liquid water):

$$Mg_5Al_2Si_3O_{10}(OH)_8 + 16H^+ = 5Mg^{2+} + 2Al^{3+} + 3H_4SiO_4 + 6H_2O. \quad (1)$$

Defining K$_1$ as the equilibrium constant for Eq. (1), assuming the activity of mineral phases and liquid water to be unity, and taking negative logarithms:

$$pK_1 = 5pMg^{2+} + 2pAl^{3+} + 3pH_4SiO_4 − 16pH. \quad (2)$$

Hence, if Al^{3+} and H$_4$SiO$_4$ are held constant, the stability of clinochlore can be expressed in terms of pMg^{2+} and pH. If pH − ⅓pAl^{3+} is controlled by gibbsite and pH$_4$SiO$_4$ is maintained at 4.0, for example, it can be seen from the last column in Table 1 that the estimated values of pH − ½pMg^{2+} in equilibrium with clinochlore range from 5.4 to 7.0. If ½pMg^{2+} in experimental samples were furthermore held to approximately 1.0 by addition of appropriate salts, the anticipated equilibrium pH of the clinoclore-gibbsite system would range from 6.4 to 8.0. It seems likely that chlorite could reach stable equilibrium in this pH range, considering that Rich and Bonnet (1975) found a swelling chlorite that appeared to form in a soil whose pH ranged from 7.6 to 8.1.

An Al-containing mineral in the soil system must maintain Al^{3+} at the same low level as other minerals if it is to be stable (Kittrick, 1969). It is instructive, therefore, to isolate the Al^{3+} variable in Eq. (2):

$$pAl^{3+} = 8pH − {}^5/_2pMg^{2+} − {}^3/_2pH_4SiO_4 + pK_1/2. \quad (3)$$

It is evident from Eq. (3) that for clinochlore to keep Al^{3+} low (high pAl^{3+}), the pH must be high and pMg^{2+} and pH$_4$SiO$_4$ must be low. To determine pK$_1$ by solubility methods, all common ion activities must be known. Unfortunately, relatively high pH values (near neutrality) are likely to present analysis problems for Al with regard to both the amount and the nature of ion species. To circumvent this problem, gibbsite of known stability (Kittrick, 1966a) can be added to the system to control pH − ⅓pAl^{3+} at constant values. Thus, at equilibrium, pAl^{3+} can be calculated from the measured pH.

If kaolinite is also added to the system, gibbsite-kaolinite equilibria should control pH$_4$SiO$_4$ (Kittrick, 1967). Measurements of pH$_4$SiO$_4$ can then be compared with predicted equilibria to provide an independent indicator of equilibrium. For the clinochlore-gibbsite-kaolinite system:

$$2Mg_5Al_2Si_3O_{10}(OH)_8 + 2Al(OH)_3 + 2OH^+ = 3Al_2Si_2O_5(OH)_4 + 10Mg^{2+} + 15H_2O. \quad (4)$$

---

[1] Scientific Paper No. SP5881. College of Agriculture Research Center, Washington State University, Pullman, Washington. Project 1885.

Table 1. Chlorite ΔG^0_f values and the level of pH $- \frac{1}{2}$pMg^{2+} in equilibrium with chlorite and gibbsite at pH$_4$SiO$_4$ of 4.0.

Chlorite	Source	ΔG^0_f kJ/mole[1]	Estimated from	pH $- \frac{1}{2}$pMg^{2+}
$Mg_5Al_2Si_3O_{10}(OH)_8$	Zen (1972)		High-temperature, high pressure data	
	Bird and Anderson (1973)	-8268	of Fawcett and Yoder (1966)	5.4
$Mg_5Al_2Si_3O_{10}(OH)_8$	Nriagu (1975)	-8190	Thermochemical model	6.8
$Mg_5Al_2Si_3O_{10}(OH)_8$	Helgeson (1969)	-8187	Seawater solubility data of MacKenzie and Garrels (1965)	6.8
$Mg_5Al_2Si_3O_{10}(OH)_8$	Tardy and Garrels (1974)	-8174	Thermochemical model	7.0
$(Si_{2.97}Al_{1.03})(Al_{1.44}Fe^{3+}_{0.07}Fe^{2+}_{0.99}Mg_{3.24})O_{10}(OH)_8$	Vermont[2]	-7793	Solubility measurements	6.5
$(Si_{2.99}Al_{1.01})(Al_{1.39}Fe^{3+}_{0.21}Fe^{2+}_{0.57}Mg_{3.52})O_{10}(OH)_8$	Quebec[2]	-7869	Solubility measurements	6.4
$(Si_{2.47}Al_{1.53})(Al_{1.60}Fe^{2+}_{3.29}Mg_{1.05})O_{10}(OH)_8$	Michigan[2]	-7290	Solubility measurements	6.3
$(Si_{2.84}Al_{1.16})(Al_{1.75}Fe^{2+}_{2.61}Mg_{1.16}Fe^{3+}_{0.12})O_{10}(OH)_8$	New Mexico[2]	-7319	Solubility measurements	6.5

[1] Literature ΔG^0_f values recalculated using $\Delta G^0_{f,Al^{3+}} = -489,400$ J/mole.
[2] Present paper.

$$pK_4 = 10pMg^{2+} - 20pH. \quad (5)$$

Thus, the system is defined by measurements of pMg^{2+} and pH.

The unfortunate reality of obtaining chlorite samples suitable in amount and purity for solubility studies is that none match the simple formula used for clinochlore. Chlorites contain both Fe^{2+} and Fe^{3+}, which makes the situation much more complicated. Following the strategy of multiple equilibria introduced by the gibbsite-kaolinite additions described above, one can add hematite to control Fe^{3+} at calculable levels. For hematite (Kittrick, 1971):

$$pFe^{3+} = 0.96 + 3pH. \quad (6)$$

From Garrels and Christ (1965, p. 196), pFe^{2+} can be calculated from pFe^{3+} and a measurement of Eh as follows:

$$pFe^{2+} = (Eh_{Fe^{2+}{-}Fe^{3+}} - 0.771)/0.0592 + pFe^{3+}. \quad (7)$$

At the extremely low levels of Fe^{3+} and Fe^{2+} anticipated in these experiments, the Fe^{3+}-Fe^{2+} couple will not control sample Eh. Unfortunately, the sample and the Fe^{3+}-Fe^{2+} couple may not even be at the same Eh (Bohn, 1968). To avoid this possibility, it was anticipated that quinhydrone could be added to the samples as an Eh buffer, to bring all portions of the sample system to the same Eh.

MATERIALS AND METHODS

Materials

Gibbsite. Commercial Alcoa hydrated alumina C-730 was obtained from the Aluminum Company of America. X-ray powder diffraction (XRD), differential thermal (DTA), and other characteristics of this material, including solubility measurements, were described by Kittrick (1966a).

Hematite. Mapico 347 was obtained from Columbia Carbon Co. The particle size range is given as 0.06 to 0.8 μm. Chemical, DTA, and XRD analyses (not reported) show this material to be very pure hematite.

Kaolinite. The English kaolinite was obtained from Hammill and Gillespie, Inc. and the "Georgia 2" kaolinite was obtained from Southern Clays, Inc. XRD and other characteristics of these kaolinites, including solubility measurements, were described by Kittrick (1966b).

Chlorite. Massive chlorites from Ward's Natural Science Establishment, Rochester, New York, were ground with an impact grinder to pass a 150 mesh sieve (<104 μm). XRD analyses of oriented samples, including various combinations of Mg and K saturation, heating, and glycerol solvation, detected no phases other than chlorite. XRD patterns before and after solution equilibria were indistinguishable. Portions of the oriented patterns are shown in Figure 1 and illustrate the characteristic relative peak intensities of high-Mg chlorites (Vermont and Quebec) and high-Fe chlorites (Michigan and New Mexico), their good crystallinity, and the lack of impurities. Random powder XRD peaks necessary for polytype identification were relatively weak, but appeared to be adequate for the task of polytype identification when compared with the data of Bailey (1975, p. 242). The Vermont, Quebec, and Ishpe-

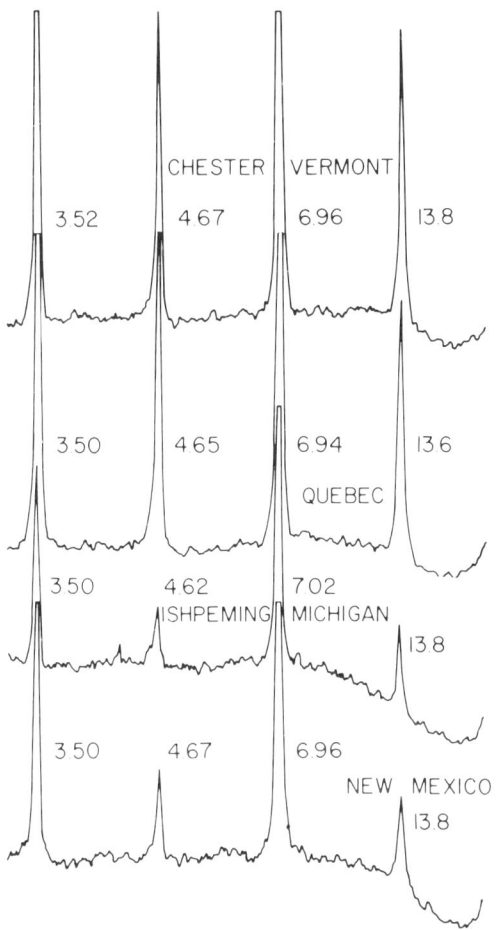

Figure 1. A portion of the chlorite X-ray diffraction patterns. Glycerated, oriented samples; Ni-filtered CuKα radiation; 1°2θ/min.

ming, Michigan, chlorites can be classified as IIb(97), whereas the chlorite from New Mexico is Ia(97).

Methods

Total chemical analysis of chlorites. Analyses for Si, Al, Ti, Fe, Mn, Ca, Mg, K, Na, and P were made by X-ray spectroscopy. Fe^{2+} and Fe^{3+} were independently determined with orthophenanthroline after HF decomposition (Roth *et al.*, 1968). Cation-exchange capacities were negligible. The unit-cell formulae were calculated according to an 18 oxygen unit cell (Jackson, 1969).

Sample preparation and equilibration. To 10.0 g of chlorite was added 10.0 g of hematite, plus 10.0 g of kaolinite or 40.0 g of gibbsite, or both. Following treatment with hot 0.50 M NaOH to remove soluble substances (after Foster, 1953), each sample was washed on a Buchner funnel with 1% NaOH, then with pH 5.0 NaOAc, and then with 0.010 M $MgCl_2$. The sample was then placed in a 250-ml polycarbonate centrifuge bottle, given several centrifuge washes, and equilibrated with 50 ml of 0.0080 M to 0.0100 M $MgCl_2$ solution. In some samples the initial pH_4SiO_4 levels were adjusted with NaOH and $Na_2SiO_3 \cdot 9H_2O$ so as to permit equilibration of the sample from both undersaturation and supersaturation with respect to variables of interest. After equilibration, the pH_4SiO_4 ranged from 3.66 to 4.40, pMg^{2+} ranged from 2.02 to 2.39, and pH ranged from 7.04 to 7.89. For some samples, a portion of the solution was removed by centrifuging for analysis after a close approach to equilibrium was indicated by successive pH measurements (a few days to a few months). For other samples, portions of the solution were removed for analysis until successive analysis indicated a close approach to equilibrium (a year or two). Equilibrated samples were sometimes centrifuge washed and equilibrated with another solution whose composition permitted a different approach to equilibrium. Quinhydrone at a concentration of 0.10 g/liter was added to some samples. Samples were agitated almost continuously in a constant temperature room at 25°C and were centrifuged in a temperature-controlled centrifuge. Room temperature during analysis was 23–25°C.

Chemical analysis of solutions. Prior to analysis, all samples were centrifuged, and an aliquot of the supernatant was further centrifuged until clear to a Tyndall beam. Si was determined colorimetrically with molybdate (APHA, 1960), Mg with an atomic absorption spectrometer, and K and Na with a flame photometer. Ion activities were computed from the extended Debye-Hückel equation. All pH measurements were with a glass combination electrode and a Corning Model 12 meter calibrated to ±0.02 units with 2 buffers. Equilibrium pH was determined with the aid of a slow-speed, strip-chart recorder. Analysis precision estimated from duplicate determinations for all analyses is approximately ±0.02 p units. Eh measurements, ranging from 0.347 to 0.388 V, were to the nearest millivolt, using a bright platinum foil electrode and a calomel reference electrode. ZoBell solution (ZoBell, 1946) was used as a standard.

EXPERIMENTAL RESULTS

Chlorite from Chester, Vermont

Chlorite-gibbsite-kaolinite-hematite system. The equilibrium of the chlorite from Vermont with its constituent ions can be depicted as follows:

$$(Si_{2.97}Al_{1.03})(Al_{1.44}Fe^{3+}_{0.07}Fe^{2+}_{0.99}Mg_{3.24})O_{10}(OH)_8 + 16.08H^+ = 2.97H_4SiO_4 + 2.47Al^{3+} + 3.24pMg^{2+} + 0.99Fe^{2+} + 0.07Fe^{3+} + 6H_2O. \quad (8)$$

$$pK_{\aleph} = 2.97pH_4SiO_4 + 2.47pAl^{3+} + 3.24pMg^{2+}$$
$$+ 0.07pFe^{3+} + 0.99pFe^{2+} - 16.08pH \quad (9)$$
$$= 2.97pH_4SiO_4 - 7.41(pH - \tfrac{1}{3}pAl^{3+})$$
$$- 6.48(pH - \tfrac{1}{2}pMg^{2+})$$
$$- 0.21(pH - \tfrac{1}{3}pFe^{3+})$$
$$- 1.98(pH - \tfrac{1}{2}pFe^{2+}). \quad (10)$$

If gibbsite is in equilibrium with the chlorite, $pH - \tfrac{1}{3}pAl^{3+}$ should be constant (Kittrick, 1980) as follows:

$$Al^{3+} + 3H_2O = 3H^+ + Al(OH)_3 \quad (11)$$
$$\tfrac{1}{3}pK_{11} = pH - \tfrac{1}{3}pAl^{3+}$$
$$= 2.68 \pm 0.07. \quad (12)$$

If gibbsite and the English kaolinite are in equilibrium with the chlorite, the pH_4SiO_4 should be constant (Kittrick, 1980) as follows:

$$Al_2Si_2O_5(OH)_4 + 5H_2O = 2Al(OH)_3 + 2H_4SiO_4 \quad (13)$$
$$\tfrac{1}{2}pK_{13} = pH_4SiO_4 = 4.5. \quad (14)$$

According to Eq. (6), $pH - \tfrac{1}{3}pFe^{3+}$ is a constant if the chlorite is in equilibrium with hematite, and $pH - pFe^{2+}$ can then be calculated from the Eh and Eq. (7).

Initial experiments involved attempts to obtain $pK_\aleph$ by measuring pH_4SiO_4, pMg^{2+}, and pH, controlling Eh with quinhydrone and calculating pFe^{2+} from Eqs. (6) and (7). For eight identical samples that had been equilibrated over a period of 479 days, however, pK values varied by 5 units or more. Furthermore, this amount of variation did not significantly diminish over time, as indicated by analyses at ten intervals during the equilibration period. The results could not be explained by analytical errors, which at most would contribute approximately ±0.5 units of variation. After much experimentation, it was discovered that the variation in pK was due to the presence of quinhydrone. Changes in pH, probably resulting from quinhydrone breakdown, prevented attainment of equilibrium. Quinhydrone was thereafter omitted from the experiments.

To determine sample equilibrium in this investigation, the three measured variables from Table 2 (pH_4SiO_4, pMg^{2+}, pH) were combined into two parameters (pH_4SiO_4 and $pH - \tfrac{1}{2}pMg^{2+}$) so that sample equilibrium could be approached from supersaturation and undersaturation with respect to each. This permits equilibrium to be approached from four different directions on a plot of pH_4SiO_4 vs. $pH - \tfrac{1}{2}pMg^{2+}$ (Figure 2). The convergence of the data for chlorite-gibbsite-kaolinite-hematite samples equilibrated for 7 to 30 days (Figure 2, triangles) indicates that a real equilibrium point was being approached, as opposed to fortuitous agreement between samples of similar composition subjected to processes having similar kinetics.

Chlorite-gibbsite-hematite system. Decreasing the number of phases by one increases the degrees of freedom of the system by one. Specifically, when this sys-

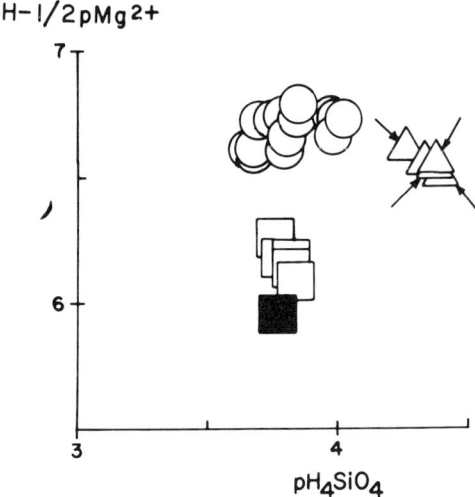

Figure 2. Solution compositions of several systems containing Vermont chlorite. Chlorite-gibbsite-kaolinite-hematite samples (△) were equilibrated for 7 to 30 days from four different directions of supersaturation and undersaturation as indicated by the arrows. Duplicates of the chlorite-gibbsite-hematite system (□) were sampled repeatedly from 98 to 534 days, with one solution obtained by immiscible displacement (■). Duplicates of the chlorite-kaolinite-hematite system (○) were sampled repeatedly from 37 to 793 days. Symbol size represents the deviations in $pH - \tfrac{1}{2}pMg^{2+}$ that could be engendered by the known variation in the gibbsite stability determination.

tem is in equilibrium, pH_4SiO_4 should not be controlled at the fixed value of Eq. (14), as when kaolinite is also present. However, the $pH - \tfrac{1}{3}pAl^{3+}$ should still be controlled by gibbsite. In Figure 2, analyses from Table 2 for duplicate samples are plotted (as squares) after equilibrating from 90 to 543 days, where there were no apparent differences due to equilibration time. Earlier analyses were undersaturated with respect to those shown. After 543 days of equilibration, the Eh of the two samples was 0.347 V and 0.355 V, respectively. Insufficient liquid remained for further analyses after regular centrifugation, so an additional set of analyses on one sample was made after immiscible displacement (Kittrick, 1980). All analyses plotted appear to be in reasonably good agreement.

Chlorite-kaolinite-hematite system. When this system is in equilibrium, the $pH - \tfrac{1}{3}Al^{3+}$ value should be controlled by kaolinite (Georgia 2 in this case) and should depend upon pH_4SiO_4 (Kittrick, 1980). As can be seen in Figure 2 (circles), analyses from Table 2 at 37 to 793 days for duplicate samples were all in reasonably good agreement. Earlier analyses were undersaturated with respect to those shown. After 793 days, the Eh of the two samples was 0.368 V and 0.377 V, respectively.

Table 2. Solution compositions of chlorite samples.

Days	pH_4SiO_4	pH	pMg^{2+}	Days	pH_4SiO_4	pH	pMg^{2+}	Days	pH_4SiO_4	pH	pMg^{2+}
Vermont chlorite, gibbsite kaolinite, hematite[1]				Vermont, chlorite, kaolinite hematite, Eh = 0.377 V				Michigan chlorite, gibbsite hematite, Eh = 0.384 V			
19	4.40	7.67	2.31	37	3.99	7.76	2.20	6	4.11	7.26	2.24
30	4.39	7.74	2.39	63	4.03	7.84	2.20	28	4.09	7.33	2.22
20	4.34	7.68	2.24	113	3.98	7.85	2.20	90	4.03	7.38	2.22
7	4.27	7.74	2.24	215	3.85	7.89	2.20	172	3.95	7.39	2.22
13	4.39	7.70	2.27	349	3.79	7.84	2.20	327	3.88	7.39	2.22
				506	3.76	7.80	2.14	544	3.82	7.38	2.20
				678	3.69	7.66	2.10	ID[2]	3.77	7.11	2.18
				793	3.70	7.76	2.08				
Vermont chlorite, gibbsite hematite, Eh = 0.347 V				Quebec chlorite, gibbsite kaolinite, hematite[1]				Michigan chlorite, gibbsite hematite, Eh = 0.387 V			
90	3.82	7.25	2.22	8	4.47	7.53	2.34	21	4.09	7.32	2.22
172	3.82	7.30	2.20	1	4.40	7.52	2.36	54	4.04	7.41	2.24
327	3.82	7.30	2.20	9	4.45	7.74	2.26	98	4.02	7.38	2.26
543	3.75	7.35	2.18	4	4.43	7.65	2.31	174	3.95	7.37	2.24
								328	3.90	7.27	2.20
								545	3.82	7.31	2.18
Vermont chlorite, gibbsite hematite, Eh = 0.355 V				Quebec chlorite, gibbsite hematite, Eh = 0.388 V				New Mexico chlorite, gibbsite, hematite, Eh = 0.382 V			
98	3.85	7.22	2.26	28	3.71	7.36	2.22	6	4.05	7.22	2.24
174	3.84	7.22	2.22	90	3.73	7.47	2.22	28	3.97	7.29	2.22
328	3.82	7.27	2.20	172	3.72	7.54	2.22	90	3.90	7.49	2.24
543	3.77	7.28	2.24	327	3.69	7.50	2.20	172	3.89	7.49	2.24
ID[2]	3.77	7.04	2.16	544	3.70	7.47	2.18	327	3.87	7.43	2.20
				ID[2]	3.66	7.23	2.18	545	3.78	7.41	2.22
								ID[2]	3.77	7.20	2.20
Vermont chlorite, gibbsite hematite, Eh = 0.368 V				Quebec chlorite, gibbsite hematite, Eh = 0.378 V				New Mexico chlorite, gibbsite, hematite, Eh = 0.384 V			
37	3.99	7.81	2.22	54	3.73	7.51	2.24	21	4.01	7.32	2.24
63	4.02	7.82	2.20	98	3.71	7.44	2.22	54	3.96	7.43	2.24
113	3.98	7.84	2.24	174	3.71	7.52	2.22	98	3.92	7.34	2.26
215	3.85	7.82	2.20	328	3.71	7.48	2.20	174	3.90	7.34	2.22
349	3.82	7.73	2.18	544	3.70	7.53	2.18	328	3.82	7.47	2.20
506	3.80	7.67	2.12					545	3.78	7.47	2.20
678	3.69	7.64	2.10								
793	3.66	7.62	2.02								

[1] Separate individual samples.
[2] ID = immiscible displacement after last analysis.

Agreement between systems. The experimental values for the three systems displayed in Figure 2 tend to lie in three separate groups. To compare them directly, it is necessary to ensure that they are displayed at the same constant values of all parameters that do not appear on the coordinates of Figure 2. Thus Eq. (10) should be solved for $pH - \frac{1}{2}pMg^{2+}$ as follows:

$$pH - \tfrac{1}{2}pMg^{2+} = 0.46pH_4SiO_4 - 1.14(pH - \tfrac{1}{3}pAl^{3+}) - 0.31(pH - \tfrac{1}{2}pFe^{2+}) - 0.03(pH - \tfrac{1}{3}pFe^{3+}) - 0.15pK_8. \quad (16)$$

It can be seen from Eq. (16) that if $pH - \frac{1}{2}pMg^{2+}$ is plotted against pH_4SiO_4 (as in Figure 2), equilibrium solution analyses for Vermont chlorite should lie along a line of slope 0.46 *if* $pH - \frac{1}{3}pAl^{3+}$, $pH - \frac{1}{2}Fe^{2+}$, and $pH - \frac{1}{3}pFe^{3+}$ are held constant. From Eq. (6), hematite should control $pH - \frac{1}{3}pFe^{3+}$ in all samples at -0.32. Where gibbsite is present, $pH - \frac{1}{3}pAl^{3+}$ should be constant at 2.68 (Kittrick, 1980). Where no gibbsite is present, the values of $pH - \frac{1}{3}pAl^{3+}$ can be obtained from the pH_4SiO_4 of the samples and the known stability of Georgia 2 kaolinite (Kittrick, 1980). From these values of $pH - \frac{1}{3}pAl^{3+}$, the experimental values of $pH - \frac{1}{2}pMg^{2+}$ can be converted to calculated values of $pH - \frac{1}{2}pMg^{2+}$ appropriate to a $pH - \frac{1}{3}pAl^{3+}$ value of 2.68 (Eq. 27, Appendix).

The calculated $pH - \frac{1}{2}pFe^{2+}$ of each sample varies with the pH and Eh of the sample (Eq. 6 and 7). The measured Eh was used in this calculation for all systems except for the chlorite-gibbsite-kaolinite-hematite system where no Eh measurements were made. The Eh of this system was assumed to be the same as that of the chlorite-gibbsite-hematite system. Values of $pH - \frac{1}{2}pMg^{2+}$ appropriate to a constant $pH - \frac{1}{2}pFe^{2+}$ of -1.0 could then be calculated from Eq. (27) in the Appendix.

Because the Fe content of the Vermont chlorite is small, the coefficient of the $pH - \frac{1}{2}pFe^{2+}$ term in Eq. (16) is small. The range in measured Eh values of about ± 0.01 volt between systems therefore had only a small impact on adjustment of experimental $pH - \frac{1}{2}pMg^{2+}$ values to a common $pH - \frac{1}{2}pFe^{2+}$ of -1.0. When adjustments for constant $pH - \frac{1}{3}pAl^{3+}$ and $pH - \frac{1}{2}pFe^{2+}$ are made, it can be seen from Figure 3 that the three independent systems agreed with each other and with a line of slope 0.46, as predicted in Eq. (16).

Calculated ΔG value. In Figure 3, the solubility of Vermont chlorite is displayed at a $pH - \frac{1}{3}pAl^{3+}$ of 2.7, a $pH - \frac{1}{3}pFe^{3+}$ of -0.3 and a $pH - \frac{1}{2}pFe^{2+}$ of -1.0. At a pH_4SiO_4 of 4.0, this chlorite can be seen to support a $pH - \frac{1}{2}pMg^{2+}$ of 6.5. From Eq. (10):

$$\begin{aligned} pK_8 &= 2.97(4.0) - 7.41(2.7) - 6.48(6.5) \\ &\quad - 0.21(-0.3) - 1.98(-1.0) \\ &= -48.2 \pm 0.5, \end{aligned}$$

where the error estimate is derived from the known variation in the gibbsite solubility determination, because it is the only term in Eq. (10) where the error is accurately known. It is thought to represent a rough estimate of *minimum* analytical error. Furthermore, from the Nernst equation, where ΔG_r is the standard free energy of reaction, and ΔG^0_f values are taken from Robie *et al.* (1978):

$$\begin{aligned} \Delta G_r &= 5.71pK = -292 \pm 3 \text{ kJ} \\ &= 2.97\Delta G^0_{f,\,H_4SiO_4} + 2.47\Delta G^0_{f,\,Al^{3+}} \\ &\quad + 3.24\Delta G^0_{f,\,Mg^{2+}} + 0.07\Delta G^0_{f,\,Fe^{3+}} \\ &\quad + 0.99\Delta G^0_{f,\,Fe^{2+}} + 6\Delta G^0_{f,\,H_2O} - \Delta G^0_{f,\,chlorite} \\ \Delta G^0_{f,\,chlorite} &= 275.2 + 2.97(-1308.0) \\ &\quad + 2.47(-489.4) + 3.24(-454.8) \\ &\quad + 0.07(-4.6) + 0.99(-78.9) \\ &\quad + 6(-237.1) \\ &= -7793 \pm 3 \text{ kJ/mole}, \end{aligned}$$

where the error estimate is the minimum due to analytical error.

Chlorite from Quebec

Chlorite-gibbsite-kaolinite-hematite system. The equilibrium of the chlorite from Quebec with its constituent ions can be given as follows:

$$\begin{aligned} &(Si_{2.99}Al_{1.01})(Al_{1.39}Fe^{3+}_{0.21}Fe^{2+}_{0.57}Mg_{3.52})O_{10}(OH)_8 \\ &\quad + 16.01H^+ = 2.99H_4SiO_4 + 2.40Al^{3+} + 3.52Mg^{2+} \\ &\quad\quad + 0.21Fe^{3+} + 0.57Fe^{2+} + 6.03H_2O \end{aligned} \quad (17)$$

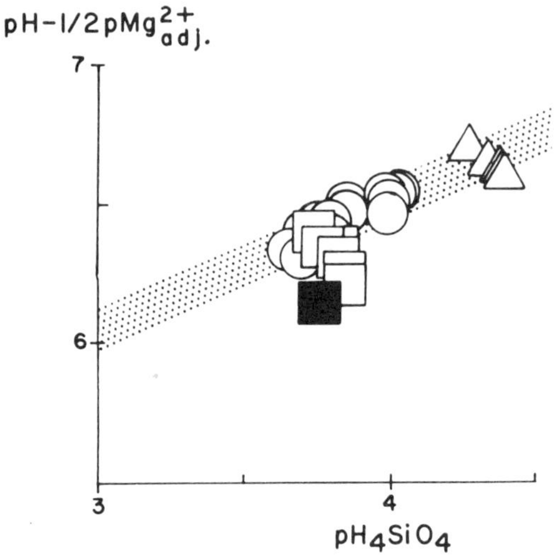

Figure 3. Adjusted solution compositions of several systems containing Vermont chlorite. Using Eq. (27), the experimental $pH - \frac{1}{2}pMg^{2+}$ values of the chlorite-kaolinite-hematite samples (○) of Figure 2 were adjusted to a $pH - \frac{1}{3}pAl^{3+}$ of 2.68. The chlorite-gibbsite-kaolinite-hematite (△) and chlorite-gibbsite-hematite (□ and ■) systems were also adjusted to a $pH - \frac{1}{2}pFe^{2+}$ of -1.0 with Eq. (27). The shaded line has a slope of 0.46. Line width and symbol size represent the deviations in $pH - \frac{1}{2}Mg^{2+}$ that could be engendered by the known variation in the gibbsite stability determination.

$$\begin{aligned} pK_{17} &= 2.99pH_4SiO_4 + 2.40pAl^{3+} + 3.52pMg^{2+} \\ &\quad + 0.57pFe^{2+} + 0.21pFe^{3+} - 16.01pH \\ &= 2.99pH_4SiO_4 - 7.20(pH - \tfrac{1}{3}pAl^{3+}) \\ &\quad - 7.04(pH - \tfrac{1}{2}pMg^{2+}) \\ &\quad - 0.63(pH - \tfrac{1}{3}pFe^{3+}) \\ &\quad - 1.14(pH - \tfrac{1}{2}pFe^{2+}). \end{aligned} \quad (18)$$

As indicated previously for this system, both $pH - \frac{1}{3}pAl^{3+}$ and $pH - \frac{1}{3}pFe^{3+}$ should be constant, controlled by gibbsite and hematite, respectively. The $pH - \frac{1}{2}pFe^{2+}$ will not necessarily be constant, but will depend upon the pH and Eh.

As for the chlorite from Vermont, the chlorite (Quebec) gibbsite-kaolinite-hematite system (Table 2) was equilibrated with solutions whose initial composition was such that the equilibrium was approached from four different directions on a plot of pH_4SiO_4 vs. $pH - \frac{1}{2}pMg^{2+}$ (Figure 4). The convergence of the samples about a line of slope 0.42 again indicates that a real equilibrium was being approached.

Chlorite-gibbsite-hematite system. Plotted in Figure 4 are analyses from Table 2 for duplicate samples equilibrated for 28 to 544 days. Earlier analyses were undersaturated with respect to those shown. After 544 days of equilibration, the Eh of the two samples was 0.388 V and 0.378 V, respectively. Insufficient liquid

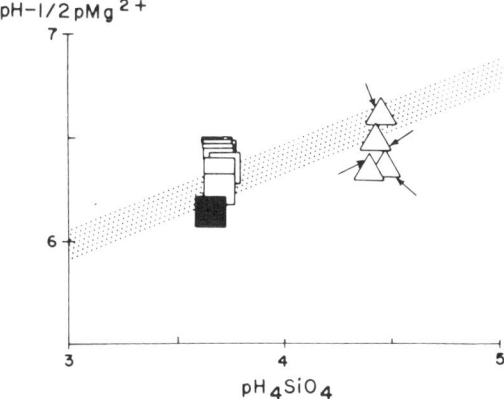

Figure 4. Solution compositions of several systems containing Quebec chlorite. The chlorite-gibbsite-kaolinite-hematite samples (△) were equilibrated for 1 to 9 days from four different types of supersaturation and undersaturation as indicated by the arrows. Samples of the chlorite-gibbsite-hematite system (□) were equilibrated for 28 to 544 days, with one solution obtained by immiscible displacement (■). The shaded line has a slope of 0.42. Line width and symbol size represent the deviations in pH $- \frac{1}{2}$pMg^{2+} that could be engendered by the known variation in the gibbsite stability determination.

remained for further analyses after regular centrifugation, so an additional set of analyses was made on one of the samples after immiscible displacement. All analyses plotted in Figure 4 for this system were very close.

Agreement between systems. To determine the theoretical slope of the chlorite solubility line on the coordinates of Figure 4, Eq. (18) must be rearranged as follows:

$$\begin{aligned}\text{pH} - \tfrac{1}{2}\text{pMg}^{2+} &= 0.42\text{pH}_4\text{SiO}_4 - 1.02(\text{pH} - \tfrac{1}{3}\text{pAl}^{3+}) \\ &\quad - 0.16(\text{pH} - \tfrac{1}{2}\text{pFe}^{2+}) \\ &\quad - 0.09(\text{pH} - \tfrac{1}{3}\text{pFe}^{3+}) \\ &\quad - 0.14\text{pK}_{17}. \end{aligned} \quad (19)$$

It can be seen from Eq. (19) that, if pH $- \frac{1}{2}$pMg^{2+} is plotted against pH$_4$SiO$_4$ (as in Figure 4), equilibrium solution analyses for Quebec chlorite should lie along a line of slope 0.42 *if* pH $- \frac{1}{3}$pAl^{3+}, pH $- \frac{1}{2}$pFe^{2+} and pH $- \frac{1}{3}$pFe^{3+} are held constant. All samples contained both gibbsite and hematite, so pH $- \frac{1}{3}$pAl^{3+} and pH $- \frac{1}{3}$pFe^{3+} should have remained constant. Eh was measured only for the chlorite-gibbsite-hematite system, however, so one cannot be sure that both systems had the same pH $- \frac{1}{2}$pFe^{2+} values.

The most reasonable assumption concerning the Eh of the samples in Figure 4 is probably that all samples had an Eh of about 0.380 V (see preceding section). This would mean that all samples had essentially the same pH $- \frac{1}{2}$pFe^{2+} value of -1.0, and no adjustment of data points would be necessary for direct comparisons. Fortunately the coefficient of the pH $- \frac{1}{2}$pFe^{2+} term in Eq. (19) is small, so only large variations in pH $- \frac{1}{2}$pFe^{2+} would have appreciable effect. Large variations in pH $- \frac{1}{2}$pFe^{2+}, at least due to the pH term, are unlikely because most samples were within 0.1 unit of pH 7.50. With no appreciable variations anticipated in pH $- \frac{1}{3}$pAl^{3+}, pH $- \frac{1}{3}$pFe^{3+}, or pH $- \frac{1}{2}$pFe^{2+} among samples in Figure 4, they may be directly compared. As can be seen in Figure 4, the two independent systems agree well with each other and with a line of slope 0.42 as predicted in Eq. (19).

Calculated ΔG^0_f value. In Figure 4, the Quebec chlorite has a pH $- \frac{1}{3}$pAl^{3+} of 2.7, a pH $- \frac{1}{3}$pFe^{3+} value of -0.3, and a pH $- \frac{1}{2}$pFe^{2+} value of -1.0. At a pH$_4$SiO$_4$ of 4.0, this chlorite can be seen to support a pH $- \frac{1}{2}$pMg^{2+} value of 6.4. From Eq. (18):

$$\begin{aligned}\text{pK}_{17} &= 2.99(4.0) - 7.20(2.7) - 7.04(6.4) \\ &\quad - 0.63(-0.3) - 1.14(-1.0) \\ &= -51.2 \pm 0.5,\end{aligned}$$

where the error estimate is derived from the known variation in the gibbsite solubility determination and is thought to represent a rough estimate of minimum analytical error. Then from the Nernst equation, where ΔG^0_f values are taken from Robie *et al.* (1978):

$$\begin{aligned}\Delta G_r &= 5.71\text{pK} = -292 \pm 3\text{kJ} \\ &= 2.99\Delta G^0_{f,\text{H}_4\text{SiO}_4} + 2.40\Delta G^0_{f,\text{Al}^{3+}} + 3.52\Delta G^0_{f,\text{Mg}^{2+}} \\ &\quad + 0.21\Delta G^0_{f,\text{Fe}^{3+}} + 0.57\Delta G^0_{f,\text{Fe}^{2+}} \\ &\quad + 6.03\Delta G^0_{f,\text{H}_2\text{O}} - \Delta G^0_{f,\text{chlorite}}. \end{aligned}$$

$$\begin{aligned}\Delta G^0_{f,\text{chlorite}} &= 292.4 + 2.99(-1308.0) + 2.40(-489.4) \\ &\quad + 3.52(-454.8) + 0.21(-4.6) \\ &\quad + 0.57(-78.9) + 6.03(-237.1) \\ &= -7869 \pm 3\text{kJ/mole},\end{aligned}$$

where the error estimate is the minimum due to analytical error.

Chlorite from Ishpeming, Michigan

Chlorite-gibbsite-hematite system. Equilibrium of chlorite from Michigan with its constituent ions can be depicted as follows:

$$\begin{aligned}(\text{Si}_{2.47}&\text{Al}_{1.53})(\text{Al}_{1.60}\text{Fe}^{2+}_{3.29}\text{Mg}_{1.05})\text{O}_{10}(\text{OH})_8 + 18.07\text{H}^+ \\ &= 2.47\text{H}_4\text{SiO}_4 + 3.13\text{Al}^{3+} + 3.29\text{Fe}^{2+} \\ &\quad + 1.05\text{Mg}^{2+} + 8.10\text{H}_2\text{O} \end{aligned} \quad (20)$$

$$\begin{aligned}\text{pK}_{20} &= 2.47\text{pH}_4\text{SiO}_4 + 3.13\text{pAl}^{3+} + 3.29\text{pFe}^{2+} \\ &\quad + 1.05\text{pMg}^{2+} - 18.07\text{pH}\end{aligned} \quad (21)$$

$$\begin{aligned}\text{pK}_{20} &= 2.47\text{pH}_4\text{SiO}_4 - 9.39(\text{pH} - \tfrac{1}{3}\text{pAl}^{3+}) \\ &\quad - 6.58(\text{pH} - \tfrac{1}{2}\text{pFe}^{2+}) \\ &\quad - 2.10(\text{pH} - \tfrac{1}{2}\text{pMg}^{2+})\end{aligned} \quad (22)$$

In Figure 5, analyses for duplicate samples (Table 2) are plotted after equilibrating from 6 to 545 days. Earlier analyses were undersaturated with respect to those shown. After 545 days of equilibration, the Eh of the two samples was 0.384 V and 0.387 V, respectively.

Insufficient liquid remained for further analyses after regular centrifugation, so an additional set of analyses on one sample was made using an immiscible displacement technique (Kittrick, 1980).

Calculated ΔG. If the analyses of Figure 5 represent equilibrium, then pH $- \frac{1}{3}$pAl^{3+} is 2.7 and pH $- \frac{1}{3}$pFe^{3+} is -0.3. From pH and Eh measurements on the 544 and 545 day samples and Eqs. (6) and (7), pH $- \frac{1}{2}$pFe^{2+} is calculated to be -0.9. From Eq. (22) we see that, if pH $- \frac{1}{3}$pAl^{3+} and pH $- \frac{1}{2}$pFe^{2+} are constant, then

$$2.10(\text{pH} - \tfrac{1}{2}\text{pMg}^{2+}) = 2.47\text{pH}_4\text{SiO}_4 + K, \text{ and}$$
$$\text{pH} - \tfrac{1}{2}\text{pMg}^{2+} = 1.18\text{pH}_4\text{SiO}_4 + K^1.$$

Thus the slope of the line in Figure 5 is 1.18. All analyses plotted in Figure 5 are in good agreement with this theoretical relationship. At a pH$_4$SiO$_4$ of 4.0 in Figure 5, the line indicates a pH $- \frac{1}{2}$pMg^{2+} of 6.3. Thus, from Eq. (22),

$$\begin{aligned}\text{pK}_{20} &= 2.47(4.0) - 9.39(2.7) \\ &\quad - 6.58(-0.9) - 2.10(6.3) \\ &= -22.8 \pm 0.5\end{aligned}$$

where the error estimate is derived from the known variation in the gibbsite solubility determination and is thought to represent a rough estimate of minimum analytical error. Then, from the Nernst equation,

$$\Delta G_r = 5.71\text{pK}_{20} = -130 \pm 3 \text{ kJ}.$$

From Eq. (20) and using ΔG^0_f values from Robie *et al.* (1978):

$$\Delta G_r = 2.47\Delta G^0_{f,\text{H}_4\text{SiO}_4} + 3.13\Delta G^0_{f,\text{Al}^{3+}}$$
$$+ 3.29\Delta G^0_{f,\text{Fe}^{2+}} + 1.05\Delta G^0_{f,\text{Mg}^{2+}}$$
$$+ 8.10\Delta G^0_{f,\text{H}_2\text{O}} - \Delta G^0_{f,\text{chlorite}}$$
$$\begin{aligned}\Delta G^0_{f,\text{chlorite}} &= 130 + 2.47(-1308.0) + 3.13(-489.4) \\ &\quad + 3.29(-78.9) + 1.05(-454.8) \\ &\quad + 8.10(-237.1) = -7290 \pm 3 \text{ kJ/mole},\end{aligned}$$

where the error estimate is the minimum due to analytical error.

Chlorite from New Mexico

Chlorite-gibbsite-hematite system. Equilibrium of the chlorite from New Mexico with its constituent ions can be depicted as follows:

$$(\text{Si}_{2.84}\text{Al}_{1.16})(\text{Al}_{1.75}\text{Fe}^{2+}{}_{2.61}\text{Mg}_{1.16}\text{Fe}^{3+}{}_{0.12})\text{O}_{10}(\text{OH})_8$$
$$+ 16.63\text{H}^+ = 2.84\text{H}_4\text{SiO}_4 + 2.91\text{Al}^{3+} + 2.61\text{Fe}^{2+}$$
$$+ 1.16\text{Mg}^{2+} + 0.12\text{Fe}^{3+} + 6.64\text{H}_2\text{O} \quad (23)$$

$$\begin{aligned}\text{pK}_{23} &= 2.84\text{pH}_4\text{SiO}_4 + 2.91\text{pAl}^{3+} \\ &\quad + 2.61\text{pFe}^2 + 1.16\text{pMg}^{2+} \\ &\quad + 0.12\text{pFe}^{3+} - 16.63\text{pH}\end{aligned} \quad (24)$$

$$\begin{aligned}\text{pK}_{23} &= 2.84\text{pH}_4\text{SiO}_4 - 8.73(\text{pH} - \tfrac{1}{3}\text{pAl}^{3+}) \\ &\quad - 5.22(\text{pH} - \tfrac{1}{2}\text{pFe}^{2+}) \\ &\quad - 2.32(\text{pH} - \tfrac{1}{2}\text{pMg}^{2+}) \\ &\quad - 0.36(\text{pH} - \tfrac{1}{3}\text{pFe}^{3+}).\end{aligned} \quad (25)$$

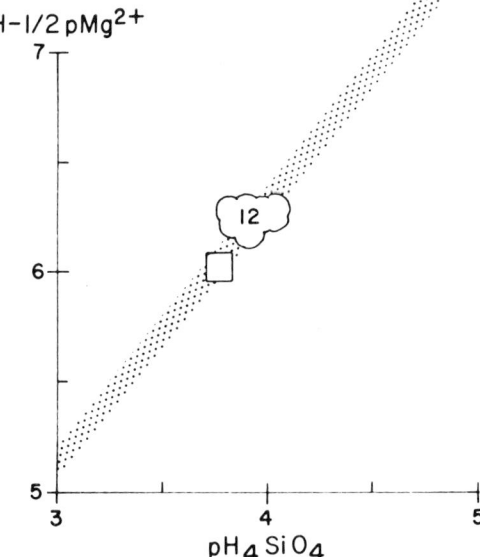

Figure 5. Solution compositions of two samples containing chlorite from Michigan, plus gibbsite and hematite (○). Duplicates were sampled repeatedly from 6 to 545 days, with one solution obtained by immiscible displacement (□). The group contains 12 analysis points. The shaded line has a slope of 1.18. Symbol size and line width represent the deviation in pH $- \frac{1}{2}$pMg^{2+} that could be engendered by the known variation in the gibbsite stability determination.

As before, with pH $- \frac{1}{3}$pAl^{3+} held constant according to Eq. (12), pH $- \frac{1}{3}$pFe^{3+} held constant according to Eq. (6), and pFe^{2+} calculated from Eq. (7), pK$_{23}$ is defined by measurements of equilibrium pH$_4$SiO$_4$, pMg^{2+}, pH, and Eh (Table 2). In Figure 6, three of these variables are plotted for duplicate samples after equilibrating for 6 to 545 days. Earlier analyses were undersaturated with respect to those shown. After 545 days of equilibration, the Eh of the two samples was 0.382 V and 0.384 V, respectively. Insufficient liquid remained for further analysis after regular centrifugation, so an additional set of analyses was made on one sample using an immiscible displacement technique.

Calculated ΔG value. If the analyses of Figure 6 represent equilibrium, then pH $- \frac{1}{3}$pAl^{3+} is 2.7 and pH $- \frac{1}{3}$pFe^{3+} is -0.3. From pH and Eh measurements on the 545 day samples and Eqs. (6) and (7), pH $- \frac{1}{2}$pFe^{2+} is calculated to be -0.9. From Eq. (25), if pH $- \frac{1}{3}$pAl^{3+}, pH $- \frac{1}{3}$pFe^{3+}, and pH $- \frac{1}{2}$pFe^{2+} are constant,

$$2.32(\text{pH} - \tfrac{1}{2}\text{pMg}^{2+}) = 2.84\text{pH}_4\text{SiO}_4 + K, \text{ and}$$
$$\text{pH} - \tfrac{1}{2}\text{pMg}^{2+} = 1.22\text{pH}_4\text{SiO}_4 + K^1.$$

Thus the slope of the line in Figure 6 is 1.22. All analyses plotted in Figure 6 are in good agreement with this

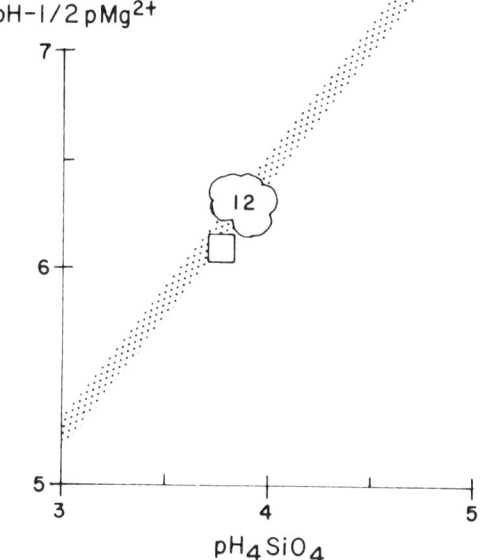

Figure 6. Solution compositions of two samples containing chlorite from New Mexico, plus gibbsite and hematite (○). Duplicates were sampled repeatedly from 6 to 545 days with one solution obtained by immiscible displacement (□). The group contains 12 analysis points. The shaded line has a slope of 1.22. Line width and symbol size represent the deviations in pH − ½pMg^{2+} that could be engendered by the known variation in the gibbsite stability determination.

relationship. At a pH$_4$SiO$_4$ of 4.0, the line indicates a pH − ½pMg^{2+} value of 6.5. Thus, from Eq. (25),

$$pK_{23} = 2.84(4.0) − 8.73(2.7) − 5.22(−0.9)$$
$$− 2.32(6.5) − 0.36(−0.3)$$
$$= 22.5 \pm 0.5,$$

where the estimate of minimum analytical error is again derived from the known variation in the gibbsite solubility determination. Similarly

$$\Delta G_r = 5.71 pK_{23} = -128 \pm 3 \text{ kJ}.$$

and from Eq. (23)

$$\Delta G_r = 2.84\Delta G^0_{f, H_4SiO_4} + 2.91\Delta G^0_{f, Al^{3+}}$$
$$+ 2.61\Delta G^0_{f, Fe^{2+}} + 1.16\Delta G^0_{f, Mg^{2+}}$$
$$+ 0.12\Delta G^0_{f, Fe^{3+}} + 6.64\Delta G^0_{f, H_2O}$$
$$- \Delta G^0_{f, chlorite}.$$

$$\Delta G^0_{f, chlorite} = 128 + 2.84(−1308.0) + 2.91(−489.4)$$
$$+ 2.61(−78.9) + 1.16(−454.8)$$
$$+ 0.12(−4.6) + 6.64(−237.1)$$
$$= −7319 \pm 3 \text{ kJ/mole}.$$

DISCUSSION AND CONCLUSIONS

Sample equilibrium

The most common flaw in mineral stability determinations by the solubility method is the lack of demonstrated sample equilibrium. In the present experiments there were five indicators of sample equilibrium for the high-Mg chlorites (Vermont and Quebec). First, good agreement was achieved between successive analyses of single samples over a long period of time. Second, there was good agreement between duplicate samples. Third, the same values were obtained for samples equilibrated from both undersaturation and supersaturation. Fourth, an independent measure of equilibrium was made involving the measured pH$_4$SiO$_4$ levels of samples containing both gibbsite and kaolinite. The pH$_4$SiO$_4$ of such samples in Figures 2 and 4 ranges from 4.3 to 4.5. These values are in good agreement with Eq. (14) and with the stability of gibbsite and kaolinite as determined by long-term solubility methods and by immiscible displacement of solutions following short-term equilibration (Kittrick, 1980). A fifth check on sample equilibrium involved a comparison of three independent systems containing chlorite, i.e., chlorite-gibbsite-kaolinite-hematite, chlorite-gibbsite-hematite, and chlorite-kaolinite-hematite. Agreement between these systems with regard to directly measured variables was good, and could be made essentially perfect, depending upon assumptions relative to sample Eh values.

The same five indicators of sample equilibrium were also applied to the high-Fe chlorites (Michigan and New Mexico), but data supporting equilibrium checks three and five are not shown. A lack of Eh measurements on the chlorite-gibbsite-kaolinite-hematite system made these data marginally useful for equilibrium constant calculations where high-Fe chlorites are involved, so they were omitted. The three equilibrium indicators shown are sufficient to indicate that the likelihood of sample equilibrium is good.

Interstratification and solid solution

A gradation exists in the brucitic layers of natural chlorites with regard to both layer completeness and the binding of adjacent smectitic units. It is, therefore, uncertain as to whether the brucitic and smectitic layers should be considered units within a single-phase mineral, or independent, regularly interstratified components. Separate brucite and talc components, for example, controlling their individual solubilities should generate solution analyses that cluster at the intersection of the brucite and talc stability lines. The pH − ½pMg^{2+} supported by brucite can be determined as follows:

brucite
$$Mg(OH)_2 + 2H^+ = Mg^{2+} + 2H_2O$$
$$\Delta G_r = \Delta G^0_{f, Mg^{2+}} + 2\Delta G^0_{f, H_2O}$$
$$- \Delta G^0_{f, Mg(OH)_2}$$
$$= −454.8 + 2(−237.1) − (−833.5)$$
$$= −95.5 \pm 0.44 \text{ kJ}.$$

There is good agreement among recent compilations (Sadiq and Lindsay, 1979; Robie et al., 1978; Parker et

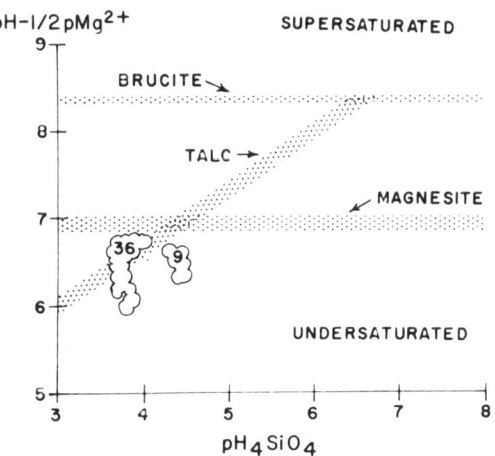

Figure 7. Solution compositions of Vermont and Quebec chlorites (from Figures 2 and 4) in relation to stability lines of brucite, talc, and magnesite. Line widths and symbol size represent estimates of experimental error. There are 36 points in one cluster and 9 in the other. The intersection of the brucite and talc lines is derived from Eq. (26).

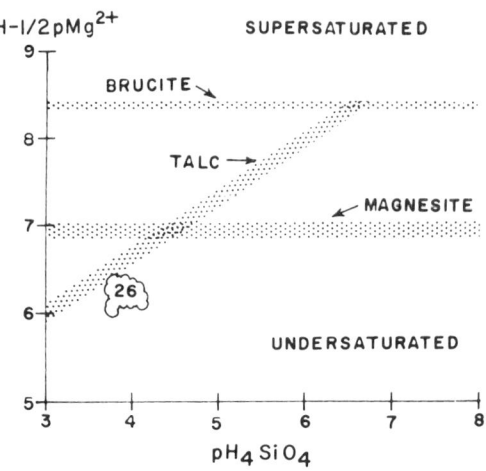

Figure 8. Solution compositions of Michigan and New Mexico chlorites (from Figures 5 and 6) in relation to stability lines for brucite, talc, and magnesite. Line widths and symbol size represent estimates of experimental error. The group contains 26 analysis points.

al., 1971) for the ΔG^0_f of brucite. The value given by Robie et al. (1978) for brucite is used above, where the error in the $\Delta G^0_{f, \text{brucite}}$ of brucite is assigned to the error in ΔG_r. Furthermore,

$$pK = \Delta G_r/5.707 = -16.73 \pm 0.08$$
$$pK = 2pMg^{2+} - 2pH$$
$$pH - \tfrac{1}{2}pMg^{2+} = -pK/2 = 8.37 \pm 0.04.$$

Thus, the stability line for brucite (as shown in Figures 7 and 8) is parallel to the pH_4SiO_4 axis, intersecting the $pH - \tfrac{1}{2}pMg^{2+}$ axis at 8.37.

If the brucitic and talcitic components of chlorite were to control solution equilibria, solution analyses should occur at the intersection of the brucite and talc stability lines. This intersection can be determined as follows:

brucite
$$Mg(OH)_2 + 2Mg^{2+} + 4H_4SiO_4$$
talc
$$= Mg_3Si_4O_{10}(OH)_2 + 4H^+ + 6H_2O \qquad (26)$$
$$\Delta G_r = \Delta G^0_{f, \text{talc}} + 6\Delta G^0_{f, H_2O} - 4\Delta G^0_{f, H_4SiO_4}$$
$$- 2\Delta G^0_{f, Mg^{2+}} - \Delta G^0_{f, \text{brucite}}$$
$$= -5525.22 + 6(-237.1) - 4(-1308.0)$$
$$- 2(-454.8) - (-833.5) = 30.84 \pm 4.35 \text{ kJ}.$$

The $\Delta G^0_{f, \text{talc}}$ selected is from Sadiq and Lindsay (1979), with the error in the ΔG_{talc} assigned to the error in ΔG_r. Then, from Eq. (26),

$$pK = 4pH^+ - 4pH_4SiO_4 - 2pMg^{2+}$$
$$= 30.84/5.707 = 5.40 \pm 0.76.$$

$$4pH_4SiO_4 = 4pH - 2pMg^{2+} - pK$$
$$pH_4SiO_4 = pH - \tfrac{1}{2}pMg - pK/4 = 8.37 - 1.84$$
$$= 6.53 \pm 0.19.$$

The intersection of the brucite and talc lines is shown in Figures 7 and 8, where it can be seen that the solution analyses are not clustered about the intersection, but rather, occur over a range in $pH - \tfrac{1}{2}pMg^{2+}$ and pH_4SiO_4 values. In particular, none lie close to the intersection of brucite and talc lines. None of the solution analyses parallel or lie close to the brucite line although some coincide with the talc stability line. It therefore appears that neither brucite individually, nor brucite and talc collectively, control solution equilibria as solid solution components. Control of solution equilibria by a talc component appears doubtful, but perhaps cannot be entirely eliminated.

When conducting mineral solubility determinations, it is always necessary to consider the possible precipitation of other mineral phases which may then control solution ion levels. Magnesite can be readily precipitated in room temperature solutions in contact with the CO_2 of the atmosphere. The shaded line in Figures 7 and 8 (Kittrick, 1973) is based upon the $\Delta G^0_{f, \text{magnesite}}$ selected from Robie et al. (1978). Because all solution analyses are undersaturated with respect to magnesite, magnesite does not appear to be controlling samples in Figures 7 and 8. Chrysotile does not ordinarily precipitate in solutions at room temperature, but again using ΔG^0_f values from Robie et al. (1978), it can be calculated that chrysotile supports a $pH - \tfrac{1}{2}pMg^{2+}$ value of 6.7 ± 0.1 at a pH_4SiO_4 of 4.0. This is somewhat more soluble than the four chlorites which range from pH − ½pMg^{2+} values 6.3 to 6.5 under those conditions (Table 1), thus chrysotile does not appear to be controlling sample solubilities either.

There seems little doubt that chlorite, with its wide and essentially continuous variation in chemical composition, is a solid solution. However, distribution of the solution analyses along lines of theoretical slope (Figures 3–6) indicate control by a single phase of bulk chlorite composition rather than control by solid solution components. The solid solution components appear to be "frozen in." The fact that the slope of solubility lines are compatible with the bulk chemical composition of the chlorites indicates that the proportion of their solid solution components does not change appreciably during the course of the equilibration. If these chlorites formed under equilibrium conditions of low temperature metamorphism, their proportion of various solid solution components would be expected to be that which is most stable for those particular conditions. That is not likely to be the same proportion of components that is most stable during room temperature solubility experiments. Because the chlorite composition showed no appreciable change during the course of those experiments, the system should be considered to be at a metastable equlibrium.

Chlorite stabilities compared

Because ΔG^0_f is an extensive variable, ΔG^0_f values of minerals of variable composition are strongly dependent upon the exact chemical composition of the individual minerals. Thus, a comparison of stabilities of minerals of variable composition cannot be obtained by comparing ΔG^0_f values (Table 1). However, their stabilities can be compared by considering their level of control of some constituent at fixed levels of other constituents. For example, at a pH$_4$SiO$_4$ of 4.0 under identical conditions of pH − ⅓pAl^{3+}, pH − ⅓pFe^{3+}, and pH − ½pFe^{2+}, the chlorite from Michigan supports a pH − ½pMg^{2+} of 6.3 compared to 6.4 for the chlorite from Quebec and 6.5 for the chlorites from Vermont and New Mexico. This can be directly compared with the estimates for clinochlore stability in Table 1, which range from 5.4 to 7.0. The estimate of 6.8 for pH − ½pMg^{2+}, as calculated from Helgeson (1969) and Nriagu (1975), is particularly close to the experimental values.

The chlorite from Michigan is the more stable (least soluble) of the four under the conditions of comparison, but one must note that the difference between the four in terms of pH − ½pMg^{2+} is approximately the width of the estimated error band for the talc stability line in Figures 7 and 8. Thus, in terms of the uncertainties involved in determining mineral stabilities, the stabilities of the Vermont, Quebec, Michigan, and New Mexico chlorites must be considered to be quite similar. They probably serve as a useable estimate of the high pH-high Mg limit in the Al$_2$O$_3$-SiO$_2$-MgO-H$_2$O-H$^+$ system and of the general stability of chlorites formed under low-temperature metamorphism. As a first approximation, a pH − ½pMg^{2+} of 6.4 at a pH$_4$SiO$_4$ of 4.0 may be used to represent a wide range of chlorites formed under low-temperature metamorphism if Al control is close to that of gibbsite. Chlorites that are formed at room temperature are presumably more stable at room temperature than those investigated here. Whether this is reflected in a measurable difference in equilibrium levels of constituents remains to be determined.

The values of pH − ½pMg^{2+} and pH$_4$SiO$_4$ in equilibrium with the four chlorites are directly determined with good accuracy and dependable precision. The values of pH − ⅓pAl^{3+} and pH − ⅓pFe^{3+} depend upon equilibrium with gibbsite and hematite respectively, which appear to be good assumptions. The values of pK$_8$, pK$_{17}$, pK$_{20}$, and pK$_{23}$ and their corresponding $\Delta G^0_{f,\ chlorite}$ are less certain because in addition to the aforementioned parameters, they also depend upon the assumption that the measured Eh of the samples is the same as that of the Fe^{2+}-Fe^{3+} couple in the samples. This assumption is important for the high-Fe chlorites. For example, for every 10 mV change in Eh, the ΔG^0_f of the Michigan chlorite changes 3.2 kJ (0.76 kcal) and the ΔG^0_f of the New Mexico chlorite changes 2.5 kJ (0.60 kcal).

The Eh of the chlorite samples is not controlled by the chlorite-hematite pair, because they support such low levels of Fe^{2+} and Fe^{3+} in solution. Quinhydrone additions controlled the Eh and probably would have ensured that the Fe^{2+}-Fe^{3+} couple was at the measured Eh. When quinhydrone was eliminated for other reasons, the Eh of the chlorite systems was left to be controlled by dissolved oxygen. This Eh was measured, but one cannot be sure that the Fe^{2+}-Fe^{3+} couple is at that Eh (Bohn, 1968). Thus, the determination of high-Fe chlorite stabilities by solubility methods can be accomplished only within the constraints of whatever uncertainty exists as to the actual Eh of the Fe^{3+}-Fe^{2+} couple.

ACKNOWLEDGMENTS

This work was supported in part by the donors of the Petroleum Research Fund, administered by the American Chemical Society, under Grant PRF5872-AC2. Appreciation is expressed to Mr. E. W. Hope and Mr. W. L. Moore who performed the solution analyses and to Dr. P. R. Hooper who analyzed the chlorites.

APPENDIX

Based upon Eq. (16), the analytically determined pH − ½pMg^{2+} for Vermont chlorite may be adjusted to a pH − ⅓pAl^{3+} of 2.68 and a pH − ½pFe^{2+} of −1.0

as follows: $pH - \frac{1}{2}pMg^{2+}_{adjusted} = pH - \frac{1}{2}pMg^{2+}_{analysis} + 1.14(pH - \frac{1}{3}pAl^{3+}_{analysis} - 2.68) + 0.31(pH - \frac{1}{2}pFe^{2+}_{analysis} + 1.0).$ (27)

REFERENCES

APHA (1960) *Standard Method for the Examination of Water and Waste-Water.* 11th ed., American Public Health Association, New York. 626 pp.

Bailey, S. W. (1975) Chlorites: *Soil Components, Vol. 2, Inorganic Components,* J. E. Gieseking, ed. Springer Verlag, New York, 191–263.

Bird, G. W. and Anderson, G. M. (1973) The free energy of formation of magnesian cordierite and phlogopite: *Amer. J. Sci.* **273**, 84–91.

Bohn, H. L. (1968) Electromotive force of inert electrodes in soil suspensions: *Soil Sci. Soc. Amer. Proc.* **32**, 211–215.

Fawcett, J. J. and Yoder, H. S. (1966) Phase relations of chlorite in the system $MgO-Al_2O_3-SiO_2-H_2O$: *Amer. Mineral.* **51**, 353–380.

Foster, M. D. (1953) Geochemical studies of clay minerals. III. The determination of free silica and free alumina in montmorillonites: *Geochim. Cosmochim. Acta* **3**, 143–154.

Garrels, R. M. and Christ, C. L. (1965) *Solutions Minerals and Equilibria:* Harper and Row, New York, 450 pp.

Helgeson, H. C. (1969) Thermodynamics of hydrothermal systems at elevated temperatures and pressures: *Amer. J. Sci.* **267**, 729–804.

Jackson, M. L. (1969) *Soil Chemical Analysis—Advanced course.* 2nd ed., Published by author. Univ. Wisconsin, Madison, Wisconsin, 895 pp.

Kittrick, J. A. (1966a) The free energy of formation of gibbsite and $Al(OH)_4^-$ from solubility measurements: *Soil Sci. Soc. Amer. Proc.* **30**, 595–598.

Kittrick, J. A. (1966b) Free energy of formation of kaolinite from solubility measurements: *Amer. Mineral.* **51**, 1457–1466.

Kittrick, J. A. (1967) Gibbsite-kaolinite equilibria: *Soil Sci. Soc. Amer. Proc.* **31**, 314–316.

Kittrick, J. A. (1969) Soil minerals in the $Al_2O_3-SiO_2-H_2O$ system and a theory of their formation: *Clays & Clay Minerals* **17**, 157–167.

Kittrick, J. A. (1971) Stability of montmorillonites. I. Belle Fourche and Clay Spur montmorillonites: *Soil Sci. Soc. Amer. Proc.* **35**, 140–145.

Kittrick, J. A. (1973) Mica-derived vermiculites as unstable intermediates: *Clays & Clay Minerals* **21**, 479–488.

Kittrick, J. A. (1980) Gibbsite and kaolinite solubilities by immiscible displacement of equilibrium solutions. *Soil Sci. Soc. Amer. J.* **44**, 139–142.

McKenzie, F. T. and Garrels, R. M. (1965) Silicates: Reactivity with sea water: *Science* **150**, 57–58.

Nriagu, J. O. (1975) Thermochemical approximations for clay minerals: *Amer. Mineral.* **60**, 834–839.

Parker, V. B., Wagman, D. D., and Evans, W. H. (1971) Selected values of chemical thermodynamic properties. Tables for the alkaline earth elements: *Nat. Bur. Stand. Tech. Note* **270-6**, 106 pp.

Rich, C. I. and Bonnet, J. A. (1975) Swelling chlorite in a soil of the Dominican Republic: *Clays & Clay Minerals* **23**, 97–102.

Robie, R. A., Hemingway, B. S., and Fisher, J. R. (1978) Thermodynamic properties of minerals and related substances at 298.15K and 1 bar (10^5 pascals) pressure and at higher temperatures: *U.S. Geol. Surv. Bull.* **1452**, 456 pp.

Roth, C. B., Jackson, M. L., Lotse, E. G., and Syers, J. K. (1968) Ferrous-ferric ratio and CEC changes on deferration of weathered micaceous vermiculite: *Israel J. Chem.* **6**, 261–273.

Sadiq, M. and Lindsay, W. L. (1979) Selection of standard free energies of formation for use in soil chemistry: *Colorado State Univ. Exp. Stat. Tech. Bull.* **134**, 1069 pp.

Tardy, Y. and Garrels, R. M. (1974) A method of estimating the Gibbs energies of formation of layer silicates. *Geochim. Cosmochim. Acta* **38**, 1101–1116.

Zen, E-an. (1972) Gibbs free energy, enthalpy, and entropy of ten rock-forming minerals: Calculations, discrepancies, implications: *Amer. Mineral.* **57**, 524–553.

ZoBell, C. E. (1946) Studies on redox potential of marine sediments. *Bull. Amer. Assoc. Petrol. Geol.* **30**, 477–513.

(Received 16 April 1981; accepted 17 September 1981)

Резюме—Высоко-Mg хлориты из Вермонта и Квебека и высоко-Fe хлориты из Мичигана и Новой Мексики уравновешивались при комнатной температуре в почти нейтральном диапазоне pH. Гиббсит, каолинит и гематит о известной стабильности были добавлены к образцам, чтобы контролировать неизмеряемые переменные на уровнях, поддающихся исчислению. Составы равновесного раствора определялись путем ненасыщения и перенасыщения. Другие индикаторы равновесия были в хорошем согласии между последующими анализами в течение длинного времени, между спаренными образцами, между независимыми системами и между независимыми измерениями равновесия. Все четыре хлориты были стабильны по отношению к бруциту и, за несколькими исключениями, по отношению к тальку в условиях проведения исследования. В случае равновесия с гиббситом величина pH − 1/2Mg^{2+} хлоритов изменялась от 6,3 до 6,5 при величине pH$_4$SiO$_4$ равной 4,0. Эти величины хорошо согласуются с прежними оценками стабильности хлоритов. Рассчитанная стандартная свободная энергия образования хлоритов зависит от растворения Fe^{2+}, вычисленного для Eh образца при предположении равновесия с гематитом, а также при допущении, что пара Fe^{2+}-Fe^{3+} имеет такое Eh как образец. [E.C.]

Resümee—Mg-reiche Chlorite von Vermont und Quebec und Fe-reiche Chlorite von Michigan und New Mexico wurden bei Raumtemperatur im neutralen pH-Bereich ins Gleichgewicht gebracht. Den Proben wurden Gibbsit, Kaolinit, und Haematit mit bekannter Stabilität hinzugefügt, um den Einfluß unmeßbarer Variablen unter Kontrolle zu halten. Die Gleichgewichtszusammensetzungen der Lösungen wurden aus der Untersättigung und Übersättigung ermittelt. Weitere Hinweise für Gleichgewicht waren eine gute Übereinstimmung von aufeinanderfolgenden Analysen übereinen langen Zeitraum zwischen Parallelproben, zwischen unabhängigen Systemen und zwischen unabhängigen Gleichgewichtsmessungen. Alle vier Chlorite waren unter den untersuchten Bedingungen in Bezug auf Brucit und—mit wenigen Ausnahmen—auch in Bezug auf Talk stabil. Im Gleichgewicht mit Gibbsit lag bei einem pH$_4$SiO$_4$ von 0,4 der pH − ½ Mg^{2+} der Chlorite zwischen 6,3 und 6,5. Diese Werte stimmen gut mit früheren Schätzungen der Chloritstabilität überein. Die berechnete Freie Standartsbildungsenergie der Chlorite hängt vom Fe^{2+}-Gehalt der Lösung ab, wie aus dem Eh der Probe berechnet wurde und sich auch aus dem Gleichgewicht mit Haematit ergab. Voraussetzung dafür ist, daß das Fe^{2+}-Fe^{3+} Paar den gleichen Eh wie die Probe hat. [U.W.]

Résumé—Des chlorites à contenu élevé en Mg du Vermont et du Québec, et des chlorites à contenu élevé en Fe de Michigan et de New Mexico ont été équilibrées à température ambiante à un pH quasiment neutre. De la gibbsite, de la kaolinite et de l'hématite de stabilité connue ont été ajoutées aux échantillons pour contrôler des variables non-mesurables à des niveaux calculables. Des compositions de solution équilibrée ont été obtenues par sousaturation et supersaturation. D'autres indicateurs d'équilibre incluent la correspondance d'analyses successives pendant une longue durée, d'échantillons répetées, de systèmes indépendants, et de mesures indépendantes d'équilibre. Les quatre chlorites étaient stables relatives à la brucite, et, avec quelques exceptions, au talc sous les conditions de l'étude. Lorsque les chlorites étaient en équilibre avec la gibbsite, leurs pH − ½Mg^{2+} s'étageaient de 6,3 à 6,5, à une valeur pour pH$_4$SiO$_4$ de 4,0. Ces valeurs s'accordent bien avec des estimations précédentes de stabilité pour les chlorites. L'énergie libre standard calculée pour la formation des chlorites dépend de la solution Fe^{2+} calculée à partir de l'échantillon Eh et de l'équilibre supposé avec l'hématite, en supposant que le couple Fe^{2+}-Fe^{3+} a le même Eh que l'échantillon. [D.J.]

16

Copyright © 1974 by Pergamon Press Ltd.
Reprinted by permission from Geochim. Cosmochim. Acta **38**:1101-1116 (1974)

A method of estimating the Gibbs energies of formation of layer silicates

YVES TARDY

Centre National de la Recherche Scientifique, Centre de Sédimentologie et Géochimie de la Surface, Strasbourg, France

and

ROBERT M. GARRELS

Department of Oceanography, University of Hawaii, Honolulu, Hawaii 96822
and Department of Geology, Northwestern University

(*Received* 10 *December* 1973; *accepted in revised form* 18 *January* 1974)

Abstract—Gibbs energies of formation (from the elements at 298·15 K) of layer and fibrous silicates can be estimated within the limits of uncertainty of the values obtained by current experimental techniques. The method of calculation is based on the assumption that each silicate can be represented by oxide and hydroxide components possessing Gibbs energies of formation within the silicate structures that are constant but may differ from the Gibbs energies assigned to the components as separate phases. Layer silicates with exchangeable ions are included in this scheme. The method permits estimation of the free energies of formation of montmorillonites, illites, chlorites and micas of complex composition.

INTRODUCTION

THE STANDARD Gibbs energies of formation of a number of layer silicates have been determined within the past several years. These compounds and their Gibbs energies of formation from the elements of 298·15 K are given in Table 1, as well as those of other species that will be used in the ensuing discussion. It has been necessary to choose 'best values' from the literature for our calculations. Table 1 is arranged and annotated so as to provide the reader with some feeling for the uncertainty in the values that are used as targets in the estimation process. If the formulae of the compounds are written in terms of oxide and hydroxide components, and the assumption is made that the free energies of formation of these components *in* the silicate matrix are independent of the silicate in which they occur, it is possible to solve for the Gibbs energy of formation of the components in the magnesium silicates, aluminum silicates, ferrous iron silicates, and in muscovite and paragonite. The rule that has been used in determining the values is to assign the magnesium in magnesium-bearing silicates to an hydroxide component, otherwise the components are expressed in oxide form. Residual magnesium is written as the oxide. It is then possible to set up a series of simultaneous equations to obtain the Gibbs energies of formation of the components in the silicate structure.

MAGNESIUM SILICATES

In the MgSiOH system, talc, chrysotile and sepiolite can be written in terms of three components. This gives three equations and three unknowns, as follows:

For chrysotile:

$$2\Delta G^\circ_{f,\text{sil}}\text{Mg(OH)}_2 + \Delta G^\circ_{f,\text{sil}}\text{MgO} + 2\Delta G^\circ_{f,\text{sil}}\text{SiO}_2 = -965 \text{ kcal.}$$

Table 1. Some free energies of formation at 298·15 K of various substances (values in kcal/mol for formation from the elements)

Substance	Formula	ΔG_f°	Author	Al corr.* (kcal)	ΔG_f° selected for this paper
Wustite	FeO (fictive)	$-60·097 \pm 0·6$	Robie and Waldbaum (1968)		$-60·1$
	FeO$_{0·947}$O	$-58·599 \pm 0·210$	Robie and Waldbaum (1968)		
	FeO (fictive?)	$-58·4$	Mel'nik (1972)		
	FeO	$-58·4$	Garrels and Christ (1965)		
Hematite	Fe$_2$O$_3$	$-177·73 \pm 0·310$	Robie and Waldbaum (1968)		$-177·7$
	Fe$_2$O$_3$	$-176·77$	Mel'nik (1972)		
	Fe$_2$O$_3$	$-177·728 \pm 0·310$	Langmuir (1969)		
	Fe$_2$O$_3$	$-177·1$	Garrels and Christ (1965)		
Goethite	FeOOHα	$-117·04$	Mel'nik (1972)		$-117·0$
	FeOOHα	$-116·375 \pm 0·160$	Langmuir (1969)		
ppt. aged Fe$_2$O$_3$	Fe$_2$O$_3$ (estimated)	$-170·0$	Estimated by authors		$-170·0$
Periclase	MgO	$-136·087 \pm 0·110$	Robie and Waldbaum (1968)		$-136·1$
	MgO	$-136·09$	Mel'nik (1972)		
	MgO	$-136·13$	Garrels and Christ (1965)		
Brucite	Mg(OH)$_2$	$-199·460 \pm 0·730$	Robie and Waldbaum (1968)		$-199·5$
	Mg(OH)$_2$	$-200·05$	Mel'nik (1972)		
	Mg(OH)$_2$	$-199·27$	Garrels and Christ (1965)		
	Mg(OH)$_2$	$-200·040 \pm 0·690$	Fisher and Zen (1971)		
Corundum	Al$_2$O$_3\alpha$	$-378·082 \pm 0·310$	Robie and Waldbaum (1968)		$-378·1$
	Al$_2$O$_3\alpha$	$-376·77$	Garrels and Christ (1965)		
Gibbsite	Al(OH)$_3$	$-273·486 \pm 0·310$	Robie and Waldbaum (1968)	$-3·5$	$-277·0$
	Al(OH)$_3$	$-277·3$	Garrels and Christ (1965)		
Potassium oxide	K$_2$O	$-76·974 \pm 0·680$	Robie and Waldbaum (1968)		$-77·0$
	K$_2$O	$-76·986$	JANAF Tables (1965)		
Potassium hydroxide	KOH	$-90·866$	JANAF Tables (1965)		$-90·9$
	KOH	$-89·5$	Garrels and Christ (1965)		
Lithium oxide	Li$_2$O	$-134·329 \pm 0·510$	Robie and Waldbaum (1968)		$-134·3$
Sodium oxide	Na$_2$O	$-90·125$	JANAF Tables (1965)		
	Na$_2$O	$-90·161 \pm 1·51$	Robie and Waldbaum (1968)		$-90·2$
	Na$_2$O	$-90·0$	Garrels and Christ (1965)		
Sodium hydroxide	NaOH	$-91·188$	JANAF Tables (1965)		
	NaOH	$-90·1$	Garrels and Christ (1965)		$-90·1$
Lime	CaO	$-144·352 \pm 0·340$	Robie and Waldbaum (1968)		$-144·4$
	CaO	$-144·4$	Garrels and Christ (1965)		
Carbon dioxide	CO$_2$ (ideal gas)	$-94·257 \pm 0·04$	Robie and Waldbaum (1968)		$-94·3$
	CO$_2$ (ideal gas)	$-94·24$	Mel'nik (1972)		
	CO$_2$ (ideal gas)	$-94·2598$	Garrels and Christ (1965)		
Water (liquid)	H$_2$O (liquid)	$-56·688 \pm 0·02$	Robie and Waldbaum (1968)		$-56·69$
Quartz	SiO$_2$	$-204·646 \pm 0·410$	Robie and Waldbaum (1968)		$-204·6$
	SiO$_2$	$-204·76$	Mel'nik (1972)		
Silica glass	SiO$_2$ (glass)	$-203·298 \pm 0·510$	Robie and Waldbaum (1968)		$-203·3$
	Amorphous hydrous gel	$-203·23$	Mel'nik (1972)		
Minnesotaite	Fe$_3^{2+}$Si$_4$O$_{10}$(OH)$_2$	$-1069·9$	Mel'nik (1972)		$-1070·0$
Talc	Mg$_3$Si$_4$O$_{10}$(OH)$_2$	$-1324·486 \pm 1·710$	Robie and Waldbaum (1968)		
	Mg$_3$Si$_4$O$_{10}$(OH)$_2$	$-1323·28$	Mel'nik (1972)		
	Mg$_3$Si$_4$O$_{10}$(OH)$_2$	$-1319·0$	Hostetler et al. (1971)		
	Mg$_3$Si$_4$O$_{10}$(OH)$_2$	$-1320·0 \pm 2·0$	Bricker et al. (1973)		$-1320·0$
Chrysotile	Mg$_3$Si$_2$O$_5$(OH)$_4$	$-965·77$	Mel'nik (1972)		
	Mg$_3$Si$_2$O$_5$(OH)$_4$	$-962·08 \pm 0·68$	Bricker et al. (1973)		
	Mg$_3$Si$_2$O$_5$(OH)$_4$	$-964·92 \pm 1·0$	Hostetler and Christ (1968)		$-965·0$
Sepiolite	[2MgO·3SiO$_2$·1·5H$_2$O]·2H$_2$O	$-1105·6$	Christ et al. (1973)		
	Mg$_2$Si$_3$O$_6$(OH)$_4$	$-1020·5$	Authors; from Christ et al. (1973)		$-1020·5$†
Pyrophyllite	Al$_2$Si$_4$O$_{10}$(OH)$_2$	$-1260·0$	Zen (set 1) (1972)		$-1260·0$
	Al$_2$Si$_4$O$_{10}$(OH)$_2$	$-1253·0$	Zen (set 2) (1972)		
Annite	KFe$_3^{2+}$AlSi$_3$O$_{10}$(OH)$_2$	$-1145·8$	Beane (1973)	$-3·5$	$-1149·3$
	KFe$_{0·3}^{2+}$Fe$_{2·7}^{3+}$AlSi$_3$O$_{12}$H$_{1·7}$	$-1148·2 \pm 6$	Zen (1973)	$-3·5$	$-1151·7$
Phlogopite	KMg$_3$AlSi$_3$O$_{10}$(OH)$_2$	$-1407·4$	Beane (1973)	$-3·5$	$-1410·9$

Table 1. (Continued)

Substance	Formula	ΔG_f°	Author	Al corr.* (kcal)	ΔG_f° selected for this paper
Kaolinite	$Al_2Si_2O_5(OH)_4$	−901·4	Karpov and Kashik (1968)		
	$Al_2Si_2O_5(OH)_4$	−902·868 ± 0·960	Robie and Waldbaum (1968)		
	$Al_2Si_2O_5(OH)_4$	−910·0	Zen (set 1) (1972)		−910·0
	$Al_2Si_2O_5(OH)_4$	−902·87	Zen (set 2) (1972)		
	$Al_2Si_2O_5(OH)_4$	−903·3	Kittrick (1971a)		
	$Al_2Si_2O_5(OH)_4$	−903·5	Kittrick (1971a)		
Muscovite	$KAl_3Si_3O_{10}(OH)_2$	−1327·5	Karpov and Kashik (1968)		
	$KAl_3Si_3O_{10}(OH)_2$	−1330·103 ± 1·320	Robie and Waldbaum (1968)	−10·5	−1340·6
	$KAl_3Si_3O_{10}(OH)_2$	−1340·0	Zen (set 1) (1972)		
	$KAl_3Si_3O_{10}(OH)_2$	−1330·0	Zen (set 2) (1972)		
Paragonite	$NaAl_3Si_3O_{10}(OH)_2$	−1326·0	Karpov and Kashik (1968)		
	$NaAl_3Si_3O_{10}(OH)_2$	−1328·0	Zen (set 1) (1972)		−1328·0
	$NaAl_3Si_3O_{10}(OH)_2$	−1318·0	Zen (set 2) (1972)		
Chlorite	$Mg_5Al_2Si_3O_{10}(OH)_8$	−1974·0	Zen (1972)		−1974·0
	$Mg_5Al_2Si_3O_{10}(OH)_8$	−1954·8	Helgeson (1969)	−7·0	−1961·8
Microcline	$KAlSi_3O_8$	−887·03	Karpov and Kashik (1968)		
	$KAlSi_3O_8$	−892·817 ± 0·970	Robie and Waldbaum (1968)	−3·5	−896·3
Potassium ion	K^+ (aq)	−67·7 ± 0·1	Robie and Waldbaum (1968)		−67·7
	K^+ (aq)	−67·466	Garrels and Christ (1965)		
Sodium ion	Na^+ (aq)	−62·539 ± 0·05	Robie and Waldbaum (1968)		−62·5
	Na^+ (aq)	−62·589	Garrels and Christ (1965)		
Magnesium ion	Mg^{2+} (aq)	−108·90 ± 0·20	Robie and Waldbaum (1968)		−108·9
	Mg^{2+} (aq)	−108·8	Mel'nik (1972)		
	Mg^{2+} (aq)	−108·99	Garrels and Christ (1965)		
Aluminum ion	Al^{3+} (aq)	−116·0 ± 0·3	Robie and Waldbaum (1968)	−3·5	−119·5
	Al^{3+} (aq)	−115·0	Garrels and Christ (1965)		
Dissolved silica	$H_4SiO_4^\circ$ (aq)	−312·72	Mel'nik (1972)		
	$H_4SiO_4^\circ$ (aq)	−312·5	Bricker et al. (1973)		−312·5
	$H_4SiO_4^\circ$ (aq)	−313·0 ± 0·34	Kittrick (1971a)		
Hydroxyl ion	OH^- (aq)	−37·594 ± 0·01	Robie and Waldbaum (1968)		−37·6
	OH^- (aq)	−37·59	Mel'nik (1972)		
Hydrogen ion	H^+ (aq)	0·0	Robie and Waldbaum (1968)		0·0
Calcium ion	Ca^{2+} (aq)	−132·180 ± 0·200	Robie and Waldbaum (1968)		−132·2
	Ca^{2+} (aq)	−132·18	Garrels and Christ (1965)		

* A correction of −3·5 kcal/mol has been made for species containing aluminum, except for those designated Zen (set 1), which are already consistent with the correction. This change was made as a result of a redetermination of ΔG_f° of gibbsite by Robie (1973).

† The ΔG_f° we derive from the value of Christ et al. (1973) is for a 'fictive' sepiolite. We assumed their K_{sp} at 298·15 K for the reaction $[2MgO \cdot 3SiO_2 \cdot 1·5H_2O] \cdot 2H_2O + 4·5H_2O_1 = 2Mg_{aq}^{2+} + 3H_4SiO_{4\ aq} + 4OH_{aq}^-$ is the same as that for the reaction

$$Mg_2Si_3O_6(OH)_4 + 6H_2O_1 = 2Mg_{aq}^{2+} + 3H_4SiO_{4\ aq} + 4OH_{aq}^-.$$

We then obtained $\Delta G_f^\circ Mg_2Si_3O_6(OH)_4$ from the second reaction. The validity of the assumption is supported in the sense that if the method of prediction developed here is used to predict ΔG_f° of the Christ et al. sepiolite formula, assuming that the 2H₂O outside the brackets have the ΔG_f° of liquid water, the value obtained is −1106·8 kcal/mol, as compared with their value of −1105·6 kcal/mol.

Note: Agreement among various investigators is generally good. However, an important selection for our purposes was the choice of Bricker et al.'s value for talc (−1320 kcal/mol), as opposed to the more negative values of Mel'nik (−1323·3 kcal/mol) and of Robie and Waldbaum (−1324·5 kcal/mol). The Bricker et al. number is close to that of Hostetler et al. (−1319 kcal/mol). Also, use of the number of Bricker et al. gives slightly better prediction than use of the more negative numbers for talc.

For talc:

$$\Delta G_{f,sil}^\circ Mg(OH)_2 + 2\Delta G_{f,sil}^\circ MgO + 4\Delta G_{f,sil}^\circ SiO_2 = -1320 \text{ kcal.}$$

For sepiolite:

$$2\Delta G_{f,sil}^\circ Mg(OH)_2 + 3\Delta G_{f,sil}^\circ SiO_2 = -1020·5.$$

Solving the simultaneous equations yields:

$$\Delta G_{f,sil}^\circ Mg(OH)_2 = -203·3 \text{ kcal/mol}$$
$$\Delta G_{f,sil}^\circ MgO = -149·2 \text{ kcal/mol}$$
$$\Delta G_{f,sil}^\circ SiO_2 = -204·6 \text{ kcal/mol,}$$

where 'sil' refers to the Gibbs energy of formation from the elements within the silicate structure.

ALUMINIUM SILICATES

In the AlSiOH system, there are two phases with measured $\Delta G_f°$ values, namely kaolinite and pyrophyllite. By carrying forward $\Delta G°_{f,sil}SiO_2$ from the MgSiOH system, values of $\Delta G°_{f,sil}$ for the components can be obtained.

For kaolinite:

$$\Delta G°_{f,sil}Al_2O_3 + 2\Delta G°_{f,sil}SiO_2 + 2\Delta G°_{f,sil}H_2O = -910 \text{ kcal.}$$

For pyrophyllite:

$$\Delta G°_{f,sil}Al_2O_3 + 4\Delta G°_{f,sil}SiO_2 + \Delta G°_{f,sil}H_2O = -1260 \text{ kcal.}$$

Solving, using $-204·6$ kcal for $\Delta G°_{f,sil}SiO_2$:

$$\Delta G°_{f,sil}Al_2O_3 = -382·4 \text{ kcal/mol}$$
$$\Delta G°_{f,sil}H_2O = -59·2 \text{ kcal/mol.}$$

It is also possible to write aluminum silicates so as to obtain a $\Delta G°_{f,sil}$ for $Al(OH)_3$, by assigning the hydroxyl to Al. For kaolinite:

$$\tfrac{4}{3}\Delta G°_{f,sil}Al(OH)_3 + \tfrac{1}{3}\Delta G°_{f,sil}Al_2O_3 + 2\Delta G°_{f,sil}SiO_2 = -910$$

$$\Delta G°_{f,sil}Al(OH)_3 = -280·0.$$

MICAS

$\Delta G_f°$ values are available for muscovite and for paragonite. Following the previous convention, we write:

For muscovite:

$$\tfrac{1}{2}\Delta G°_{f,sil}K_2O + \tfrac{3}{2}\Delta G°_{f,sil}Al_2O_3 + 3\Delta G°_{f,sil}SiO_2 + \Delta G°_{f,sil}H_2O = -1340·5.$$

Solving, using preceding values:

$$\Delta G°_{f,sil}K_2O = -188·0 \text{ kcal/mol.}$$

For paragonite:

$$\tfrac{1}{2}\Delta G°_{f,sil}Na_2O + \tfrac{3}{2}\Delta G°_{f,sil}Al_2O_3 + 3\Delta G°_{f,sil}SiO_2 + \Delta G°_{f,sil}H_2O = -1328.$$

Solving:

$$\Delta G°_{f,sil}Na_2O = -162·8 \text{ kcal/mol.}$$

It is of interest to calculate the micas with the alternative components KOH or NaOH, using up the excess hydroxyl as $Al(OH)_3$. The calculation yields:

$$\Delta G°_{f,sil}KOH = -123·6 \text{ kcal/mol}$$
$$\Delta G°_{f,sil}NaOH = -111·0 \text{ kcal/mol.}$$

FERROUS IRON SILICATES

Gibbs energy data are available only for minnesotaite, the ferrous equivalent of talc. As before,

$$3\Delta G°_{f,sil}FeO + 4\Delta G°_{f,sil}SiO_2 + \Delta G°_{f,sil}H_2O = -1070.$$

Substituting preceding values for $\Delta G^\circ_{f,\text{sil}}\text{SiO}_2$ and $\Delta G^\circ_{f,\text{sil}}\text{H}_2\text{O}$,

$$\Delta G^\circ_{f,\text{sil}}\text{FeO} = -64\cdot 1 \text{ kcal.}$$

Comparison of $\Delta G^\circ_{f,\text{sil}}$ and ΔG°_f of Components

It is now possible to compare ΔG°_f of the components of the silicates as individual phases with ΔG°_f of the components in the silicate structure (Table 2). Table 2 also contains $\Delta G^\circ_{f,\text{sil}}$ values that will be derived later.

Table 2. Free energies of formation from the elements at 298·15 K of components of layer silicates as individual phases and as 'silicated' phases (values in kcal/mol)

Component	A $\Delta G^\circ_{f\text{ free}}$	B $\Delta G^\circ_{f(\text{silicated})}$	B − A $\Delta G^\circ_{\text{silication}}$
K_2O	−77·0	−188·0	−111·0
Na_2O	−90·2	−162·8	−72·6
MgO	−136·1	−149·2	−13·1
FeO	−60·1	−64·1	−4·0
Fe_2O_3	−177·7	−177·7	0·0
Al_2O_3	−378·1	−382·4	−4·3
SiO_2	−204·6	−204·6	0·0
H_2O	−56·69	−59·2	−2·5
$Mg(OH)_2$	−199·5	−203·3	−3·8
$Al(OH)_3$	−277·0	−280·0	−3·0
$FeOOH$	−117·0	−117·0	0·0
KOH	−90·9	−123·6	−32·7
$NaOH$	−90·1	−111·0	−20·9
K_2O	−77·0	−188·0 ex*	−111·0
CaO	−144·4	−182·8 ex	−38·4
Na_2O	−90·2	−175·4 ex	−85·2
MgO	−136·1	−159·5 ex	−23·4
H_2O	−56·69	−58·6 ex	−1·9
Li_2O	−134·3	−190·6 ex	−56·3

* ex refers to the ΔG°_f value for ions in exchange sites, treating them as oxides.

Figure 1 is a plot of the values of $\Delta G^\circ_{\text{silication}}$ of oxides ($\Delta G^\circ_{f,\text{sil}} - \Delta G^\circ_{f\text{·free}}$) from Table 1 vs the electronegativity of the cations. Various scales of electronegativity are available, but the differences between them are slight. The values used here are from Gordy and Thomas (1956). The 'silication' energies of the alkali metal oxides are large; it is clear that metal ions with large electronegativity values have oxides with small silication energies. The graph is similar to that obtained by what might be termed the 'hydroxylation' energy of these oxides. Figure 2 shows values of ΔG°_f of some hydroxides minus ΔG°_f of the corresponding oxides, again plotting against electronegativity. The similarity of the relations for 'silication' in Fig. 1 and 'hydroxylation' in Fig. 2 suggests that 'silication' energies for other oxides could be predicted fairly closely.

Ferric Iron Silicates

Because of the absence of ΔG°_f data on ferric iron silicates, Fig. 1 must act as a guide to the silication energy of Fe_2O_3. The electronegativity of Fe^{3+} is the same

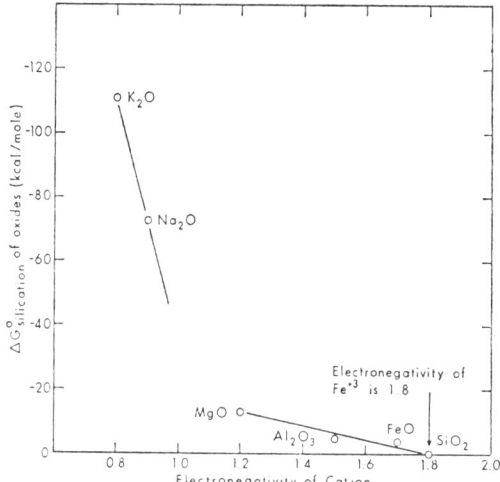

Fig. 1. Difference of Gibbs energy of formation from the elements in kcal/mol of various oxides within silicate structures ($\Delta G°_{f,\text{sil}}$) and free energy of formation of oxides as free phases, plotted against electronegativities of cations.

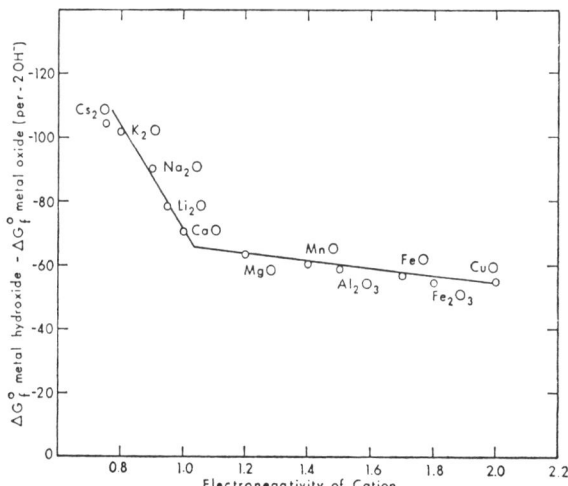

Fig. 2. Difference in Gibbs energies of formation from the elements in kcal/mol of metal hydroxides and energy of formation of corresponding oxides (per 2 OH$^-$) plotted against electronegativities of the cations.

as that of Si^{4+} (1·8), so the implication is clear that $\Delta G°_{\text{silication}}$ must be close to zero for Fe$_2$O$_3$. This conclusion is borne out by the behavior of the iron oxides. A controversy has continued for years as to whether Fe$_2$O$_3$ or FeOOH is the stable phase at room temperature in water, so that $\Delta G°$ for the hydroxylation of Fe$_2$O$_3$ is certainly close to $-56·7$ kcal/mol.

It is concluded that little error will be made by assuming that $\Delta G°_{\text{silication}}$ of Fe$_2$O$_3$ is zero. However, the pure ferric silicates, ferric minnesotaite and ferric greenalite, written using the components Fe$_2$O$_3$, SiO$_2$ and H$_2$O, should be slightly favored with respect to breakdown to hematite, quartz and water, a conclusion not in harmony with the absence of such phases in nature. On the other hand, natural montmorillonites with as many as 2/3 of the octahedral positions occupied by Fe^{3+} have deen described (WILDMAN et al., 1968), so the end-members containing only Fe_2O_3 should not be far from equilibrium with their breakdown products.

In summary, the $\Delta G°_{silication}$ value for Fe_2O_3 is that predicted from Fig. 1, and it appears that the error in the values is small, probably less than a kilocalorie per mol.

EXCHANGE ENERGIES

One of the steps in determining the free energy of formation of a compound is an evaluation of the energy associated with the interlayer cations. In most exchange studies, reactions are written as,

$$Na_2X + 2K^+ = K_2X + 2Na^+,$$

where $X^{(2-)}$ denotes the entire negative framework. For our purposes, we convert the formula of a typical montmorillonite from the usual form, i.e.

$$Mg_{0.4}[(Al_{1.5}Fe^{3+}_{0.2}Mg_{0.3})(Al_{0.5}Si_{3.5})O_{10}(OH)_2]$$

to

$$0.4Mg^{2+}O^{2-}_{ex} \cdot Al_2O_3 \cdot 0.1Fe_2O_3 \cdot 0.3Mg(OH)_2 \cdot 3.5SiO_2 \cdot 0.7H_2O.$$

Writing the exchangeable cation as an oxide, even though it is under the influence of the net negative charge of the entire framework, is in a sense a device for substituting an 'equivalent O^{2-}' for the negative framework.

Differences in standard $\Delta G_f°$ of 'exchangeable oxides' can be determined from the abundant ion exchange data in the literature. Most of the studies that have been made involve only two cations. The values for exchange constants are about the same for illites and montmorillonites (TRUESDELL and CHRIST, 1968). In Table 3, column B, the differences in $\Delta G_f°$ of H_2O_{ex}, Na_2O_{ex}, CaO_{ex}, K_2O_{ex} and Li_2O_{ex}, compared to MgO_{ex} as an arbitrary zero, are listed from our estimates of the best values in the literature. Both DREVER (1971) and RUSSELL (1970) measured the cations occupying the exchangeable sites of their marine clays. Column A shows the differences in Gibbs energies of the exchange oxides, using their data for the ratios of cations in the exchange positions and the composition of standard seawater to calculate the exchange constants. Activity coefficients for the seawater species were taken from GARRELS and CHRIST (1965). The correspondence between columns A and B is remarkable, for it shows that nearly the same $\Delta G°_{f,ex}$ values are obtained whether a simple 2-ion exchange is being studied, or whether the exchange sites have significant percentages of as many as 4 cations. It might have been anticipated that multiple occupation of the exchange sites would result in marked non-ideality of behavior for some of the cations.

Column C of Table 3 is based on the assumption that $\Delta G°_{ex}K_2O$ is identical to $\Delta G°_{sil}K_2O$ derived from muscovite, and hence permits assigning $\Delta G°_{ex}$ values for exchangeable species by using exchange constants between K_2O_{ex} and the other cations. The validity of this assumption rests upon a small value of ΔG for the reaction:

$$K^+_{hydrated} \text{ [silicate framework]} = K^+_{dry} \text{ [silicate framework]} + nH_2O. \qquad (1)$$

Table 3. Differences in $\Delta G_f°$ of layer silicates as a function of occupancy of negative sites by various cations. All values are in kilocalories per 2 equivalents of exchange sites. In Columns A and B occupancy of sites by Mg^{2+} is arbitrarily assigned at zero. Column C gives the free energies of formation of various oxides *in exchange sites* ($\Delta G_{ex}°$), based on the values of Column B and the assumption that $\Delta G_{sil}° K_2O = \Delta G_{ex}° K_2O$.

	A* Marine clays	B† Bi-ionic systems	C $\Delta G_{ex}°$
H_2O_{ex}	?	+100.9	−58.6
MgO_{ex}	0.0	0.0	−159.5
Na_2O_{ex}	−14.5	15.9	−175.4
CaO_{ex}	−23.5	−23.3	−182.8
K_2O_{ex}	−28.5	−28.5	−188.0
Li_2O_{ex}	?	−31.1	−190.6

* Average values from 22 samples by DREVER (1971) and from a composite of samples by RUSSELL (1970).

† Based on equilibrium constants:
$\log[K^+]/[H^+] = 2.2$ TRUESDELL and CHRIST (1968)
$\log[K^+]/[Na^+] = -0.8$ TRUESDELL and CHRIST (1968)
$\log[K^+]/[Li^+] = -0.9$ TRUESDELL and CHRIST (1968)
and MARTIN and LAUDELOUT (1963)
$\log[K^+]^2/[Ca^{2+}] = -1.45$ LAUDELOUT et al. (1968)
$\log[K^+]^2/[Mg^{2+}] = -1.45$ DEIST and TALIBUDEEN (1967).

Exchange constants are determined for reactions such as:

$$K^+_{hydrated} \text{[silicate framework]} + Na^+_{hydrated} =$$
$$Na^+_{hydrated} \text{[silicate framework]} + K^+_{hydrated}.$$

When both ions retain all or part (?) of their hydration shells when in the interlayer positions of montmorillonites, exchange is rapid. As interlayer charge increases, potassium tends to become 'fixed,' presumably as a result of a reaction like that suggested by (1) above. Such 'fixed' potassium is slow to exchange. However, muscovite reacts with pure water to release K^+ and fix H^+ (GARRELS and HOWARD, 1957) and despite the fact that the reaction is not entirely reversible, an exchange constant for $\log[K^+]/[H^+]$ of about 10^4 is indicated. Consequently, H^+ is more effective in displacing K^+ from mica than it is in displacing K^+ from illites and montmorillonites. For montmorillonite K^+–H^+ exchange, $\log[K^+]/[H^+] = 2.2$ (Table 3). The overall implication is that fixation of potassium involves a small energy change, perhaps of the order of 1 or 2 kcal, but that the loss of its bound water decreases the interlayer spacing, and makes displacement of 'dry' K^+ by other 'wet' cations kinetically difficult. In other words, at low total charge of the layer structure, reaction (1) is displaced to the left (montmorillonites); whereas with increasing charge (illites) it goes to the right, but the driving force in either direction is small.

The use of different $\Delta G_f°$ values (except for K_2O) for interlayer 'oxides,' as opposed to the values for the same oxides in other structural positions, has as a precedent the extensive use of the same device by organic chemists, who have assigned different $\Delta G_f°$ values to organic units identical compositionally but different structurally (cf. Cox and Pilcher, 1970).

Estimation of $\Delta G_f°$ Values for Montmorillonites and Illites

A check on the numbers derived for the $\Delta G°_{sil}$ of various components in layer silicate structures can now be made by applying them to the values of $\Delta G_f°$ for these species that have been derived from solubility studies by various investigators. Kittrick (1971a, b), Routson and Kittrick (1971) and Weaver et al. (1971) have published solubility data for montmorillonites and illites, employing various aqueous media for their studies.

Routson and Kittrick (*op. cit.*) worked with illites in alkaline solutions, in which K^+ was unquestionably the exchange ion; Kittrick (1971a, b) used acid solutions, in which our analysis of the exchange ion occupancy indicates that it was H^+. Among their experiments, Weaver et al. (*op. cit.*) used two solutions spiked with magnesium. Our computations for the exchange cations for their experiments, using the values of Table 3, indicates occupation of the exchange sites by hydrogen ion rather than magnesium, in disagreement with their assumption.

We recalculated the solubility data of these authors using the $\Delta G_f°$ values of the species listed in Table 1. We also made the assumption that the iron released from the montmorillonites and illites was in equilibrium with a ferric oxide with a $\Delta G_f°$ of about -170 kcal, rather than the $-177\cdot7$ kcal/mol for precipitated Fe_2O_3 (the value for coarsely crystallized hematite) used by all these investigators.

Langmuir and Whittemore (1971) have shown that ferric oxides age slowly after precipitation, showing a range of stability increase of about 10 kcal/mol as a maximum. The experiments of the various investigators were carried out over various time intervals; we have assumed that the precipitated Fe_2O_3 (or FeOOH ?) was considerably less stable than coarsely crystalline hematite.

A sample calculation shows how the $\Delta G_f°$ values of illites and montmorillonites were obtained from solubility data. The data of Weaver et al. (1971) for the Colony, Wyoming montmorillonite are used. Our method is essentially identical to theirs, but because we have made some changes in $\Delta G_f°$ values of the species they used and also decided that their montmorillonite was in the H^+ form, the calculation bears repetition.

For the dissociation into solution of the montmorillonite:

$$H^+_{0\cdot40}[(Al_{1\cdot52}Fe^{3+}_{0\cdot22}Mg_{0\cdot29})(Al_{0\cdot19}Si_{3\cdot81})O_{10}(OH)_2]_c + 5\cdot71H^+_{aq} + 3\cdot57H_2O_l$$
$$= 1\cdot71Al^{3+}_{aq} + 3\cdot81H_4SiO_{4aq} + 0\cdot11\ Fe_2O_{3\ poorly\ crystallized} + 0\cdot29Mg^{2+}_{aq}.$$

The equilibrium constant is

$$K_{eq} = \frac{[Mg^{2+}]^{0\cdot49}[Al^{3+}]^{1\cdot71}[H_4SiO_4]^{3\cdot81}}{[H^+]^{5\cdot71}}$$

assuming the activities of montmorillonite, poorly crystallized Fe_2O_3, and H_2O are unity. Substituting the values of Weaver et al. (1971, p. 827) for the activities of the dissolved species:

$$\log K_{eq} = 0.991; \quad \Delta G_r^\circ = -1.352 \text{ kcal.}$$

$$\Delta G_r^\circ = -1.35 = 0.29 \Delta G_f^\circ Mg^{2+} + 1.71 \Delta G_f^\circ Al^{3+} + 3.81 \Delta G_f^\circ H_4SiO_4$$
$$+ 0.11 \Delta G_f^\circ Fe_2O_3 - \Delta G_f^\circ \text{ montmorillonite} - 3.57 \Delta G_f^\circ H_2O$$

(values from Table 1)

$$\Delta G_f^\circ \text{ montmorillonite} = -1241.5 \text{ kcal/mol.}$$

The other values listed in the table were calculated similarly.

Predicted Free Energies of Formation from Oxide Components

It is now possible to compare the ΔG_f° values calculated from solubility measurements with those predicted by summing the oxide components. A representative calculation for a montmorillonite, by summing the oxide components, is illustrated in Table 5 for Colony montmorillonite of Table 4. Table 6 shows the Fithian illite

Table 4. ΔG_f° of montmorillonites and illites recalculated from solubility data, using the ΔG_f° values of species listed in Table 1

Mineral	Structural formula	ΔG_f° (kcal/mol)	References*
Illites			
Fithian	$K_{0.64}^+[(Al_{1.54}Fe_{0.29}^{3+}Mg_{0.19})(Al_{0.49}Si_{3.51})O_{10}(OH)_2]$	−1277.7	Routson and Kittrick (1971)
Beavers Bend	$K_{0.53}^+[(Al_{1.66}Fe_{0.20}^{3+}Mg_{0.18})(Al_{0.38}Si_{3.62})O_{10}(OH)_2]$	−1274.7	Routson and Kittrick (1971)
Goose Lake	$K_{0.59}^+[(Al_{1.58}Fe_{0.24}^{3+}Mg_{0.15})(Al_{0.35}Si_{3.65})O_{10}(OH)_2]$	−1272.1	Routson and Kittrick (1971)
Montmorillonites			
Colony	$K_{0.40}^+[(Al_{1.52}Fe_{0.22}^{3+}Mg_{0.29})(Al_{0.19}Si_{3.81})O_{10}(OH)_2]$	−1241.5	Weaver et al. (1971)
Aberdeen	$H_{0.415}^+[(Al_{1.29}Fe_{0.335}^{3+}Mg_{0.445})(Al_{0.18}Si_{3.82})O_{10}(OH)_2]$	−1230.6	Kittrick (1971b)
Bell Fourche	$H_{0.28}^+[(Al_{1.515}Fe_{0.225}^{3+}Mg_{0.29})(Al_{0.065}Si_{3.935})O_{10}(OH)_2]$	−1240.6	Kittrick (1971a)

* Original solubilities used in calculations, but ΔG_f° values of species are taken from Table 1.

rewritten in terms of oxides. Table 7 shows that the method of estimation of Gibbs energies of layer silicates gives numbers within the uncertainty of the measured values.

Table 5. Calculation of ΔG_f° of colony montmorillonite from oxide components (values used are from Column B, Table 2)

Component	Moles	ΔG_{sil}°/mol	ΔG°	ΔG_f° from solubility
H_2O_{ex}	0.20	−58.6	−11.7	
$SiO_{2\ sil}$	3.81	−204.6	−779.5	
$Al_2O_{3\ sil}$	0.86	−382.4	−328.9	
$Mg(OH)_{2\ sil}$	0.29	−203.3	−59.0	
$Fe_2O_{3\ sil}$	0.11	−177.7	−19.5	
$H_2O_{\ sil}$	0.71	−59.2	−42.0	
			−1240.6	−1241.5

Table 6. Calculation of ΔG_f° of Fithian illite from oxide components
(values used are from Column B, Table 2)

Component	Moles	ΔG_f°/mol	ΔG_f°	ΔG_f° from solubility
K_2O_{ex}	0·32	−188·0	−60·0	
$SiO_{2\ sil}$	3·51	−204·6	−718·1	
$Al_2O_{3\ sil}$	1·015	−382·4	−388·1	
$Mg(OH)_{2\ sil}$	0·19	−203·3	−38·6	
H_2O_{sil}	0·81	−59·2	−48·0	
$Fe_2O_{3\ sil}$	0·145	−177·7	−25·8	
			−1278·8	−1277·7

A calculation based on observed mineral relations may serve as further second order evidence that the calculated ΔG_f° values give useful numbers. Illite, K-spar and quartz appear to be a stable assemblage in ancient shales. On the other hand, montmorillonite and K-spar occur frequently in marine sediments and seem to persist indefinitely (up to 10^8 yr) with no suggestion of increase of illitic layers in sediments that contain interstitial waters supersaturated with respect to quartz. COUTURE (1973) determined that montmorillonitic deep-sea sediments average about 30 ppm of dissolved SiO_2 in their interstitial waters.

The formulae and free energies of formation of 'average' montmorillonite and 'end-member' illite (GARRELS and MACKENZIE, in press) are given in Table 7. For the reaction

average montmorillonite + K-spar = end member illite + dissolved silica

$K_{0·40}[(Mg_{0·34}Fe^{3+}_{0·17}Fe^{2+}_{0·04}Al_{1·50})(Al_{0·17}Si_{3·83})O_{10}(OH)_2] + 0·4KAlSi_3O_8$
$= K_{0·80}[(Mg_{0·34}Fe^{3+}_{0·17}Fe^{2+}_{0·04}Al_{1·50})(Al_{0·57}Si_{3·43})O_{10}(OH)_2] + 1·6H_4SiO_{4aq}.$

The calculated equilibrium silica value is 18 ppm as SiO_2. Thus the calculation shows that montmorillonite–K-spar should be stable in systems with pore waters above 18 ppm, but if quartz precipitates, and lowers the silica below 18 ppm (equilibrium value for quartz is 6 ppm), illite interlayers should form at the expense of K-spar. PERRY and HOWER (1970) demonstrated the formation of illite from montmorillonite at the expense of K-spar in Gulf Coast sediments with increasing depth.

DISCUSSION

The validity of the method proposed for calculating the Gibbs energies of layer and fibrous silicates rests chiefly at the moment on its success in predicting the experimental values for illites and montmorillonites, as well as on the good agreement between the calculated value and those for annite (Table 7). A number of points need discussion and clarification.

Grain size and degree of crystallinity

The ΔG_f° values for the end-members used in establishing ΔG_{sil}° for their components are presumably all values for well-crystallized materials. Use of these

Table 7. Comparison of measured $\Delta G_f°$ 298·15 K values for minerals with those estimated from assignment of constant values of $\Delta G°_{sil}$ 298·15 K of their components. $\Delta G_f°$ measured from Tables 1 and 4, $\Delta G°_{sil}$ for components from Table 2. $\Delta G_f°$ values in kcal/mol from the elements

Mineral	Composition	$\Delta G_f°$ measured or calculated from experiments	Uncertainty (estimated by authors)	$\Delta G_f°$ estimated
A. Minerals used to obtain $\Delta G°_{sil}$ of components				
Talc	$Mg_3Si_4O_{10}(OH)_2$	−1320·0	±5	Same as measured
Chrysotile	$Mg_3Si_2O_5(OH)_4$	−965·0	±1	Same as measured
Sepiolite	$Mg_2Si_3O_6(OH)_4$	−1020·5	±1	Same as measured
Kaolinite	$Al_2Si_2O_5(OH)_4$	−910·0	±1	Same as measured
Pyrophyllite	$Al_2Si_4O_{10}(OH)_2$	−1260·0	±2	Same as measured
Muscovite	$KAl_3Si_3O_{10}(OH)_2$	−1340·5	±2	Same as measured
Paragonite	$NaAl_3Si_3O_{10}(OH)_2$	−1328·0	±2	Same as measured
Minnesotaite	$Fe_3Si_4O_{10}(OH)_2$	−1070·0	±5	Same as measured
B. Measured and predicted values				
Fithian illite		−1277·7	±2	−1278·8
Beavers Bend illite		−1274·7	±2	−1276·2
Goose Lake illite		−1272·1	±2	−1273·4
Aberdeen montmorillonite	See Table 4	−1230·6	±2	−1228·3
Belle Fourche montmorillonite		−1240·6	±2	−1236·4
Colony montmorillonite		−1241·5	±2	−1240·6
Annite	$KFe_3^{2+}AlSi_3O_{10}(OH)_2$	−1149·3	±5	−1150·2
	$KFe_{0.3}^{3+}Fe_{2.7}^{2+}AlSi_3O_{12}H_{1.7}$	−1151·7	±6	−1149·1
Clinochlore	$Mg_5Al_2Si_3O_{10}(OH)_8$	−1961·8 (Helgeson)	±?	−1958·7
		−1974·0 (Zen)	±?	
Phlogopite	$KMg_3AlSi_3O_{10}(OH)_2$	−1410·9	±?	−1400·7
C. Predicted values				
Greenalite	$Fe_3^{2+}Si_2O_5(OH)_4$			−720·0
'Ferric' minnesotaite	$Fe_2^{3+}Si_4O_{10}(OH)_2$			−1055·0
'Average' montmorillonite*	$K_{0.40}[(Mg_{0.34}Fe_{0.17}^{3+}Fe_{0.04}^{2+}Al_{1.50})(Al_{0.17}Si_{3.83})O_{10}(OH)_2]$			−1266·5
'End member' illite*	$K_{0.80}[(Mg_{0.34}Fe_{0.17}^{3+}Fe_{0.04}^{2+}Al_{1.50})(Al_{0.57}Si_{3.43})O_{10}(OH)_2]$			−1299·0

* GARRELS and MACKENZIE, In S.E.P.M. Special Publication 20, *Studies in Paleo-oceanography*.

members to predict $\Delta G_f°$ values for other phases should give values applicable only to similarly well-crystallized materials. Furthermore, the composition of the phase to be predicted must be known accurately. Poorly crystallized silicates may have $\Delta G_f°$ values several kilocalories more positive than well-crystallized ones, and when grain sizes for most minerals drop below about 1 μm, surface energies also become important in the total stability of the substance (cf. KITTRICK's 1966 study of kaolinites). In fact, the agreement between predicted and experimental values for montmorillonites and illites is 'too good'; we expected to find much larger differences than those shown in Table 7.

We assumed that grain size and degree of crystallinity differences of the experimental materials might cause differences of several kilocalories between calculated and measured values, and that there might be additional discrepancies caused by analytical errors in determining the compositions of the phases. For example, if the moles of SiO_2 reported in a montmorillonite were 3·79 instead of 3·82, the difference in $\Delta G_f°$ calculated could be of the order of 5 kcal.

Experimental determination of $\Delta G_f°$ from solubilities

Another factor that made the good agreement between calculated and measured values for montmorillonites and illites suspect was the experimental problem of being certain that the solutions used by the experimenters to calculate $\Delta G_f°$ values were truly in equilibrium with the starting phases, and not simply in equilibrium between solution and some intermediate phase formed incongruently and metastably during the experiment. Kittrick, in particular, was apparently well aware of this problem, and we refer the reader to his studies to see how he treated this question. However, there is a residual problem in many solubility studies of a possible change in composition between original and final composition of experimental minerals. For example, a potassium montmorillonite placed in a strong $MgCl_2$ solution would be expected to take Mg^{2+} immediately into the exchange positions, and eventually to respond further by increase of octahedral Mg by some complicated mechanism, perhaps involving loss of Fe^{3+} as Fe_2O_3 with concomitant increase in total charge and change in gross composition. At any rate, the final phase equilibrated with the solution would not have the same composition as the initial phase, and Gibbs energies calculated on the basis of initial composition would be changed.

Also, there is serious question as to whether montmorillonites retain H^+ in their exchange positions for more than a short time, presumably being replaced by Mg and/or Al displaced from the negative framework. Despite these considerations, we found good agreement with experimental values assuming exchangeable H^+.

Structural considerations

Just as there is a difference between $\Delta G°_{sil}MgO$ (octahedral Mg), and $\Delta G°_{ex}MgO$ (Mg on interlayer sites) for the components of the exchangeable silicates, there should be other differences related to structural positions; e.g. between aluminum in tetrahedral sites and aluminum in octahedral positions. As far as we can tell, these differences are not determinable without more accurate values for measured $\Delta G_f°$ values of the various minerals. Furthermore, if the difference between the predicted value for phlogopite and Beane's value (Table 7) is correct, the implication is that the value of $\Delta G°_{sil}Mg(OH)_2$ that we have used is too positive when brucite interlayers are present. In fact, the behavior of Mg has puzzled us throughout our calculations: we have no physical explanation of why the rule of assigning as much Mg as possible to the component $Mg(OH)_{2\,sil}$ seems to work. It is our hope, that as more accurate and consistent values become available for the phases we have listed, and that as values for other phases are measured, a more sophisticated method of calculation, based on structural considerations, can be developed. SLAUGHTER (1966a, b, c) has shown that calculations based entirely on bonding considerations can give approximate values for $\Delta G_f°$ of silicate minerals.

The success of the method we have proposed, limited as it is in terms of good checks on predictions, suggests that a relatively simple method, more soundly based on structural considerations, can be developed to yield calculated Gibbs energies of formation of high accuracy.

SUMMARY

A method of estimating the Gibbs energies of formation of layer silicates has been developed. It is based on the assumption that the oxide and hydroxide components of these silicates have fixed Gibbs energies of formation within the silicates. The general rule is that all components are treated as oxides, with the exception of Mg, which is treated insofar as possible as $Mg(OH)_2$ in all silicates.

Acknowledgements—We are deeply grateful to E-AN ZEN, C. L. CHRIST and R. ROBIE, U.S. Geological Survey, D. WALDBAUM, Princeton University, C. WEAVER and K. BECK, Georgia Institute of Technology, J. I. DREVER and W. GUNTER, University of Wyoming, J. A. KITTRICK and S. V. MATTIGOD, Washington State University, R. G. BURNS, Massachusetts Institute of Technology, and R. SPEED, Northwestern University, for reading the manuscript and making many helpful comments and criticisms, as well as drawing our attention to sources of data we had missed. We have done our best to incorporate their suggestions into the final manuscript. Dr. ROBERT EDDY of Tufts University was particularly helpful in getting us on the track of using oxide and hydroxide components as the basis for estimation.

This work was supported by funds from the University of Hawaii and the Petroleum Research Fund of the American Chemical Society (RMG), and by GRANT GA 29207 of the National Science Foundation (RMG and YT), and by a NATO Fellowship (YT). The many typings and much of the bibliographic work were done by CYNTHIA GARRELS.

REFERENCES

BEANE R. (1973) Personal communication.
BRICKER O. P., NESBITT H. W. and GUNTER W. D. (1973) The stability of talc. *Amer. Mineral.* **58**, 64–72.
CHRIST C. L., HOSTETLER P. B. and SIEBERT R. M. (1973) Studies in the system $MgO-SiO_2CO_2-H_2O$ (III): the activity-product constant of sepiolite. *Amer. J. Sci.* **273**, 65–83.
COUTURE R. (1973) Personal communication.
COX J. D. and PILCHER G. (1970) *Thermochemistry of Organic and Organometallic Compounds*, 648 pp. Academic Press.
DEIST J. and TALIBUDEEN O. (1967) Thermodynamics of K–Ca ion exchange in soils. *J. Soil Sci.* **18**, 138–148.
DREVER J. I. (1971) Early diagenesis of clay minerals, Rio Ameca Basin, Mexico. *J. Sediment. Petrol.* **41**, 982–994.
FISHER J. R. and ZEN E-AN (1971) Thermochemical calculations from hydrothermal phase equilibrium data and the free energy of H_2O. *Amer. J. Sci.* **270**, 297–314.
GARRELS R. M. and CHRIST C. L. (1965) *Solutions, Minerals and Equilibria*, 450 pp. Harper & Row.
GARRELS R. M. and HOWARD P. (1957) Reactions of feldspar and mica with water at low temperature and pressure. *Proc. 6th Nat. Conf. Clays and Clay Minerals*, pp. 68–88. Pergamon Press.
GARRELS R. M. and MACKENZIE F. T. Chemical history of the oceans as deduced from postdepositional history of sedimentary rocks. Paper given at S.E.P.M. Ann. Conv. Symposium on History of the Oceans, Houston, Texas, 1971. In S.E.P.M. Special Publication 20, *Studies in Paleo-oceanography*.

GORDY W. and THOMAS W. J. O. (1956) Electronegativities of the elements. *J. Phys. Chem.* **24**, 439–444.
HELGESON H. C. (1969) Thermodynamics of hydrothermal systems at elevated temperatures and pressures. *Amer. J. Sci.* **267**, 729–804.
HOSTETLER P. B. and CHRIST C. L. (1968) Studies in the system $MgO-SiO_2-CO_2-H_2O$ (I): the activity-product constant of chrysotile. *Geochim. Cosmochim. Acta* **32**, 482–497.
HOSTETLER P. B., HEMLEY J. J., CHRIST C. L. and MONTOYA J. W. (1971) Talc–chrysotile equilibrium in aqueous solutions. *Geol. Soc. Amer. Abstr.* **3**, 605.
JANAF (1965) (Joint Army–Navy–Air Force Thermochemical Panel) *JANAF Thermochemical Tables.* Sponsored by Project Principia of the Advanced Research Projects Agency through Air Force contract AF 04(611)-7554, by D. R. STULL, project director (and others) at The Thermal Research Laboratory, Dow Chemical Co., Midland, Michigan. Distributed by Clearinghouse for Federal Scientific and Technical Information.
KARPOV I. K. and KASHIK S. A. (1968) Computer calculation of standard isobaric–isothermal potentials of silicates by multiple regression from a crystallochemical classification. *Geochem. Int.* **5**, 706–713.
KITTRICK J. A. (1966) Free energy of formation of kaolinite from solubility measurements. *Amer. Mineral.* **51**, 1457–1466.
KITTRICK J. A. (1971a) Stability of montmorillonites: I. Belle Fourche and Clay Spur montmorillonites. *Soil Sci. Soc. Amer. Proc.* **35**, 140–145.
KITTRICK J. A. (1971b) Stability of montmorillonites: II. Aberdeen montmorillonite. *Soil Sci. Soc. Amer. Proc.* **35**, 820–823.
LANGMUIR D. (1969) The Gibbs free energies of substances in the system $Fe-O_2-H_2O-CO_2$ at 25°C. *U.S. Geol. Surv. Prof. Paper* **650-B**, B180–B184.
LANGMUIR D. and WHITTEMORE D. O. (1971) Variations in the stability of precipitated ferric oxyhydroxides. In *Advances in Chemistry Series*, No. 106, *Nonequilibrium Systems in Natural Water Chemistry.* Amer. Chem. Soc.
LAUDELOUT H., VAN BLADEL R., BOLT G. H. and PAGE A. L. (1968) Thermodynamics of heterovalent cation exchange reactions in a montmorillonite clay. *Trans. Faraday Soc.* **64**, 1477–1488.
MARTIN H. and LAUDELOUT H. (1963) Thermodynamique de l'échange des cations alcalins dans les argiles. *J. Chim. Phys.* **60**, 1086–1099.
MEL'NIK Y. P. (1972) *Thermodynamic Constants for the Analysis of Conditions of Formation of Iron Ores*, 193 pp. Institute of the Geochemistry and Physics of Minerals, Academy of Sciences, Ukrainian S.S.R., Kiev. (In Russian.)
PERRY E. A., JR. and HOWER J. (1970) Burial diagenesis in Gulf Coast pelitic sediments. *Clays Clay Mineral.* **18**, 165–177.
ROBIE R. A. (1973) Personal communication.
ROBIE R. A. and WALDBAUM D. R. (1968) Thermodynamic properties of minerals and related substances at 298·15°K (25°C) and one atmosphere (1·013 bars) pressure and at higher temperatures. *U.S. Geol. Surv. Bull.* **1259**, 256 pp.
ROUTSON R. C. and KITTRICK J. A. (1971) Illite solubility. *Soil Sci. Soc. Amer. Proc.* **35**, 714–718.
RUSSELL K. L. (1970) Geochemistry and halmyrolysis of clay minerals. *Geochim. Cosmochim. Acta* **34**, 893–908.
SLAUGHTER M. (1966a) Chemical binding in the silicate minerals. Part I. Models for determining crystal-chemical properties. *Geochim. Cosmochim. Acta* **30**, 299–313.
SLAUGHTER M. (1966b) Chemical binding in the silicate minerals. Part II. Computational methods and approximations for the binding energy of complex silicates. *Geochim. Cosmochim. Acta* **30**, 315–322.
SLAUGHTER M. (1966c) Chemical binding in the silicate minerals. Part III. Application of energy calculations to the prediction of silicate mineral stability. *Geochim. Cosmochim. Acta* **30**, 323–339.
TRUESDELL A. H. and CHRIST C. L. (1968) Cation exchange in clays interpreted by regular solution theory. *Amer. J. Sci.* **266**, 402–412.
WEAVER R. M., JACKSON M. L. and SYERS J. K. (1971) Magnesium and silicon activities in

matrix solutions of montmorillonite-containing soils in relation to clay mineral stability. *Soil Sci. Soc. Amer. Proc.* **35,** 823–830.

WILDMAN, W. E., JACKSON M. L. and WHITTIG L. D. (1968) Iron-rich montmorillonite formation in soils derived from serpentinite. *Soil Sci. Soc. Amer. Proc.* **32,** 787–794.

ZEN E-AN (1972) Gibbs free energy, enthalpy and entropy of ten rock-forming minerals: Calculations, discrepancies, implications. *Amer. Mineral.* **57,** 524–553.

ZEN E-AN (1973) Thermochemical parameters of minerals from oxygen-buffered hydrothermal equilibrium data: method, application to annite and almandine. *Contrib. Mineral. Petrol.* **39,** 65–80.

17

Copyright © 1969 by Microforms International Marketing Corporation, as exclusive copyright licensee of Pergamon Press journal back files
Reprinted from Geochim. Cosmochim. Acta **33**:455–481 (1969)

Evaluation of irreversible reactions in geochemical processes involving minerals and aqueous solutions—II. Applications

Harold C. Helgeson, Robert M. Garrels and Fred T. Mackenzie

Department of Geology, Northwestern University, Evanston, Illinois 60201

(*Received* 19 *June* 1968; *accepted in revised form* 4 *October* 1968)

Abstract—Equilibrium relations among common rock-forming minerals and aqueous solutions over a range of temperatures and pressures are known experimentally for a number of systems and can be calculated for others. This information permits prediction of the mass transfer involved in chemical reactions characteristic of geochemical processes. Calculations of this kind are used to examine various chemical and geologic implications of irreversibility in idealized models of weathering, evaporative concentration, diagenesis, hydrothermal rock alteration, and ore deposition.

Introduction

Part I of this two-part contribution (Helgeson, 1968b) consists of a summary of thermodynamic principles, definitions, and relations pertinent to evaluation of irreversible chemical reactions in geologic systems involving aqueous solutions. In the present communication these principles are employed to evaluate quantitatively the consequences of irreversibility in a variety of geochemical processes.

Equilibrium calculations define limiting conditions for real processes. In recent years, sufficient thermodynamic and chemical data have become available to permit calculation of the free energy changes associated with a large number of reversible reactions in geologic systems (Helgeson, 1969). Despite uncertainties in these data, equilibrium diagrams can now be constructed for a substantial number of systems involving aqueous solutions at temperatures to 300°C. The chemical composition of the aqueous phase in these systems often can be assessed from analytical data for natural solutions such as river water, sea water, ground water, pore water, fluid inclusions, and thermal springs. The analytical data can be converted into activities of components with the aid of appropriate activity coefficients, which are now predictable for high temperatures and high concentrations of electrolytes (Helgeson, 1967b, 1969; Helgeson and James, 1968). The framework of equilibrium relations defined by these data makes it possible to compare natural systems with idealized equilibrium models, and to evaluate the compositional changes involved in a variety of irreversible processes of geologic interest.

Various chemical aspects of irreversibility in weathering, evaporative concentration, diagenesis, hydrothermal rock alteration, and ore deposition are considered in the following pages. *In all cases, the chemical models presented are necessarily hypothetical*; i.e. the assumption is made that partial equilibrium (Barton *et al.* 1963; Helgeson, 1968b) is maintained among the aqueous species and product minerals in the geochemical processes being considered. Further, all original reactant minerals are taken to be present in excess and various metastable phases that might form in the actual process are not included in the calculations. Nevertheless, the results of the calculations suggest that theoretical consideration of irreversibility may contribute significantly to a better understanding of the chemistry of geologic processes.

The assumption that partial equilibrium is maintained among the product minerals and the aqueous phase makes it possible to predict quantitatively the mass transfer and distribution of species in geochemical processes without explicit provision for the kinetics of reaction, except specification of relative reaction rates for multiple reactant minerals. The calculations are carried out for a given starting composition of an aqueous phase in a system undergoing change. Activity coefficients are required to define partial or overall equilibrium states, and specific provision must be included for ions and complexes in the aqueous solution, the appearance and disappearance of stable and metastable minerals, and the effects of solid solution on the changes in the composition of the aqueous phase. However, where solid solubility among the minerals is slight, omitting provision for solid solution usually introduces negligible errors in calculations of the mass transfer involved in geochemical processes. Most of the calculations presented below were carried out with a computer to provide simultaneously for the variables discussed above. The activity of H_2O was taken as unity in the calculations because uncertainties introduced by this assumption are considered insignificant (HELGESON, 1967b, 1969).

WEATHERING

The processes of weathering an igneous rock can be regarded as a combination of hydrolysis reactions on the one hand, and acid attack by CO_2-charged soil waters on the other. For granitic rocks, the changes in composition of the aqueous phase involved in such reactions are represented in the diagrams depicted in Figs. 1–3, 5 and 6. Equilibrium phase relations in the systems K_2O–Al_2O_3–SiO_2–H_2O and Na_2O–Al_2O_3–SiO_2–H_2O at 25°C and 1 atm are shown in Figs. 2 and 3, and the changes in the number of moles or mass of minerals produced and destroyed/1000 grams of H_2O during the hydrolysis of K-feldspar and coexisting K-feldspar and albite are depicted in Figs. 1, 4 and 5.

Because hydrolytic reactions in the weathering process are almost always accompanied by acid attack, weathering in most environments cannot be represented by a simple hydrolysis model. Such a model constitutes a limiting approximation of weathering which is probably applicable only in areas of high rainfall, little vegetation, and good drainage. Nevertheless, it is of interest to compute the mass transfer attending hydrolysis of minerals in order to provide a basis for comparison with calculations for more complicated models. For this purpose, we shall first consider the hydrolysis of K-feldspar and albite separately, and then examine hydrolytic reactions for the two coexisting minerals. To define limiting cases, we shall regard these minerals as pure phases and consider the temperature and pressure to be 25°C and 1 atm, respectively.

Hydrolysis of K-feldspar

Disregarding the exchange of H^+ for K^+ on the surface of the reacting mineral (CORRENS and VON ENGLEHARDT, 1938), the initial hydrolysis of K-feldspar can be represented by

$$KAlSi_3O_8 + 8H_2O \rightarrow K^+ + Al(OH)_4^- + 3H_4SiO_4 \tag{1}$$

and the mole transfer contributing to a given state in the progress of the reaction can be calculated from a truncated Taylor's expansion such as

$$\Delta m_{\bar{s}} = \bar{n}_{\bar{s}} \Delta \xi + \frac{\bar{n}_{\bar{s}}'(\Delta \xi)^2}{2!} \qquad (2)$$

or

$$\Delta \bar{x}_{\phi} = \bar{n}_{\phi} \Delta \xi + \frac{\bar{n}_{\phi}'(\Delta \xi)^2}{2!} \qquad (3)$$

where $m_{\bar{s}}$ and $\bar{n}_{\bar{s}}$ are the molality and reaction coefficient (per 1000 g of H_2O) for the $\bar{s}$th aqueous species in the irreversible reaction, $\bar{x}_{\phi}$ is the number of moles/1000 g of H_2O of the ϕth mineral in the system, $\bar{n}_{\phi}$ refers to the subscripted reaction coefficient/1000 g of H_2O of the mineral in the reaction, $\bar{n}_{\bar{s}}'$ and $\bar{n}_{\phi}'$ are the derivatives of $\bar{n}_{\bar{s}}$ and $\bar{n}_{\phi}$ with respect to ξ, the progress variable for the reaction, and $\Delta \xi$ refers to a small increment of reaction progress (HELGESON, 1968b).

Reaction (1) is an approximation of the actual hydrolysis of K-feldspar. It is written in terms of the predominant species in the aqueous phase; however, a number of partial equilibrium states can be assumed to hold in the solution, and these equilibria affect the reaction coefficients in the actual hydrolysis reaction. The important partial equilibrium states are

$$Al(OH)^{2+} \rightleftarrows Al^{3+} + OH^- \qquad (4)$$

$$Al(OH)_4^- \rightleftarrows Al^{3+} + 4OH^- \qquad (5)$$

$$H_4SiO_4 \rightleftarrows H_3SiO_4^- + H^+ \qquad (6)$$

$$H_2O \rightleftarrows H^+ + OH^-. \qquad (7)$$

By combining differential equations derived from statements of the Law of Mass Action for these reactions with mass balance relations for the elements involved in reaction (1), nonsingular matrices of linear equations can be set up to compute values of the reaction coefficients ($\bar{n}$) and their derivatives ($\bar{n}'$) for the actual hydrolysis reaction (HELGESON, 1968b). Computer evaluation of the matrix equations together with mass transfer calculations using equations (2) and (3) result in the changes in solution composition depicted in Fig. 1. The changes in composition shown in Fig. 1 were calculated for successive small increments of progress in the hydrolysis of K-feldspar. The identities of the minerals with which the solution becomes saturated as the irreversible reaction proceeds is determined by comparing activity products for all minerals in the system computed from activity coefficients and the molalities of aqueous species with those calculated from thermodynamic data.

As reaction (1) proceeds and the solution composition approaches that represented by line A in Fig. 1, the aqueous phase approaches saturation with respect to gibbsite. Equilibrium between gibbsite and the aqueous phase can be described by writing

$$Al(OH)_3 \rightleftarrows Al^{3+} + 3OH^- \qquad (8)$$

for which

$$a_{Al^{3+}} a_{OH^-}^3 = K_8 \qquad (9)$$

where K_8 is the equilibrium constant for reaction (8). The point at which the solution becomes saturated with gibbsite is determined by computing (equation 9) hypothetical values of K_8 from activity coefficients and the molalities of Al^{3+} and OH^- in solution after each increment of ξ in reaction (1). Each hypothetical value of K_8 is compared with the activity product constant for gibbsite based on experimental data ($10^{-34.03}$, KITTRICK, 1966) to define the point at which the solution becomes saturated. The exact point of saturation is found by invoking an interval halving procedure. Because in this case the aqueous phase is dilute, Debye–Hückel activity coefficients can be used to convert molalities to activities in the calculations.

In most geochemical systems involving aqueous solutions the activities of aluminum species in the aqueous phase are so low that even large relative changes in their activities do not appreciably affect the mass transfer among the solids in the system. For this reason, such a process can be viewed as one in which aluminum is essentially conserved among the solid phases (HELGESON, 1968b). With aluminum conserved, the reaction of K-feldspar with solution A (Fig. 1) can be represented as

$$KAlSi_3O_8 + 8H_2O \rightarrow Al(OH)_3 + K^+ + OH^- + 3H_4SiO_4. \quad (10)$$

Of course, in the actual reaction, a small amount of aluminum is transferred to the aqueous solution to maintain the partial equilibrium states represented by reactions (4) and (5), as well as that between gibbsite and the aqueous phase (equation 9). In a manner analogous to that described above for reaction (1), evaluation of equations (2) and (3) together with matrix equations describing the conservation of mass and partial equilibrium states in the system defines the changes in solution composition as the reaction represented by reaction 10 proceeds (interval AB in Fig. 1). It can be seen in Fig. 1 that the molality of H_4SiO_4 increases in solution from A to B, which causes equilibrium to be established between kaolinite and the solution at B. If all partial equilibrium states are maintained and K-feldspar is present in excess, the gibbsite produced in the interval AB will react at B with K-feldspar while the activity of H_4SiO_4 in solution remains constant. As this reaction proceeds, the solution, which is in equilibrium with both gibbsite and kaolinite, changes composition from B to C in Fig. 1. With aluminum conserved, the reaction taking place in the interval BC can be represented by writing

$$2KAlSi_3O_8 + 4Al(OH)_3 + H_2O \rightarrow 3Al_2Si_2O_5(OH)_4 + 2K^+ + 2OH^-. \quad (11)$$

The coefficients in reaction (11) are defined by the requirement for constant activity of H_4SiO_4 to maintain equilibrium between the solution, gibbsite and kaolinite.

Reaction (11) continues until all of the gibbsite produced between B and C in Fig. 1 has been consumed. At C (defined by equation (3) when $\Delta \bar{x}_{gibbsite} = 0$), kaolinite begins to form as a product of the continued reaction of the solution with K-feldspar. With aluminum again conserved, this reaction can be written as

$$KAlSi_3O_8 + \tfrac{11}{2}H_2O \rightarrow \tfrac{1}{2}Al_2Si_2O_5(OH)_4 + K^+ + OH^- + 2H_4SiO_4. \quad (12)$$

As a result of reaction (12), the solution changes composition from C to D in Fig. 1. At D, K-mica forms at the expense of kaolinite and K-feldspar, and the solution changes composition to E, where reaction ceases and overall equilibrium is established among K-feldspar, K-mica, kaolinite and the aqueous phase. Although quartz may

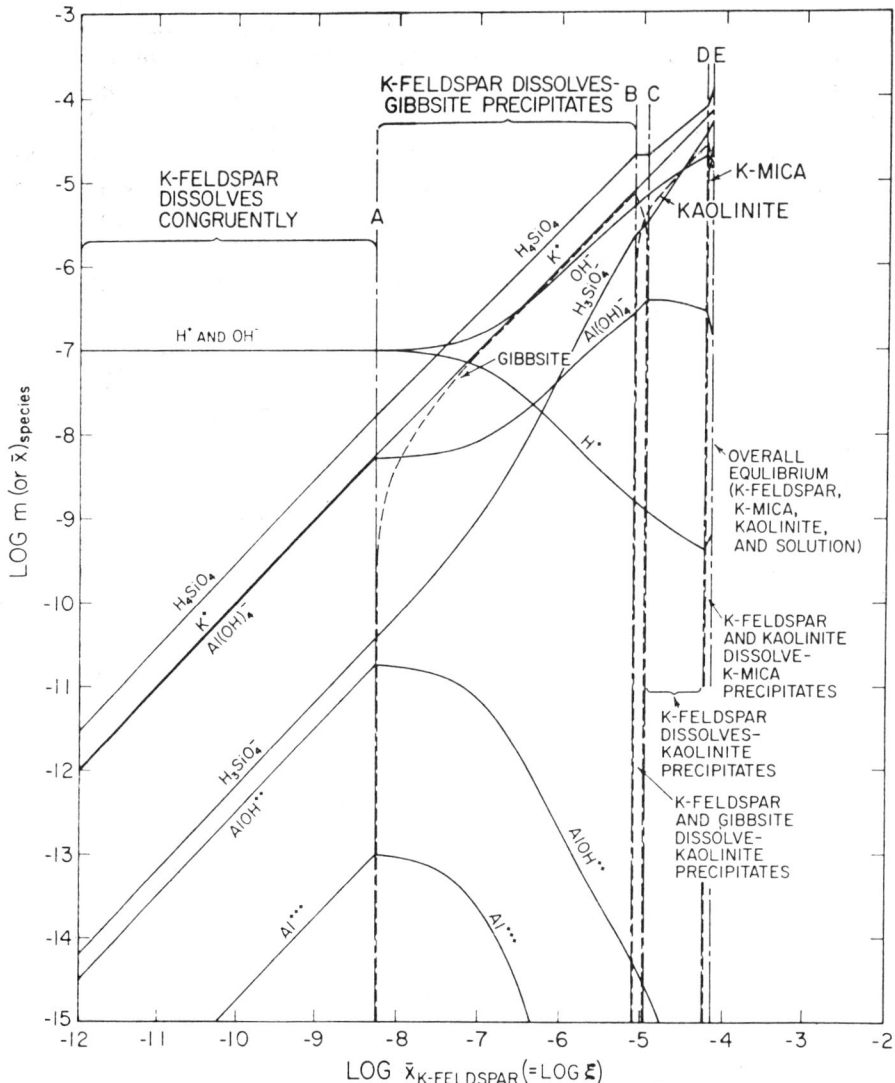

Fig. 1. Molalities (m) of species (—) in the aqueous phase and moles of minerals/1000 g of H_2O ($\bar{x}$) produced and destroyed (– – –) during the hydrolysis of K-feldspar at 25°C and 1 atm. The abscissa represents the number of moles of K-feldspar destroyed/1000 g of H_2O, which corresponds to the progress variable (ξ) for the hydrolysis reaction. The equilibrium constants employed in the calculations were taken from the sources given in the captions of Figs. 2 and 6; activity coefficients were computed from the Debye–Hückel equation (GARRELS and CHRIST, 1965). The letters designate points on reaction path ABCDE in Fig. 2.

also be produced by the reaction taking place in the interval DE, this partial equilibrium state is rarely established in the weathering process; consequently, the solution is allowed to become super-saturated with respect to quartz.

As written, reactions (10)–(12) are only approximations of the actual irreversible reaction involved in the hydrolytic process. As indicated above, the coefficients in the written reactions are not actually constants owing to the changes in the molalities of aqueous species that do not appear in the reactions as written but nevertheless contribute to the actual reaction coefficients. When these changes become large and more than one species involving a given ion is present in significant concentrations, as is the case in the interval DE in Fig. 1, no single reaction can be written to represent the compositional changes in the hydrolytic process. However, if all important partial equilibrium states are included, the influence of all such species is defined accurately in the computer calculations. The sequence of incongruent events in the hydrolysis of K-feldspar computed above appear to be consistent with those encountered experimentally by Wollast (1967).

It can be seen in Fig. 1 that the only aluminum species present in significant concentrations in the aqueous phase during the hydrolysis of K-feldspar is $Al(OH)_4^-$. In contrast, the concentration of $H_3SiO_4^-$ approaches that of H_4SiO_4 at the end of the reaction process. The species $H_3SiO_4^-$ accounts for the bulk of the charge balance required by K^+ in the interval DE in Fig. 1. It can be seen in Fig. 2 that partial equilibrium between K-mica and kaolinite requires a_{K^+}/a_{H^+} to be constant in the aqueous phase. Consequently, the pH of the solution in the interval DE in Fig. 1 decreases, which is an opposite trend to the pH change attending the preceding hydrolysis reactions. The solution composition depicted in Fig. 1 is represented in the logarithmic activity diagram in Fig. 2 by reaction path ABCDE. The stages of reaction progress are labeled with the same letters in the two figures.

The extent to which minerals are produced and destroyed by the hydrolysis of K-feldspar is shown in Fig. 1. In mass units, the overall process results in the destruction of 0·03 g of K-feldspar and production of 0·0005 g of gibbsite, 0·007 g of kaolinite, and 0·01 g of K-mica/1000 g of H_2O. In the process, all of the gibbsite and 0·005 g of the kaolinite/1000 g of H_2O produced in the early stages of hydrolysis are subsequently destroyed. These figures apply to hydrolysis in a closed system where reaction progress reaches point E in Fig. 2. In an open system, the total mass transfer would be restricted to that taking place among the phases involved in the early partial equilibrium states, whereas phases involved in later partial equilibrium states might not appear as reaction products. In contrast to the closed system, the original reactant mineral may be completely destroyed in an open system.

Hydrolysis of albite

The hydrolysis of albite is represented in Fig. 3 by reaction path ABCDEF, which is based on computer calculations like those reported above. The changes in the molalities of the aqueous species along this reaction path are similar to those accompanying the hydrolysis of K-feldspar shown in Fig. 1. It can be seen in Fig. 3 that the partial equilibrium states established among the minerals during the process are identical to those encountered in the hydrolysis of K-feldspar, except that the solution becomes saturated with respect to Na-montmorillonite instead of K-mica before overall equilibrium is achieved. The extent to which minerals are produced and/or destroyed/1000 g of H_2O along reaction path ABCDEF in Fig. 3 is illustrated in Fig. 4.

It can be seen in Fig. 4 that the hydrolysis of albite results in destruction of 0·3 g of albite/1000 g of H_2O. In the process, 0·0005 g of gibbsite and 0·02 g of

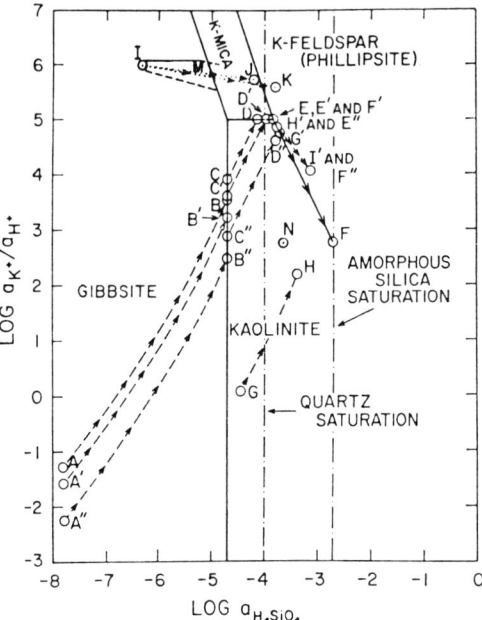

Fig. 2. Activity diagram for the system K_2O–Al_2O_3–SiO_2–H_2O at 25°C, unit activity of water, and 1 atm. The positions of the stability field boundaries shown in the diagram were calculated from thermodynamic data (BARANY, 1964; WELLER and KING, 1963; BARANY and KELLEY, 1961; KING and WELLER, 1961; WICKS and BLOCK, 1963; KELLEY and KING, 1961; WAGMAN et al., 1968; KITTRICK, 1966; LATIMER, 1952; ROSSINI et al., 1952; ROBIE, 1966; HELGESON, 1969). Where necessary, the thermodynamic data for the silicates were corrected to be consistent with $-217,650$ cal mole^{-1} for the enthalpy of formation from the elements of quartz (WISE and MARGRAVE, 1963; GOOD et al., 1964). The solubilities of amorphous silica and quartz at 25°C were taken as 115 ppm and 6 ppm respectively (MOREY et al., 1962, 1964). WALDBAUM'S (1966) values of $-946,000$ and $-937,300$ cal. mole^{-1} for the enthalpy of formation from the elements respectively of low microcline and low albite were used in the calculations along with the Third Law entropy of adularia (to represent K-feldspar) given by KELLEY et al. (1953). The free energy of gibbsite was taken to be 274,200 cal. mole^{-1} (KITTRICK, 1966). The calculations involved in constructing the diagram have been summarized elsewhere (GARRELS and CHRIST, 1965). Irreversible reaction paths (dotted and dashed lines) are shown in the diagram for the hydrolysis of K-feldspar (ABCDE) and coexisting K-feldspar and albite with relative reaction rates of 1:1 (A'B'C'D'E'F'G'H'I') and 0·1:1 (A"B"C"D"E"F"), titration of CO_2 with the equilibrium assemblage, K-feldspar + kaolinite + solution (EF), weathering of Sierra Nevada rocks (GH), and reaction of clay minerals with Bermuda sea water (IJ and IK)—see text. The area labeled M designates the compositional range of surface sea water and point N represents the *average* composition of world streams.

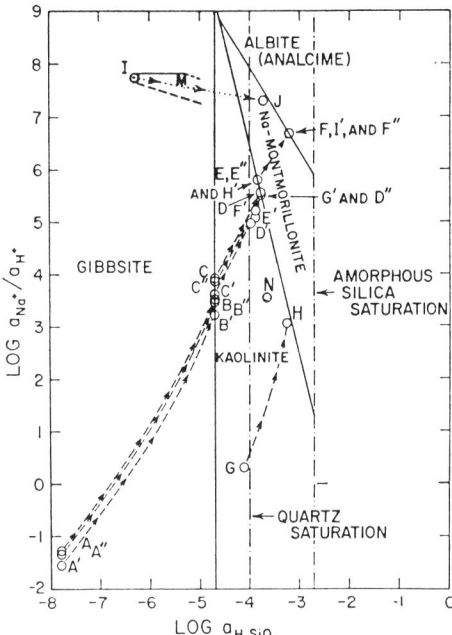

Fig. 3. Activity diagram for the system Na_2O–Al_2O_3–SiO_2–H_2O at 25°C, unit activity of water, and 1 atm. The sources of data used in constructing the diagram are given in the caption of Fig. 2. The stability field boundaries shown for montmorillonite are thermodynamically consistent with one another, but they are based on analyses of waters issuing from sediments that reportedly contain coexisting montmorillonite and kaolinite (ALTSCHULER et al., 1963; HEM, 1966; WHITE et al., 1963; BRICKER and GARRELS, 1965; GARRELS and MACKENZIE, 1967). Irreversible reaction paths (dotted and dashed lines) are shown in the diagram for the hydrolysis of albite (ABCDEF) and coexisting K-feldspar and albite with relative reaction rates of 1:1 (A′B′C′D′E′F′G′H′I′) and 0·1:1 (A″B″C″D″E″F″), weathering of Sierra Nevada rocks (GH), and reaction of clay minerals with Bermuda sea water (IJ)—see text. The area labeled M designates the composition of surface sea water and point N represents the *average* composition of world streams.

kaolinite/1000 g of H_2O are made and subsequently destroyed, and 0·2 g of montmorillonite/1000 g of H_2O are produced. This mass transfer is substantially greater than that attending the hydrolysis of K-feldspar presented above. In fact, the amount of albite destroyed and Na-montmorillonite produced/1000 g of H_2O is more than ten times the amount of K-feldspar destroyed and K-mica produced in the hydrolysis of K-feldspar.

Hydrolysis of coexisting K-feldspar and albite

To compute the extent to which the presence of one mineral affects the hydrolysis of another requires designation of the relative rate at which equal amounts of the two minerals react with the aqueous phase. The relative reaction rate can be defined as

$$\psi_{(\phi_r/\Phi_r)} = \frac{d\bar{x}_{\phi_r}/dt}{d\bar{x}_{\Phi_r}/dt} \qquad (13)$$

where $\psi_{(\phi_r/\Phi_r)}$ represents the reaction rate per cm² of surface area for the ϕ_rth reactant mineral ($\phi_r = 1, 2, 3, \ldots, \Phi_r$) relative to that of the Φ_rth reactant mineral, and t refers to time. The reaction coefficients for equal amounts of the reactant minerals in the overall reaction are then related by

$$\bar{n}_{\phi_r} = \psi_{(\phi_r/\Phi_r)} \bar{n}_{\Phi_r}. \tag{14}$$

To examine the hydrolysis of coexisting albite and K-feldspar, we shall assign alternate hypothetical values of 1·0 and 0·1 to $\psi_{\text{(K-feldspar/albite)}}$, which defines two comparative cases of possible geologic significance.

With relative reaction rates for albite and coexisting K-feldspar equal, we can represent the initial hydrolytic reaction as

$$KAlSi_3O_8 + NaAlSi_3O_8 + 16H_2O \rightarrow 2Al(OH)_4^- + K^+ + Na^+ + 6H_4SiO_4. \tag{15}$$

In the case of $\psi_{\text{(K-feldspar/albite)}} = 0\cdot 1$, reaction (15) becomes

$$0\cdot 1 KAlSi_3O_8 + NaAlSi_3O_8 + 8\cdot 8 H_2O \rightarrow$$
$$1\cdot 1 Al(OH)_4^- + 0\cdot 1 K^+ + Na^+ + 3\cdot 3 H_4SiO_4. \tag{16}$$

The complete hydrolysis of albite and coexisting K-feldspar is represented in Figs. 2 and 3 by reaction paths A'B'C'D'E'F'G'H'I' for the case of equal reaction rates, and by reaction path A"B"C"D"E"F"G" for $\psi_{\text{(K-feldspar/albite)}} = 0\cdot 1$. The extent to which minerals are produced and/or destroyed/1000 g of H_2O is depicted (with the same letter annotations) in Figs. 4 and 5. Figure 5 is a schematic "paragenesis" diagram for the case of equal reaction rates of K-feldspar and albite.

As might be predicted, the presence of albite has a dramatic effect on the behavior of K-feldspar in the hydrolytic process. It can be seen in Figs. 4 and 5 that K-feldspar equilibrates with the aqueous phase long before equilibrium is established with albite; therefore, K-feldspar becomes a reaction product during the further hydrolysis of albite. In the case of equal reaction rates, more than 0·01 g of authigenic K-feldspar/1000 g of H_2O is produced in the later stages of the hydrolytic process, which means a net destruction of only 0·0006 g of K-feldspar/1000 g of H_2O over the whole reaction path. If the reaction rate of K-feldspar is one tenth that of albite, 0·005 g of K-feldspar/1000 g of H_2O are produced in the later stages of reaction progress, but the net destruction of K-feldspar remains 0·0006 g/1000 g of H_2O. In both instances, 0·3 g of albite/1000 g of H_2O are destroyed, which is identical to the amount destroyed in the hydrolysis of albite alone, and 0·0005 g of gibbsite/1000 g of H_2O are produced and destroyed, which is the same as that produced and destroyed in the separate hydrolysis of K-feldspar and albite. The amount of kaolinite produced and destroyed is the same for the two cases of coexisting K-feldspar and albite, and it is equal to that attending the hydrolysis of albite alone (0·02 g/1000 g of H_2O). Along both reaction paths for the coexisting feldspars, 0·2 g of montmorillonite/1000 g of H_2O are produced, which is also equal to that produced by the separate hydrolysis of albite. K-mica is not a reaction product when $\psi_{\text{(K-feldspar/albite)}}$ is 0·1, but 0·002 grams of K-mica are formed and subsequently dissolved when the relative reaction rate is unity. The amount of K-mica produced and destroyed during the hydrolysis of coexisting K-feldspar and albite is only 1/5th that produced in the hydrolysis of K-feldspar alone. In contrast, three times the amount of kaolinite produced by the hydrolysis

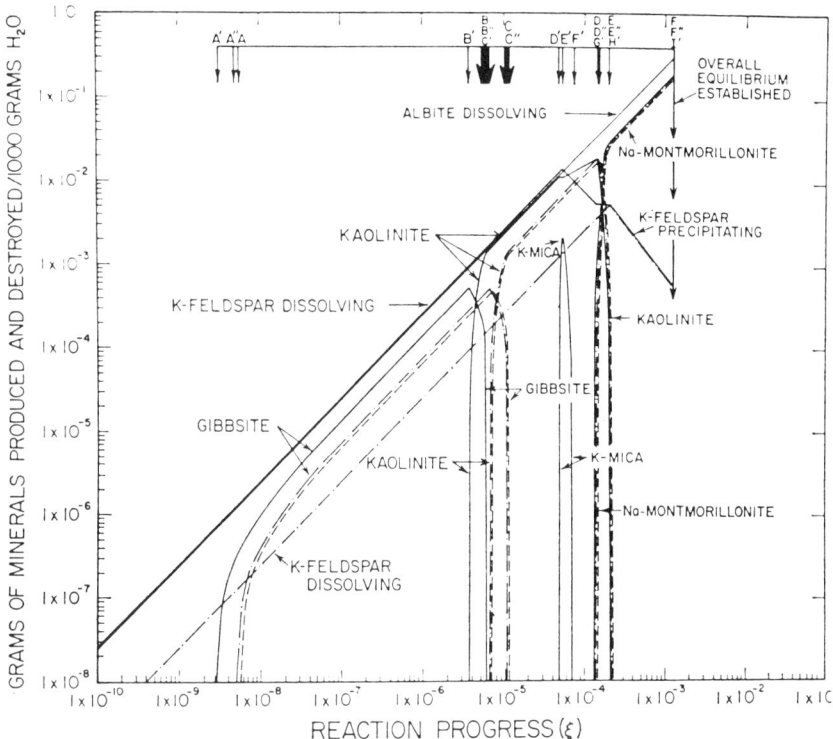

Fig. 4. Grams of minerals produced and destroyed/1000 g of H_2O during the hydrolysis of albite (– – –), and K-feldspar and coexisting albite with relative reaction rates of 1:1 (—) and 0·1:1 (— · —) as a function of the progress variable, ξ, at 25°C and 1 atm—see text. The arrows and letter annotations at the top of the diagram designate stages of reaction progress corresponding to those shown in Figs. 2 (primed letters only) and 3. The equilibrium constants employed in the mass transfer calculations were taken from the sources cited in the captions of Figs. 2, 3, and 6.

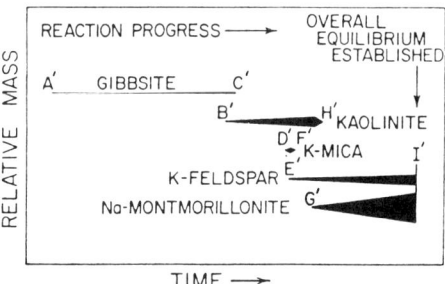

Fig. 5. Schematic diagram showing the paragenesis and relative mass of authigenic minerals produced by the hydrolysis of coexisting K-feldspar and albite with a relative reaction rate of 1:1. The letters indicate the stages of reaction progress designated in Figs. 2, 3, and 4.

237

of K-feldspar is produced and destroyed in the hydrolysis of albite or coexisting albite and K-feldspar.

Several generalizations appear justified on the basis of the mass transfer calculations summarized above:

(1) The relative amounts of original coexisting reactant minerals destroyed in irreversible reactions are not necessarily indicative of the relative rates of reaction per cm^2 of surface area of the minerals, nor of their relative abundance in the original rock.

(2) The last original reactant mineral to equilibrate controls the extent of the mass transfer among the phases, but not necessarily the sequence of phases produced in the reaction process. Thus in the cases examined above, the presence of K-feldspar has little influence on the amount of gibbsite, kaolinite, and montmorillonite produced and destroyed by the hydrolysis of albite, but it may cause the temporary appearance of K-mica, depending on the relative rates at which the reactant minerals hydrolyze.

(3) The relative amounts of the authigenic phases produced does not depend on the relative reaction rates or abundances (provided the minerals are present in excess) of the original reactant minerals, but only on the partial equilibrium states established in the system and the extent to which the last reactant mineral to equilibrate is destroyed in the reaction process.

(4) The nature of the authigenic minerals and the relative extent to which they are present in a rock can be interpreted in terms of the reaction process responsible for the mineralogic and chemical characteristics of the rock.

The changing molalities and activities of the species in the aqueous phase during the hydrolysis of K-feldspar and albite result in a final solution pH of 9·8 in all of the cases examined above, except that of K-feldspar hydrolyzing alone. In the latter instance, the final pH achieved is 9·2. The total molality of aluminum, which is almost entirely present as $Al(OH)_4^-$, in the final solution is $1·3 \times 10^{-7}$ in the case of K-feldspar hydrolysis, but $2·5 \times 10^{-8}$ in the other cases. The difference is the result of the partial extraction of aluminum from the aqueous phase to form montmorillonite. In all but one of the cases examined above, the total concentration of SiO_2 in the aqueous phase reaches 100 ppm at the end of the reaction process, 43 per cent of which is present as H_4SiO_4; in contrast, K-feldspar hydrolysis leads to a total silica concentration in the final solution of 11 ppm, 26 per cent of which is present as $H_3SiO_4^-$.

In an open system where the aqueous phase exits from the weathering profile before overall equilibrium is achieved, the total mass transfer among the phases will be correspondingly different in extent than that computed above for the closed systems. However, the authigenic mineral assemblages produced by the hydrolysis of K-feldspar and albite, and the reaction paths in Figs. 2 and 3 apply to open as well as closed systems, and they are typical of those observed in weathering profiles. Nevertheless, to characterize adequately the mass transfer involved in the actual weathering process, we must take into account the role played by CO_2 in the soil zone.

CO_2 titration

The effect of CO_2 on the hydrolytic process discussed above can be described in part by evaluating the mass transfer attending addition of CO_2 to a solution initially in equilibrium with K-feldspar and kaolinite at point E in Fig. 2. For the purpose of this calculation we shall assume the initial solution composition and amount of kaolinite present at the outset to be that produced along reaction path ABCDE in Figs. 1 and 2, and consider K-mica to be absent from the reactant mineral assemblage. If aluminum is conserved and equilibrium is maintained between kaolinite and the solution, the CO_2 titration reaction can be described in terms of the predominant species in the aqueous phase by writing

$$2KAlSi_3O_8 + \bar{n}_{CO_{2(g)}}CO_{2(g)} + (\bar{n}_{CO_{2(g)}} + 9)H_2O \rightarrow$$
$$2K^+ + 2HCO_3^- + Al_2Si_2O_5(OH)_4 + 4H_4SiO_4 + (\bar{n}_{CO_{2(g)}} - 2)H_2CO_3 \quad (17)$$

where $\bar{n}_{CO_{2(g)}}$ is the reaction coefficient for $CO_{2(g)}$ with the limit, $\bar{n}_{CO_{2(g)}} \geq 2$.

The reaction coefficient for HCO_3^- in reaction (17) is fixed by the requirement for charge balance of K^+. Because the pH of the solution decreases along the reaction path, the reaction coefficients for $CO_{2(g)}$, H_2O, and H_2CO_3 change during the titration process. For example, the average value of $\bar{n}_{CO_{2(g)}}$ in reaction (17) is 3·0 as ξ increases from 0 to $1·0 \times 10^{-4}$, but it becomes 4·6 for $\xi = 1·0 \times 10^{-4}$ to $2·0 \times 10^{-4}$, 8·2 for $\xi = 2·0 \times 10^{-4}$ to $3·0 \times 10^{-4}$, and 12·4 for $\xi = 3·0 \times 10^{-4}$ to $5·0 \times 10^{-4}$. On the other hand, $H_3SiO_4^-$ contributes negligibly to the reaction process so that $\bar{n}_{H_4SiO_4}$ remains essentially constant.

Simultaneous evaluation of equation (2) and the respective Law of Mass Action equations for the dissociation of the complexes in the aqueous phase involved in reaction (17) yields the changes in molalities of K^+, H_4SiO_4, HCO_3^-, and other species illustrated in Fig. 6. With K-feldspar present in excess, the composition of the solution changes from E to F in Figs. 2 and 6 as $\Sigma\ CO_2$ increases in solution; at point F, amorphous silica begins to precipitate from the aqueous phase. Although aluminum was conserved in balancing reaction (17), it can be seen in Fig. 6 that the molalities of Al^{3+}, $A(OH)^{2+}$, and $Al(OH)_4^-$ in solution change as the reaction progresses. However, the extent of this change is insignificant with respect to the total mass transfer of aluminum between K-feldspar and kaolinite. The overall CO_2 titration reaction depicted in Figs. 2 and 6 results in the destruction of 0·3 g of K-feldspar and formation of 0·1 g of kaolinite/1000 g of H_2O. The relative importance of acid attack compared to hydrolysis in the weathering process is reflected by the fact that this mass transfer is approximately ten times that computed above for the hydrolysis of K-feldspar.

The region to the right of line QQ' in Fig. 6 represents chemical environments typical of soil profiles in which $CO_{2(g)}$ is a significant constituent of the soil atmosphere. The important role played by atmospheric CO_2 in the weathering process is also evident in Fig. 6. It can be deduced by inspection of the curves to the left of line QQ' that atmospheric CO_2 increases the mass transfer resulting from hydrolysis of K-feldspar by a factor of three in the production of kaolinite.

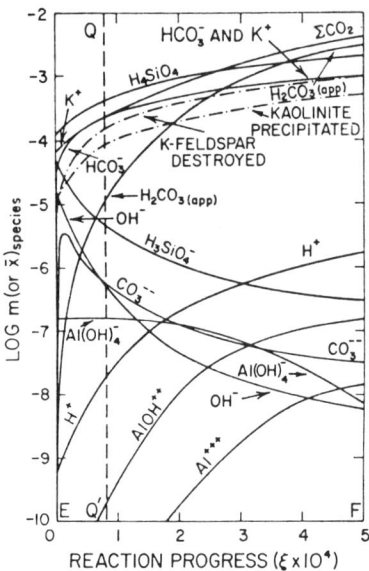

Fig. 6. Concentrations of species in an aqueous phase in equilibrium with K-feldspar and kaolinite as CO_2 is added to the solution at 25°C and one atmosphere. The variable ξ on the abscissa is the progress variable for the overall titration reaction—see text. The left and right extremities of the diagram correspond to E in Figs. 1 and 2, and F in Fig. 2, respectively; i.e. the initial solution and moles of kaolinite/1000 g of H_2O shown on the ordinate are those produced by the hydrolysis of K-feldspar represented in Figs. 1 and 2. Because the aqueous phase is dilute, unit activity coefficients were used in the calculations. The equilibrium constants required for construction of the diagram were taken from (or computed from data given by) SILLÉN and MARTELL (1964), COBBLE (1964), HEM and ROBERSON (1967), KITTRICK (1966), LATIMER (1952), WAGMAN et al. (1968), and HELGESON (1969). The vertical dashed line labeled QQ' designates the composition of the solution that would be in equilibrium with the Earth's atmosphere ($P_{CO_2} = 10^{-3.5}$ atm). The dash–dot curves define the number of moles/1000 g of H_2O of K-feldspar destroyed and kaolinite produced by the addition of CO_2 to the solution.

Weathering in the Sierra Nevada

The mass transfer involved in the hydrolysis of K-feldspar, albite, and K-feldspar + albite, and that resulting from the CO_2 titration discussed above can now be compared with the compositional changes in a more general model of the weathering process. For this purpose, we shall consider the system K_2O–Na_2O–CaO–MgO–Al_2O_3–SiO_2–CO_2–H_2O and recast the results of the mass balance calculations presented by GARRELS and MACKENZIE (1967) for spring waters in the Sierra Nevada Mountains. These calculations define the mass transfer arising from reaction of CO_2-charged soil waters with an igneous mineral assemblage representing Sierra Nevada rocks.

To evaluate the weathering process, we shall assume the pH of the water to be 4·7 prior to reaction and let the water be charged initially with 10^{-3} moles of CO_2/1000 g of H_2O (which is consistent with a P_{CO_2} in the soil of $10^{-1.5}$ atm) as a result of bacterial oxidation of organic material in the soil profile. These values appear to be reasonable approximations for the average pH and CO_2 concentration in the fresh soil waters of the Sierra Nevada (FETH et al., 1964). The igneous rock in the Sierra Nevada can be represented by the assemblage K-feldspar + andesine ($Na_{0.62}Ca_{0.38}Al_{1.38}Si_{2.62}O_8$) + biotite [idealized as $KMg_3AlSi_3O_{10}(OH)_2$]. The composition of the aqueous phase resulting from weathering of this rock is that of the perennial spring waters in the Sierra Nevada (FETH et al., 1964; GARRELS and MACKENZIE, 1967). The pH of these waters is 6·8, and 8.95×10^{-4} moles of CO_2/1000 g of H_2O are present in solution as bicarbonate. The weathering process in the Sierra Nevada takes place in an open system, and the composition of the perennial spring waters is not an equilibrium composition for the complete mineral assemblage in the weathered rock. It is, however, consistent with partial equilibrium in the weathering profile.

Two segments of reaction paths representing changes in the composition of the solution during the weathering process in the Sierra Nevada are shown in Figs. 2 and 3 (paths GH). Because the solution is dilute, the activity ratios shown in these figures were computed from the molalities of the ions using Debye–Hückel activity coefficients. Owing to the fact that CaO and MgO are not components in the systems portrayed in Figs. 2 and 3, the remainder of the reaction paths for the weathering process cannot be shown on the diagrams. The detailed reactions and the concentrations of species in solution along the reaction paths have been presented elsewhere (GARRELS and MACKENZIE, 1967).

Mass transfer calculations indicate that apparent *average* relative reaction rates (on a mole basis) of 27:1·6:1 are required for the destruction of andesine:biotite:K-feldspar to account for the composition of the perennial spring waters reported by FETH et al. (1964). [The apparent *average* relative reaction rate is defined as the relative number of moles of reactant minerals destroyed per unit time without regard to surface area, which should not be confused with the relative reaction rate per cm^2 of surface area defined by equation (13).] The calculations reveal that the reaction of the soil waters with the igneous mineral assemblage results in the destruction of 0·005 g of K-feldspar, 0·008 g of biotite, and 0·1 g of andesine/1000 g of H_2O. In the process 0·03 g of calcium montmorillonite (idealized as $Ca_{0.17}Al_{2.33}Si_{3.67}O_{10}(OH)_2$) and 0·04 g of kaolinite/1000 g of H_2O are produced. The total amount of igneous rock destroyed in the process is thus 0·113 g/1000 g of H_2O. Of this, the product minerals account for 0·07 g/1000 g of H_2O and the remainder, 0·043 g/1000 g of H_2O (43 ppm), is carried off by the aqueous phase. The total concentration of K^+, Na^+, Ca^{2+}, Mg^{2+}, and SiO_2 observed in the perennial spring waters of the Sierra Nevada is approximately 43 ppm (FETH et al., 1964), and the calculated concentrations of the individual constituents are in close agreement with available analytical data (GARRELS and MACKENZIE, 1967).

Mass transfer calculations such as those discussed above provide a basis for assessing chemical weathering rates. This can be done by estimating infiltration rates for soil waters. For example, if we assume an average infiltration rate of 50 cm^3 of water/cm^2 of land surface/year in the Sierra Nevada (FETH et al., 1964) and take the area of the Sierra Nevada to be 50,000 km^2, the calculations summarized above indicate that the weathering process will destroy 1.2×10^5 metric tons of K-feldspar, 2.0×10^5 tons of biotite, and 2.5×10^6 tons of plagioclase/year. As a result, 7.5×10^5 metric tons of montmorillonite and 1.0×10^6 tons of kaolinite will be produced per year with a net release of 6.6×10^5 tons of silica to the aqueous phase. If the original igneous rock contains 33 per cent plagioclase by weight and has a density of 2.7 g/cm^3, this mass transfer is consistent with chemical weathering of the igneous terrain in the Sierra Nevada to a depth of 56 m in one million years. The calculated concentration of silica released by the weathering process is approximately 26 ppm, which agrees with the analytical mean of the concentration of dissolved silica reported by FETH et al. (1964) for stream waters in the Sierra Nevada.

EVAPORATIVE CONCENTRATION

Chemical sedimentation is particularly well suited for simple mass transfer calculations. As an example we shall examine the precipitation of sepiolite, calcite, and amorphous silica in an evaporating lake.

An equilibrium saturation diagram for the system $CaO-MgO-SiO_2-CO_2-H_2O$ at

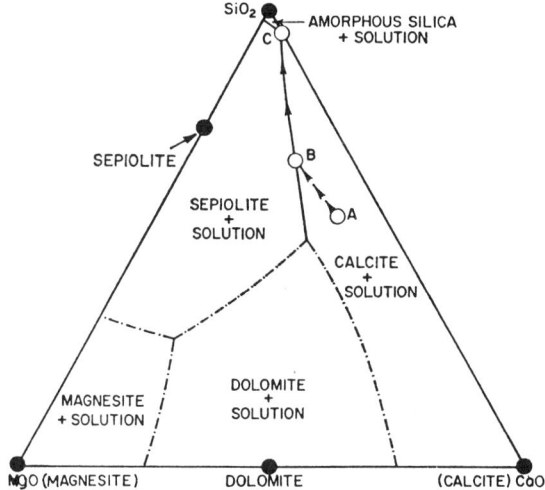

Fig. 7. Saturation diagram for the system $CaO-MgO-CO_2-SiO_2-H_2O$ at 25°C and 1 atm. Approximate positions of cotectic field boundaries are shown by dot–dash lines. Reaction path ABC represents the path of changing solution composition attending progressive evaporation of an aqueous phase with an initial composition at point A—see text and Fig. 8.

25°C and 1 atm is depicted in Fig. 7. The field boundaries (cotectics) shown in this diagram represent the intersections of saturation surfaces projected through H_2O (imagined as a tetrahedral point above the plane of the paper). The chemical composition of the perennial spring waters in the Sierra Nevada Mountains (FETH et al., 1964) is represented by point A in Fig. 7. It has been demonstrated from theoretical considerations (GARRELS and MACKENZIE, 1967) that progressive evaporation of these waters should lead first to precipitation of calcite along path AB in Fig. 7. At point B, sepiolite begins to coprecipitate with calcite according to

$$3H_4SiO_4 + Mg^{2+} \to MgSi_3O_6(OH)_2 + 2H^+ + 4H_2O. \tag{18}$$

As long as equilibrium is maintained between sepiolite, calcite, and solution, the evaporation process causes the composition of the solution to change along the cotectic BC in Fig. 7. At point C, the solution becomes saturated with respect to amorphous silica and, with further evaporation, calcite, sepiolite, and amorphous silica precipitate (in equilibrium with each other and with the aqueous phase) while the solution composition remains at point C in Fig. 7.

The changes in the molalities of the species in the aqueous phase (calculated from mass balance equations and statements of the Law of Mass Action for the complexes in solution) along the reaction path described above are illustrated in Fig. 8. The vertical dashed lines in Fig. 8 represent (from left to right) points A, B, and C in Fig. 7. The distribution of species during the "drying up" process at point C in Fig. 7 is shown in the interval between C and D in Fig. 8.

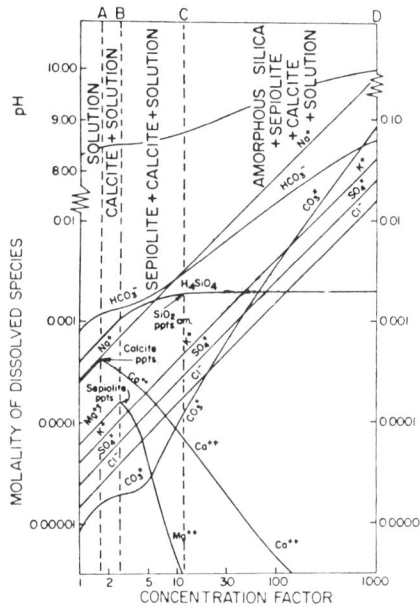

Fig. 8. Concentrations of species in the aqueous phase along evaporation path ABC in Fig. 7 (revised from GARRELS and MACKENZIE, 1967, and reproduced with permission from American Chemical Society Publications, Inc.). The vertical dashed lines denote points ABC in Fig. 7, and the curves shown in the interval CD depict changes in the molalities of aqueous species resulting from the "drying-up" process at point C in Fig. 7. The activity coefficients and equilibrium constants used to construct the diagram were taken directly, or computed from data taken from the sources cited in the captions of previous figures. In the case of sepiolite, solubility data (GARRELS and MACKENZIE, 1967) were used to calculate an approximate free energy of formation at 25°C.

Mass transfer calculations indicate that 0·026 g of calcite, 0·017 g of sepiolite, and 0·025 g of silica are precipitated from solution by evaporation of 1000 g of H_2O along path ABCD in Fig. 8. These calculations provide a basis for predicting chemical sedimentation rates for evaporating lakes in arid climates. For the purpose of making such a prediction we shall assume a steady-state flux of water through a lake with a surface area of 25 km² and consider the water influx and evaporation rate to be 90 cm/year (BRADLEY, 1929; STRAHLER, 1963). The water flux is then

2·25 × 10¹³ g/yr. We shall also assume that the composition of the waters entering the lake is that of the perennial spring waters in the Sierra Nevada (point A, Fig. 7) and that all of the evaporation takes place in the summer months so that each summer the lake behaves as though it were receiving no influx of water. Under these conditions, and if the mean annual lake temperature is 25°C, the evaporation process will result in an annual precipitation of 560 metric tons of amorphous silica, 600 metric tons of calcite, and 380 metric tons of sepiolite. The total chemical sedimentation rate (for these phases) is thus greater than 1500 metric tons/yr, which corresponds to 6 tons/cm²/10⁶ yr. This is approximately equivalent to a total depth accumulation of calcite, amorphous silica, and sepiolite in a weight ratio of 1·6:1·5:1·0 at the rate of 30 m in a million years.

DIAGENESIS

Various equilibrium models have been proposed for the world ocean (SILLÉN, 1961, 1963, 1967a,b; KRAMER, 1965). In general, these models fail to take into account differences in the chemical potentials of components from place to place in the ocean system, and by definition, they contain no provision for irreversible reactions between sea water and the sediments contained in the ocean. To be more realistic, a model of the world ocean should provide for different conditions of partial equilibrium in the various parts of the ocean system, which as a whole is almost certainly not in internal equilibrium.

It can be seen in Figs. 9 and 10 that sea water cannot be regarded as a homogeneous part of the ocean system when generalizations are made about stable mineral assemblages in sediments. Equilibrium phase relations in the systems $K_2O-Al_2O_3-SiO_2-H_2O$ and $Na_2O-Al_2O_3-SiO_2-H_2O$ at 0°C and 1 atm are depicted on activity diagrams in these figures. The areas labeled A at the ends of the shaded parallelograms represent the approximate compositional range of deep ocean waters (MOORE *et al.*, 1962; ARMSTRONG, 1965; GARRELS and THOMPSON, 1962; BERNER, 1965). In contrast, interstitial waters in modern deep sea and shelf sediments have compositions (SIEVER *et al.*, 1965; WILDE, 1966) that fall within the shaded areas labeled B, and surface sea water has a compositional range in area M. Inspection of Figs. 9 and 10 reveals that in some parts of the deep ocean, sea water contains concentrations of K^+, Na^+, H^+, and H_4SiO_4 that correspond to equilibrium activities with respect to the assemblages, K-mica (illite?) + kaolinite + Na-montmorillonite + quartz, K-mica (illite?) + Na-montmorillonite + K-feldspar + quartz, or kaolinite + Na-montmorillonite + K-feldspar. Idealized compositions of minerals are used here for the purpose of discussion, but this is not intended to imply that all equilibrium assemblages in deep ocean sediments can be described in these terms. An obvious example is montmorillonite, which may contain substantial amounts of calcium and/or magnesium. The equilibrium composition of sea water for any one of the assemblages described above differs from that of the others only by small differences in the activities of H_4SiO_4 and/or H^+. In parts of the deep ocean (area A in Figs. 9 and 10) where sea water is rich in dissolved silica, the equilibrium assemblage consists of K-feldspar (phillipsite?) + montmorillonite. It should be emphasized that the positions of the stability field boundaries shown in Figs. 9 and 10 are based on thermodynamic data and *not* on the water compositions represented by the shaded areas in the diagrams.

Although provision has not been included in Figs. 9 and 10 for solid solution and for all of the phases that occur in these systems at 0°C (and 25°C in the case of Figs. 2 and 3), the phase relations depicted in these figures define limiting conditions for provisional evaluation of the mass transfer involved in diagenetic reactions. Chlorite, illite, montmorillonite, phillipsite, and analcime are all common constituents of deep sea sediments, but the paucity of reliable thermodynamic data for these phases precludes other than qualitative provision for them in the

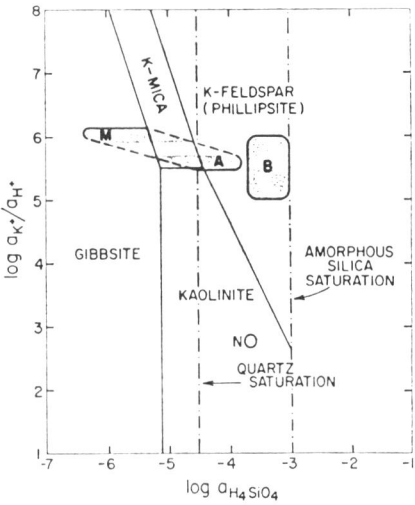

Fig. 9. Activity diagram for the system K_2O–Al_2O_3–SiO_2–H_2O at 0°C, unit activity of water, and one atmosphere. The shaded areas labeled M and A represent the compositional range of surface and deep ocean waters, respectively, and those annotated N and B designate, respectively, the *average* composition of world streams and the compositional range of interstitial waters in deep sea and shelf sediments—see text. In addition to the data cited in the caption of Fig. 2, heat capacities taken from CRISS and COBBLE (1964), KELLEY (1960, 1962), and PANKRATZ (1961) were used to calculate the positions of the stability field boundaries. Estimated heat capacities of the ions and H_4SiO_4 at 0°C were used in the calculations. The methods and uncertainties involved in calculating equilibrium constants for this system (and those represented in Figs. 10 and 11) at temperatures other than 25°C have been discussed elsewhere (GARRELS and CHRIST, 1965; HELGESON, 1967b, 1969).

present analysis. The first of these phases could be included in the diagrams by making the assumption that the chlorite in the ocean is magnesian chlorite ($Mg_5Al_2Si_3O_{10}(OH)_8$) and constructing a log $a_{Mg^{2+}}/a^2_{H^+}$ axis perpendicular to the plane of the paper in Figs. 2, 3, 9, and 10. A chlorite stability "volume" would then overlie all of the stability fields shown in these figures. It can be demonstrated that the boundaries of this chlorite stability volume (with respect to all other phases in the system) would traverse a relatively short range of $a_{Mg^{2+}}/a^2_{H^+}$ values that is probably not greatly different from the $a_{Mg^{2+}}/a^2_{H^+}$ range in sea water. Approximate locations for the stability fields of phillipsite and analcime are indicated in Figs. 9

and 10. Although not shown in the diagrams, the illite stability field occurs in the vicinity of the K-mica + K-feldspar + kaolinite assemblage, cutting off the association K-mica — kaolinite. Potassium montmorillonite occupies a stability field in Fig. 9 analogous to that for sodium montmorillonite in Fig. 10.

In accord with HOLLAND's (1965), HESS' (1966), and SILLÉN's (1967a,b) observations, there is no evidence from the equilibrium diagrams presented here that sea water is not in partial equilibrium with the mineral assemblages comprising the

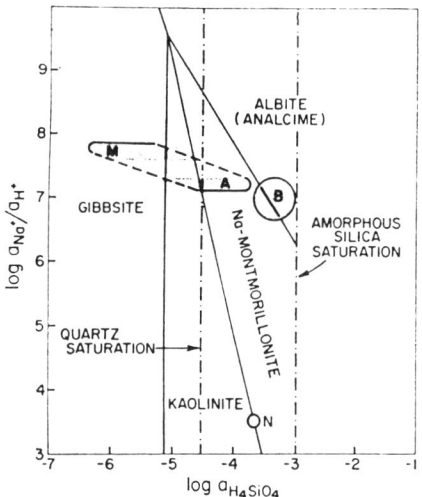

Fig. 10. Activity diagram for the system $Na_2O-Al_2O_3-SiO_2-H_2O$ at 0°C, unit activity of water, and 1 atm. The shaded areas labeled M and A represent the compositional range of surface and deep ocean waters, respectively, and those annotated N and B designate, respectively, the *average* composition of world streams and the compositional range of interstitial waters in deep sea and shelf sediments—see text. The source of data and methods used in calculating the positions of the stability field boundaries are indicated in the captions of previous figures. The standard enthalpy of formation and heat capacity of sodium montmorillonite at 25°C used in the calculations were estimated, partly on the basis of high temperature solubility data (HEMLEY et al., 1961).

bulk of deep sea sediments. Most sediments, including wackes, sands, silts, arkoses, etc., contain mineral associations for which there is an equilibrium composition of sea water or interstitial water in areas A and B in Figs. 9 and 10. However, there is no single equilibrium composition of sea water for all of the assemblages in these sediments. The major variable is the activity of silica. Region A appears to be the locus for a large number of irreversible reaction paths in the deep ocean system. It probably reflects local differences in the composition of deep ocean water that result at least in part from reactions between sea water and various silicate sediments supplied to the different parts of the ocean.

Irreversible reactions

Various checks and balances seem to operate in the deep ocean environment (in a manner analogous to the role played by carbonate equilibria in the shallow ocean) to prevent extreme local variations in the composition of sea water. Aside from ocean currents, perhaps the most important of these is the effect of reactions involving silicates on the activities of silica and hydrogen ion in sea water (SILLÉN, 1961; DEFFEYES, 1963; GARRELS, 1965; MACKENZIE and GARRELS, 1965, 1966a,b; MACKENZIE et al., 1967; SILLÉN, 1967a,b). For example, in areas of the ocean where the activity of silica is high owing to local river water contributions or the presence of submarine volcanics, kaolinite and/or illite (in the sediments supplied to that part of the ocean) should react with the high-silica sea water to produce montmorillonite or zeolites. These reactions, which consume silica, serve as a check on the local build-up of silica in sea water. In the absence of reactive clastic material, amorphous silica saturation (~60 ppm at 0°C) serves as the upper limit (assuming no supersaturation) of silica activity in sea water. The lower limit in deep ocean water is probably controlled by the presence of excess gibbsite + kaolinite and (or) K-mica (illite?) in the clastic sediments.

It can be seen in Figs. 9 and 10 that the compositions of interstitial waters in the deep sea sediments sampled to date (SIEVER et al., 1965) are distinctly different (primarily with respect to the activity of silica) from those of deep ocean waters. This is believed to be due in part to the silica released to interstitial waters by accumulations of expired diatoms incorporated in the sediments. It can be seen in Figs. 9 and 10 that interstitial waters in deep ocean sediments have compositions (in the system $Na_2O-K_2O-Al_2O_3-SiO_2-H_2O$) consistent with the equilibrium assemblage, montmorillonite + K-feldspar (phillipsite?), or albite (analcime?) + K-feldspar (phillipsite?), depending on the activity of H_4SiO_4. The latter assemblage may or may not include amorphous silica.

During diagenesis of deep sea sediments, sea water trapped in pore spaces changes composition from area A to area B in Figs. 9 and 10. As a result, kaolinite and K-mica (but not necessarily illite) in the original sediment should react with the pore water to form montmorillonite during diagenesis. Although the process is almost certainly complex in detail, the relations portrayed in the diagrams suggest that the general tendency during diagenesis is to destroy kaolinite. It appears likely that montmorillonite-illite solid solutions form at the expense of kaolinite with burial. If silica is high in the pore water, zeolites may form at the expense of the clay minerals, which should in turn produce feldspars with further burial (MACKENZIE and GARRELS, 1966b).

The shaded areas labeled M in Figs. 9 and 10 indicate the compositional range of surface sea water, and those annotated N represent the *average* composition of world stream waters (LIVINGSTONE, 1963). Because the temperature of these waters is close to 25°C, the shaded areas M and N are also plotted in Figs. 2 and 3. If Figs. 2 and 3 are respectively superimposed over Figs. 9 and 10 (along an imaginary temperature axis perpendicular to the plane of the paper), the compositional range of sea water in the world ocean can be imagined as a parallelopiped connecting the shaded regions M and A in the superimposed diagrams. It can be seen from the positions of the shaded regions labeled M and N that equilibrium mineral assemblages in brack water and neritic environments (which lie between regions A, M, and N in terms of water composition) differ from those in deep ocean environments. Kaolinite and K-mica (illite?) are the stable phases in these environments. Near river mouths,

K-mica (illite) and (or) montmorillonite may react with brack water to form kaolinite. However, the composition of river waters is so variable that specific equilibrium mineral assemblages for any given river vary considerably.

Mass transfer

The extent to which components in the ocean system are redistributed as a result of diagenetic reactions taking place between a particular sedimentary mineral assemblage and a given water in region A or B of Figs. 9 and 10 can be calculated from the relations depicted in the activity diagrams. Rough calculations of this kind indicate that the total mass transfer attending any one of these reactions may exceed 0·05 g/1000 g of H_2O. Although this mass transfer is sufficient to affect significantly the composition of sea water, it is consistent with conversion of only a few per cent of the detrital material in the ocean to authigenic minerals.

Recent experimental work involving surface sea water (MACKENZIE et al., 1967) is probably indicative of the extent to which mass transfer takes place in the upper part of the ocean. The reaction paths labeled IJ and IK in Figs. 2 and 3 represent the observed changes in the composition of Bermuda sea water ($H_4SiO_4 = 0.03$ ppm initially) as it reacts with kaolinite (path IJ, Fig. 2), K-mica (path IK, Fig. 2) and montmorillonite (path IJ, Fig. 3) at 25°C. Although the reaction products are not known experimentally, it appears that all of the reactions produced a magnesian aluminosilicate, probably chlorite. The mass transfer attending these reactions resulted in the destruction of 0·01 g of kaolinite, 0·02 g of K-mica, and 0·02 g of montmorillonite/1000 g of H_2O.

HYDROTHERMAL ROCK ALTERATION AND ORE DEPOSITION

Spatial and genetic associations of sulfides with altered wall rocks is common in hydrothermal ore deposits. The geochemical processes responsible for these associations can be characterized by evaluating the mass transfer involved in irreversible reactions between igneous or sedimentary mineral assemblages and vein solutions.

Origin of solutions

Depending on the composition of the reactant rock and the aqueous phase involved, metamorphism of sediments may have a substantial effect on the composition of interstitial solutions (ELLIS and MAHON, 1964, 1967; HELGESON, 1967b,c). In fact, the solution may take on exotic compositional characteristics such as those of the Salton Sea geothermal brines (HELGESON, 1968a). This is true in spite of the fact that the metamorphic processes may affect the sediments only slightly in a closed system. For example, if a sediment contains 20 wt. % K-feldspar and has a porosity of 20 per cent, a liter of interstitial solution would occupy the pore spaces in 5000 cm³ of rock, which would contain ∼2,200 g of K-feldspar. If the system is closed and the interstitial solution consists of sea water, reaction of the pore solution with the K-feldspar in the rock at 300°C would result in the destruction of ∼40 g of K-feldspar/1000 g of H_2O (HELGESON, 1967b), or 1·8 per cent of the total K-feldspar in the rock. To detect the effects of the process on any single grain of feldspar would be difficult without a microscope, despite the fact that effects on the chemistry of the aqueous phase are substantial. If the system were open, the effects on the rock would be severe, possibly resulting in complete metamorphism of the original sediment.

As a consequence of the large mass transfer attending metamorphism of sediments, ore-forming metals present in trace concentrations in silicate minerals are released to the aqueous phase. The metasomatic process may thus render the interstitial solution a potential ore-forming fluid. Mass transfer calculations for the reaction of interstitial sea water with an arkosic sediment containing lead-bearing K-feldspar at elevated temperatures indicate that the aqueous solution may derive concentrations of lead in excess of a part per million as a result of the metamorphic process (HELGESON, 1967c). If such a solution is carried by a vein system to a different chemical, thermal, and mineralogic environment, hydrothermal rock alteration and ore deposition may occur.

Zonal deposition of sulfides is a possible consequence of changes in the composition of hydrothermal solutions caused by irreversible reactions between the solutions and their wall rocks. The extent to which sulfides may be precipitated in this way can be evaluated in terms of the change in the hydrogen ion activity resulting from the alteration process. For this purpose, we shall consider a hydrothermal solution originating from a source rock reservoir such as a metamorphosed sediment at 300°C, and assume that a given increment of the solution has traveled through fissures at a high flow rate far enough to permit the temperature to drop from 300°C to 200°C, but also at a high enough velocity to prevent significant interaction with the wall rocks en route. The flow rates required for this are probably not greater than those exhibited in the Salton Sea geothermal wells (HELGESON, 1967c, 1968a). We shall assume that eventually (at a temperature of 200°C) the solution encounters a brecciated and contorted part of the fissure system in a granitic rock where the flow velocity decreases substantially, which allows the solution to react with its environment.

Rock alteration

Equilibrium phase relations in the system $Na_2O-K_2O-Al_2O_3-SiO_2-H_2O$ at 200°C are depicted in the activity diagram in Fig. 11. Point A in Fig. 11 represents the composition of the solution in the source rock, which corresponds to an equilibrium composition for coexisting K-feldspar, albite, quartz, and solution at 300°C (HELGESON, 1967b). Point A thus represents the composition of the solution when it arrives in the depositional environment. Reaction of this solution with a granitic wall rock at 200°C will result in the precipitation of K-mica and quartz at the expense of albite and K-feldspar. If we assume that the rate at which albite is destroyed relative to the rate of reaction of K-feldspar per cm² of surface area is 2:1 on a mole basis, and we conserve aluminum, this reaction can be written as

$$2NaAlSi_3O_8 + KAlSi_3O_8 + 2H^+ \rightarrow KAl_3Si_3O_{10}(OH)_2 + 2Na^+ + 6SiO_{2(quartz)}. \quad (19)$$

Taking into account the dissociation constant for NaCl (PEARSON *et al.*, 1963) and the general composition of fluid inclusions, we shall assume the activities of Na^+ and Cl^- in the solution are unity. Under these conditions the pH of the solution at point A is 4·9 and the activity of K⁺ is 0·11. As reaction (19) takes place, the solution composition changes along path AB in Fig. 11. At point B, further reaction between albite and the aqueous phase causes the composition of the solution to change along the stability field boundary, BC. At point C, equilibrium is established between K-feldspar, albite, K-mica, quartz, and solution. Mole transfer calculations indicate that as a result of the reactions, the pH of the solution increases along the reaction path from 4·9 at A to 5·3 at B and 5·5 at C.

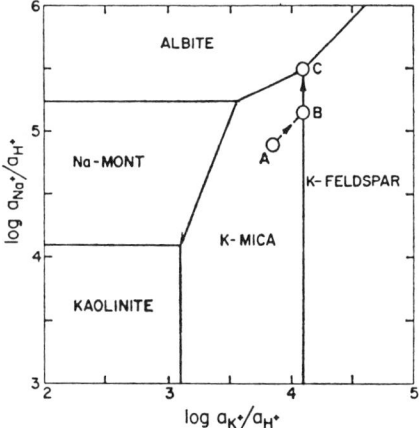

Fig. 11. Activity diagram depicting equilibrium relations among the phases (in the presence of quartz) in the system K_2O–Na_2O–Al_2O_3–SiO_2–H_2O at 200°C, unit activity of water, and 1 atm. The dashed arrows and letter annotations describe the reaction path for a hydrothermal solution reacting with K-feldspar and albite in the wall rocks of a vein system—see text. The positions of the stability field boundaries were calculated from thermodynamic data taken from sources cited in the captions of preceding figures. Estimates based in part on high temperature solubility data (HEMLEY et al., 1961), were made for the thermodynamic properties of sodium montmorillonite.

Because the solution in our model is moving through the fracture, the compositional changes represented by path ABC in Fig. 11 may occur over a long distance, depending on the diffusion rates, reaction rates, flow velocity, and flow rate obtaining in the system. We are concerned here only with the mass transfer resulting from reaction of a given increment of solution with its wall rock as it travels through the potential environment for ore deposition; we are not concerned with the spatial distribution of the reaction products.

Isothermal precipitation of sulfides

If the molality of total sulfide in the initial solution at point A in Fig. 11 is 10^{-4} (3 ppm) and the solution is initially saturated with respect to both sphalerite and pyrrhotite (1·2 ppm total zinc and 4·2 ppm total ferrous iron), the pH change resulting from the alteration reactions along path ABC in Fig. 11 should cause precipitation of 1·1 ppm zinc, 3·3 ppm iron, and 2·5 ppm sulfide as sphalerite and pyrrhotite. These figures are based on evaluation of mass transfer equations using thermodynamic data for the aqueous sulfide species, metal ions, and chloride complexes (COBBLE, 1964; HELGESON, 1964; SILLÉN and MARTELL, 1964; BARNES et al., 1966; HELGESON, 1969) to compute equilibrium constants for 200°C with the aid of predictive equations (HELGESON, 1967a, 1969).

The amount of sphalerite precipitated/1000 g of H_2O as a result of the interaction of a hydrothermal solution with its wall rock in the model examined above is more than sufficient to cause the formation of major ore deposits. The quantity of solution required to produce a $30,000,000 orebody ($\sim 2 \times 10^5$ short tons of sphalerite) would be less than that contained in 100 cubic miles of source rock sediments having a porosity of 25 per cent. Under different conditions of pH, total sulfide, total chloride, initial concentrations of metals, and/or temperature, ore concentrations of gold (HELGESON and GARRELS, 1968), silver, lead, copper, and other ore-forming metals may be precipitated in a similar manner, and the amounts of solution required to form an orebody will be different. If partial equilibrium is maintained throughout the depositional process, the resulting ore deposits will exhibit a zonal distribution of sulfides with predictable cotectic-peritectic (replacement) relations and paragenetic sequences. The zonal pattern exhibited by an ore deposit is a consequence (in part) of the fact that the activity product constants for the sulfides, and the extent to which the various metals are complexed in solution are not respectively equivalent. The order and nature of precipitation, and the amount of each sulfide deposited at any one time as a result of the rock alteration process is controlled in part by these variables. There can be little doubt that a zoned ore deposit containing a variety of metals can be formed from a single solution. The mass transfer involved is highly temperature dependent; in fact, *metal ratios in ore deposits formed in this way constitute geothermometers.*

Any process that decreases the activity of hydrogen ion in "acid" hydrothermal solutions (which is probably the general case—HELGESON, 1964, 1967b,c, 1969) tends to cause precipitation of sulfides. This mechanism is almost certainly responsible in part for lead and zinc precipitation in limestones and dolomites. For example, mass transfer calculations indicate that reaction between a hydrothermal solution with an initial pH of 4 and a limestone at 150°C may cause precipitation of more than eight ppm zinc and one ppm lead as sphalerite and galena. These calculations were carried out with provision for chloride and sulfide complexing in a 3·0 m NaCl solution containing an initial concentration of 10^{-4} m total sulfide using thermodynamic data presented elsewhere (HELGESON, 1969). The spatial distribution of the sulfides precipitated in limestones and dolomites reflects the extent to which equilibrium is achieved at any given point in the vein system, which depends on the velocity and rate of flow of the solution and the surface area available for reaction.

Acknowledgments—We are indebted to C. L. CHRIST, J. J. HEMLEY, H. J. GREENWOOD and E-AN ZEN for their helpful suggestions and constructive criticism of this paper in various stages of its development. We are also grateful to R. O. FOURNIER and H. L. BARNES for reviewing the manuscript. The work was supported by the National Science Foundation (NSF grants GP-4140, GA-828, GE-9758, and GU-1700), the Petroleum Research Fund of the American Chemical Society, and research funds made available by the Office of Research Coordination, Northwestern University. The assistance of ANDREW NIGRINI and T. A. JONES in carrying out computer calculations is also acknowledged with thanks.

REFERENCES

ALTSCHULER Z. S., DWORNIK E. J. and KRAMER H. (1963) Transformation of montmorillonite to kaolinite during weathering. *Science* **141**, 148–152.

ARMSTRONG F. A. J. (1965) Silicon. In *Chemical Oceanography* (editors J. P. Riley and G. Skirrow) Vol. 1, pp. 7–22. Academic Press.

BARANY R. (1964) Heat and free energy of formation of muscovite. *U.S. Bur. Mines Rep. Invest.* 6356.

BARANY R. and KELLEY K. K. (1961) Heats and free energies of formation of gibbsite, kaolinite, halloysite, and dickite. *U.S. Bur. Mines Rep. Invest.* 5825.

BARNES H. L., HELGESON H. C. and ELLIS A. J. (1966) Ionization constants in aqueous solutions. In *Handbook of Physical Constants* (Revised edition), (editor S. P. Clark, Jr.). *Geol. Soc. Amer. Mem.* **97**, 401–414.

BARTON P. L., BETHKE P. M. and TOULMIN P. 3rd (1963) Equilibrium in ore deposits. *Min. Soc. Amer. Spec. Paper* **1**, 171–185.

BERNER R. A. (1965) Activity coefficients of bicarbonate, carbonate, and calcium ions in sea water. *Geochim. Cosmochim. Acta* **29**, 947–965.

BRADLEY W. H. (1929) The varves and climate of the Green River epoch. *U.S. Geol. Surv. Prof. Paper* 158.

BRICKER O. P. and GARRELS R. M. (1965) Mineralogical factors in natural water equilibria. In *Proc. 4th Ann. Rudolphs. Conf.*, pp. 449–469. Rutgers Univ., New Brunswick, New Jersey.

COBBLE J. W. (1964) The thermodynamic properties of high temperature aqueous solutions. VI. Applications of entropy corresponding to thermodynamics and kinetics. *J. Amer. Chem. Soc.* **86**, 5394–5401.

CORRENS C. W. and VON ENGELHARDT W. (1938) Neue Untersuchungen über die Verwitterung des Kalifeldspates. *Chem. Erde* **12**, 1–22.

CRISS C. M. and COBBLE J. W. (1964) The thermodynamic properties of high temperature aqueous solutions. V. The calculations of ionic heat capacities up to 200°. Entropies and heat capacities above 200°. *J. Amer. Chem. Soc.* **86**, 5390–5393.

DEFFEYES K. S. (1963) Role of oceanic carbonate and silicate sedimentation in determining atmospheric carbon dioxide pressure (abstract). *Geol. Soc. Amer. Spec. Paper* **76**, 39–40.

ELLIS A. J. and MAHON W. A. J. (1964) Natural hydrothermal systems and experimental hot-water/rock interactions. *Geochim. Cosmochim. Acta* **28**, 1323–1357.

ELLIS A. and MAHON W. A. J. (1967) Natural hydrothermal systems and experimental hot-water/rock interactions (Part II). *Geochim. Cosmochim. Acta* **31**, 519–538.

FETH J. H., ROBERSON C. E. and POLZER W. L. (1964) Sources of mineral constituents in water from granitic rocks, Sierra Nevada, California, and Nevada. *U.S. Geol. Surv. Water-supply Paper* 1535-I, 170 pp.

GARRELS R. M. (1965) The role of silica in the buffering of natural waters. *Science* **148**, 69.

GARRELS R. M. and THOMPSON M. E. (1962) A chemical model for sea water at 25°C and one atmosphere total pressure. *Amer. J. Sci.* **260**, 57–66.

GARRELS R. M. and CHRIST C. L. (1965) *Solutions, Minerals, and Equilibria*, 450 pp. Harper & Row.

GARRELS R. M. and MACKENZIE F. T. (1967) Origin of the chemical compositions of some springs and lakes. In *Equilibrium Concepts in Natural Water Systems*, pp. 222–242 (editor R. F. Gould). *Adv. Chem. Ser.* 67. Amer. Chem. Soc.

GOOD W. D., LACINA J. L., DEPRATER B. L. and MCCULLOUGH J. P. (1964) A new approach to the combustion calorimetry of silicon and organo-silicon compounds. Heats of formation of quartz, fluorosilicic acid, and hexamethyldisiloxane. *J. Phys. Chem.* **68**, 579–587.

HELGESON H. C. (1964) *Complexing and Hydrothermal Ore Deposition*, 128 pp. Pergamon.

HELGESON H. C. (1967a) Thermodynamics of complex dissociation in aqueous solutions at elevated temperatures. *J. Phys. Chem.* **71**, 3121–3136.

HELGESON H. C. (1967b) Solution chemistry and metamorphism. In *Researches in Geochemistry*, (editor P. H. Abelson), Vol. 2, pp. 362–402. Wiley.

HELGESON H. C. (1967c) Silicate metamorphism in sediments and the genesis of hydrothermal ore solutions. In *Genesis of Stratiform Lead–Zinc–Barite–Fluorite Deposits*, (editor J. S. Brown). *Econ. Geol. Monogr.* **3**, 333–342.

HELGESON H. C. (1968a) Geologic and thermodynamic characteristics of the Salton Sea Geothermal System. *Amer. J. Sci.* **266**, 129–166.

HELGESON H. C. (1968b) Evaluation of irreversible reactions in geochemical processes involving minerals and aqueous solutions—I. Thermodynamic relations. *Geochim. Cosmochim. Acta* **32**, 853–877.

HELGESON H. C. (1969) Thermodynamics of hydrothermal systems at elevated temperatures and pressures. *Amer. J. Sci.* **267**, in press.

HELGESON H. C. and GARRELS R. M. (1968) Hydrothermal transport and deposition of gold. *Econ. Geol.* **63**, 622–635.

HELGESON H. C. and JAMES W. R. (1968) Activity coefficients in concentrated electrolyte solutions at elevated temperatures, an abstract. *Abstracts of Papers*, 155th Nat. Meeting, Amer. Chem. Soc. S-130, San Francisco, California, April, 1968.

HEM J. D. (1966) Personal communication.

HEM J. D. and ROBERSON C. E. (1967) Form and stability of aluminium hydroxide complexes in dilute solution. *U.S. Geol. Surv. Water-Supply Paper* 1827-A.

HEMLEY J. J., MEYER C. and RICHTER D. H. (1961) Some alteration reactions in the system Na_2O–Al_2O_3–SiO_2–H_2O. *U.S. Geol. Surv. Prof. Paper* **424D**, 338–340.

HESS P. C. (1966) Phase equilibria of some minerals in the K_2O–Na_2O–Al_2O_3–SiO_2–H_2O system at 25°C and 1 atmosphere. *Amer. J. Sci.* **264**, 289–309.

HOLLAND H. D. (1965) The history of ocean water and its effect on the chemistry of the atmosphere. *Proc. Nat. Acad. Sci.* **53**, 1173–1182.

KELLEY K. K. (1960) Contributions to the data on theoretical metallurgy. XIII. High temperature, heat content, heat capacity, and entropy data for the elements and inorganic compounds. *U.S. Bur. Mines Bull.* 584.

KELLEY K. K. (1962) Heats and free energies of formation of anhydrous silicates. *U.S. Bur. Mines Rep. Invest.* 5901.

KELLEY K. K., TODD S. S., ORR R. L., KING E. G. and BONNICKSON K. R. (1953) Thermodynamic properties of sodium–aluminum and potassium–aluminum silicates. *U.S. Bur. Mines Rep. Invest.* 4955.

KELLEY K. K. and KING E. G. (1961) Contributions to the data on theoretical metallurgy. XIV. Entropies of the elements and inorganic compounds. *U.S. Bur. Mines Bull.* 592.

KING E. G. and WELLER W. W. (1961) Low temperature heat capacities and entropies at 298·15° of diaspore, kaolinite, dickite, and halloysite. *U.S. Bur. Mines Rep. Invest.* 5810.

KITTRICK J. A. (1966) The free energy of formation of gibbsite and $Al(OH)_4^-$ from solubility measurements. *Proc. Soil Sci. Soc. Amer.* **30**, 595–601.

KRAMER J. R. (1965) History of sea water. Constant temperature–pressure equilibrium models compared to liquid inclusion analyses. *Geochim. Cosmochim. Acta* **29**, 921–946.

LATIMER W. M. (1952) *Oxidation Potentials* (2nd edition), 375 pp. Prentice-Hall.

LIVINGSTONE D. A. (1963) Chemical composition of rivers and lakes. *U.S. Geol. Surv. Prof. Paper* 440-G.

MACKENZIE F. T. and GARRELS R. M. (1965) Silicates-reactivity with sea water. *Science* **150**, 57–58.

MACKENZIE F. T. and GARRELS R. M. (1966a) Chemical mass balance between rivers and oceans. *Am. J. Sci.* **264**, 507–525.

MACKENZIE F. T. and GARRELS R. M. (1966b) Silica–bicarbonate balance in the ocean and early diagenesis. *J. Sediment Petrol.* **36**, 1075–1084.

MACKENZIE F. T., GARRELS R. M., BRICKER O. P. and BICKLEY F. (1967) Silica in sea water: control by silicate minerals. *Science* **155**, 1404–1405

MOORE G. W., ROBERSON C. E. and NYGREN H. D. (1962) Electrode determination of the carbon dioxide content of sea water and deep sea sediment. *U.S. Geol. Surv. Prof. Paper* **450-B**, 83–86.

Morey G. W., Fournier R. O. and Rowe J. J. (1962) The solubility of quartz in water in the temperature interval from 25° to 300°C. *Geochim. Cosmochim. Acta* **26**, 1029–1043.

Morey G. W. Fournier R. O. and Rowe J. J. (1964) The solubility of amorphous silica at 25°C. *J. Geophys. Res.* **69**, 1995–2002.

Pankratz L. B. (1964) High-temperature heat contents and entropies of muscovite and dehydrated muscovite. *U.S. Bur. Mines Rep. Invest.* 6371.

Pearson D., Copeland C. S. and Benson S. W. (1963) The electrical conductance of aqueous sodium chloride in the range 300 to 383°. *J. Amer. Chem. Soc.* **85**, 1044–1047.

Robie R. A. (1966) Thermodynamic properties of minerals. In *Handbook of Physical Constants* (Revised edition), (editor S. P. Clark, Jr.). *Geol. Soc. Amer. Mem.* **97**, 438–458.

Rossini R. D., Wagman D. D., Evans W. H., Levine S. and Jaffe I. (1952) Selected values of chemical thermodynamic properties. *Nat. Bur. Stand. Circ.* 500.

Siever R., Beck K. C. and Berner R. A. (1965) Composition of interstitial waters of modern sediments. *J. Geol.* **73**, 39–73.

Sillén L. G. (1961) The physical chemistry of sea water. In *Oceanography*, (editor M. Sears). *Amer. Assoc. Adv. Sci. Publ.* **67**, 549–581.

Sillén L. G. (1963) How has sea water got its present composition? *Svensk Kem. Tidskr.* **75**, 161–177.

Sillén L. G. (1967a) The ocean as a chemical system. *Science* **156**, 1189–1197.

Sillén L. G. (1967b) Gibbs phase rule and marine sediments. In *Equilibrium Concepts in Natural Water Systems*, (editor R. F. Gould), pp. 57–69. *Adv. Chem. Ser.* 67. Amer. Chem. Soc.

Sillén L. G. and Martell A. E. (1964) *Stability Constants of Metal–Ion Complexes. Chem. Soc., London, Spec. Publ.* No. 17, 754 pp.

Strahler A. N. (1963) *The Earth Sciences*, 681 pp. Harper & Row.

Wagman D. D., Evans W. H., Harlow I., Parker V. B., Bailey S. M. and Schumm R. H. (1968) Selected values of chemical thermodynamic properties, Parts I and II. *Nat. Bur. Stand. Tech. Note* 270-3.

Waldbaum D. R. (1966) Calorimetric investigations of the alkali feldspars. Ph.D. Thesis, Harvard University.

Weller W. W. and King E. G. (1963) Low temperature heat capacity and entropy at 298·15°K of muscovite. *U.S. Bur. Mines Rep. Invest.* 6281.

White D. E., Hem J. D. and Waring G. A. (1963) Chemical composition of subsurface waters. *U.S. Geol. Sur. Prof. Paper* 440-F.

Wicks C. E. and Block F. E. (1963) Thermodynamic properties of 65 elements—their oxides, halides, carbides, and nitrides. *U.S. Bur. Mines Bull.* 605.

Wilde P. (1966) pH of deep-sea sediments (abstract). *Program, 1966 Ann. Meeting, Geol. Soc. Amer.*, pp. 240–241.

Wise S. S. and Margrave J. L. (1963) Fluorine bomb calorimetry. V. The heats of formation of silicon tetrafluoride and silica. *J. Phys. Chem.* **67**, 815–821.

Wollast R. (1967) Kinetics of the alteration of K-feldspar in buffered solutions at low temperature. *Geochim. Cosmochim. Acta* **31**, 635–648.

SOLUBILITY DIAGRAMS FOR EXPLAINING ZONE SEQUENCES IN BAUXITE, KAOLIN AND PYROPHYLLITE–DIASPORE DEPOSITS

Yoshiro Tsuzuki

Department of Earth Sciences, Faculty of Science, Nagoya University, Chikusa-ku, Nagoya, Japan

Abstract—Solubility diagrams defined by log $[Al^{3+}]$ and log $[H_4SiO_4]$ are given for hydrous alumina or aluminum silicate minerals which appear in bauxite, kaolin and pyrophyllite–diapore deposits. They are constructed based on thermodynamic data of relevant reactions both at the room temperature and at elevated temperatures.

An aqueous solution reacts to a mineral, in this case K-feldspar, and, by dissolving it, becomes saturated with respect to a certain mineral. This mineral begins to be precipitated and the solution changes its composition as a result of the precipitation as well as further dissolution of the original mineral. Then, it attains saturation with respect to another mineral, which is precipitated thereafter. Thus, different minerals are precipitated in turn.

The sequences of precipitation of minerals can be shown on the diagrams under different conditions. A sequence, aluminum hydroxide → kaolinite or pyrophyllite → silica mineral plus kaolinite or pyrophyllite is expected in a weakly acid solution. In contrast, a sequence, silica mineral → silica mineral plus kaolinite or pyrophyllite is expected in a strongly acid solution. The possibility of application of the sequence of precipitation thus expected to alteration zoning is also discussed.

INTRODUCTION

Bauxite, kaolin, and pyrophyllite–diaspore deposits are formed by weathering or hydrothermal alteration, both of which are caused by aqueous solutions permeating rocks; and zoning of alteration products is a common feature of these deposits.

Solubility diagrams have been proved useful in interpreting natural mineral assemblages or chemical reactions occurring naturally. Log $[Al^{3+}]$ or log $[AlO_2^-]$–pH–log $[H_4SiO_4]$ diagram by Garrels and Christ (1965) and by Curtis and Spears (1971); (pH–$\frac{1}{3}$ pAl^{3+})–pH$_4$SiO$_4$ diagram by Kittrick (1969), and log (total dissolved Al)–log (total dissolved Si) diagram by Gardner (1970) are examples.

The writer constructed similar solubility diagrams both at room temperature and at elevated temperatures. In this paper, these diagrams are given and zone sequences expected from them are shown.

SOLUBILITY DIAGRAMS

Construction

Solubility diagrams with coordinates of log $[Al^{3+}]$ and log $[H_4SiO_4]$ were constructed at fixed activities of H^+ and K^+ and at fixed temperatures, and are shown in Figure 1 and Figures 4–9. The following reactions were considered in their construction:

$$KAlSi_3O_8 + 4H^+ + 4H_2O = K^+ + Al^{3+} + 3H_4SiO_4, \quad (1)$$
K-feldspar(microcline)

$$KAl_3Si_3O_{10}(OH)_2 + 10H^+ = K^+ + 3Al^{3+} + 3H_4SiO_4, \quad (2)$$
muscovite

$$Al_2Si_4O_{10}(OH)_2 + 6H^+ + 4H_2O = 2Al^{3+} + 4H_4SiO_4, \quad (3)$$
pyrophyllite

$$Al_2Si_2O_5(OH)_4 + 6H^+ = 2Al^{3+} + H_2O + 2H_4SiO_4, \quad (4)$$
kaolinite

$$AlO(OH) + 3H^+ = Al^{3+} + 2H_2O, \quad (5)$$
diaspore

$$AlO(OH) + 3H^+ = Al^{3+} + 2H_2O, \quad (6)$$
boehmite

$$Al(OH)_3 + 3H^+ = Al^{3+} + 3H_2O, \quad (7)$$
(crystalline) gibbsite

$$SiO_2 + 2H_2O = H_4SiO_4. \quad (8)$$
α-quartz

The equilibrium constants of these equations at various temperatures were given by Helgeson (1969) except for (3), (5) and (6), for which they were computed by employing the equations proposed by Helgeson (1969) and the thermodynamic data for pyrophyllite (Haas and Holdaway, 1973), diaspore (Haas, 1972) and boehmite (Barin and Knacke, 1973).

Using the equilibrium constants of the above reactions, the solubility lines on which a solid mineral coexists with solution were drawn for various minerals on each diagram. For reaction (1), as an example, equilibrium constant K_1 is expressed by

$$K_1 = \frac{[K^+][Al^{3+}][H_4SiO_4]^3}{[H^+]^4}, \quad (9)$$

or

$$\log K_1 = \log[K^+] + \log[Al^{3+}] +$$
$$3\log[H_4SiO_4] - 4\log[H^+]. \quad (10)$$

If $\log[K^+]$ and $\log[H^+]$ have fixed values, this equation is represented by a straight line on a $\log[Al^{3+}]$ vs. $\log[H_4SiO_4]$ diagram. Lines F_1 and F_2 in Figure 1 are such lines.

The solubility line of amorphous silica was drawn after Kennedy (1950), when the solubility of amorphous silica was important.

Although many workers (e.g. Frink and Sawhney, 1967; Parks, 1972; Huang and Keller, 1972) have discussed the state of Al in aqueous solution, dominant species in this study are considered to be Al^{3+}, $Al(OH)^{2+}$ and $Al(OH)_4^-$, following Helgeson (1968) and Helgeson et al. (1969). Assuming the activity coefficients of the ions to be unity, the total dissolved Al can be given from the activity of Al^{3+} and the pH using the equilibrium constants of the following equations, and is indicated as the ordinate on the righthand side of the diagrams.

$$Al(OH)_4^- = Al^{3+} + 4OH^-, \quad (11)$$

$$Al(OH)^{2+} = Al^{3+} + OH^-, \quad (12)$$

$$H_2O = H^+ + OH^-. \quad (13)$$

Although H_4SiO_4 dissociates into H^+ and $H_3SiO_4^-$, the ratio of $H_3SiO_4^-$ to H_4SiO_4 is negligibly small in an acid or nearly neutral solution.

Interpretation

Here, the chemical reactions which follow the congruent dissolution of K-feldspar into an aqueous solution are assumed. These reactions at pH = 5, $[K^+] = 0.001$, and temperature = 25°C may be predicted based on the solubility diagram, Figure 1. The dissolution of K-feldspar causes the total concentration of silica, aluminum and potassium in the solution to increase, keeping the relative proportions of 3:1:1. Therefore, the concentration of aluminum and silica produced by the dissolution of K-feldspar should fall on the straight line, I, which passes the point log (total dissolved Al) = 0 ($m_{Al} = 1$) and log $[H_4SiO_4] = 0.477$ ($m_{H_4SiO_4} = 3$) and inclines 45 to the ordinate. As the dissolution of K-feldspar proceeds, the concentration of aluminum and silica change from lower left to upper right along this line, which is called the initial path.

When the initial path reaches the gibbsite line at point a, gibbsite begins to be precipitated. The continuing dissolution of K-feldspar brings about the increase of dissolved silica along the gibbsite line, because the precipitation of gibbsite keeps the Al^{3+} concentration constant.

When the path reaches point b, the intersection of the gibbsite and kaolinite lines, gibbsite begins to be dissolved and kaolinite begins to be precipitated. The material balance at this point is expressed by

$$2KAlSi_3O_8 + 4Al(OH)_3 + 2H^+$$
$$= 3Al_2Si_2O_5(OH)_4 + 2K^+ + H_2O. \quad (14)$$

After gibbsite is totally consumed, the path turns to the lower right along the kaolinite line, on which the H_4SiO_4 concentration increases with the precipitation of kaolinite.

When the path reaches point c, the intersection of the kaolinite and amorphous silica lines, amorphous silica begins to be precipitated together with kaolinite. The material balance at this point is expressed by

$$2KAlSi_3O_8 + H_2O + 2H^+ = 4SiO_2$$
$$+ Al_2Si_2O_5(OH)_4 + 2K^+. \quad (15)$$

The path does not proceed further along the silica line. If the path proceeded downward along the silica line, the dissolution of K-feldspar would not take place. Proceeding in this direction means a decrease of the dissolved Al^{3+} without precipitating Al-bearing mineral, which is contradictory to the dissolution of K-feldspar. The reactions occurring in this system are illustrated in Figure 2.

Evaluation of basic assumptions

Before looking at the application of solubility diagrams to zoning, the validity of basic assumptions involved will be examined.

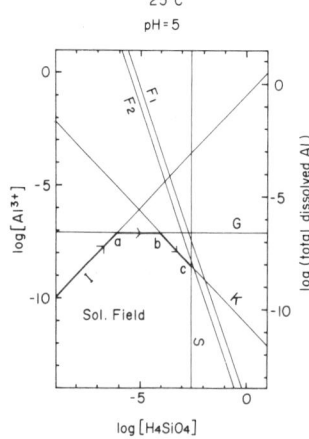

Figure 1. An example of solubility diagrams for K-feldspar alteration. G—gibbsite, K—kaolinite, S—amorphous silica, F_1—K-feldspar (microcline) at $[K^+] = 0.001$, F_2—K-feldspar (microcline) at $[K^+] = 0.01$. I—initial path for K-feldspar dissolution.

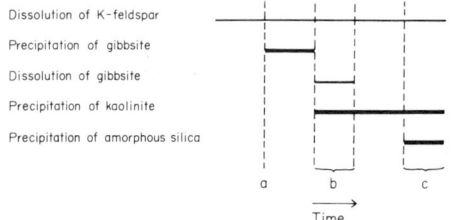

Figure 2. Schematic diagram showing the sequence of chemical reactions which follow the congruent dissolution of K-feldspar at pH = 5, $[K^+] = 0.001$, temperature = 25°C.

(1) *Thermodynamic data employed.* As mentioned above, the thermodynamic data used for constructing solubility diagrams were derived mainly from Helgeson (1969). For room temperature, more recent data were given by Curtis and Spears (1971), Huang and Keller (1973) and others. These data produce slightly different diagrams, but essential features of the diagrams are not affected.

(2) *Dissolution and precipitation mechanism.* In this study, the dissolution of K-feldspar followed by the precipitation of other minerals is considered instead of the replacement of K-feldspar by other minerals. O'Neil and Taylor (1967), Wilson *et al.* (1971), Keller *et al.* (1971), Nakagawa *et al.* (1973), and Tsuzuki *et al.* (1973, 1974) suggested that the alteration process of a mineral in a solution can be regarded as composed of the dissolution of this mineral, followed by the precipitation of another mineral.

(3) *Congruent dissolution.* The initial stage of the dissolution of K-feldspar may be the exchange reaction of K^+ for H^+ (Wollast, 1967). Experiments by Correns (1961) and Morey and Chen (1955), however, show that K-feldspar is dissolved almost congruently in the main stage.

(4) *Dissolution of K-feldspar.* Natural rocks have, of course, more complex composition than monomineral K-feldspar. If the most soluble mineral in a rock is Na-feldspar instead of K-feldspar, the sequence of reactions shown in Figure 2 remains unchanged, because the positions of the initial path and the solubility lines are the same as in Figure 1. Because these two minerals are widely distributed and liable to be dissolved, the present interpretation of solubility diagrams may be applied to natural zoning in various rocks.

(5) *Partial equilibrium.* In the interpretation of solubility diagrams, the partial equilibrium was assumed following Helgeson (1968) and Helgeson *et al.* (1969). That is, it was assumed that the precipitation of minerals occurs, maintaining equilibrium, while K-feldspar is being dissolved continuously into a solution, i.e. equilibrium is not yet attained. This assumption may be allowed, because the dissolution of K-feldspar is a slow reaction. The incompatible association of K-feldspar and kaolinite is commonly observed in altered granitic rocks (Meyer and Hemley, 1959).

(6) *Change in concentration of K^+ and H^+.* Although the solubility diagrams were constructed for fixed K^+ and H^+ concentration, these concentrations vary with the dissolution and precipitation reactions in question. This change in K^+ and H^+ concentration can be estimated quantitatively by the method of Helgeson (1968) and Helgeson *et al.* (1969), but, for the purpose of the present study, a simpler method of estimation is sufficient. The change in pH during the reaction has no important effects on sequences of precipitation, because the mutual relations between the solubility lines of pyrophyllite, kaolinite, diaspore, boehmite and gibbsite are not affected by the change in pH. The effect of changing K^+ concentration can be estimated by drawing solubility lines of K-feldspar and sericite at different K^+ concentrations, like F_1 and F_2 in Figure 1.

At point c in Figure 1, K^+ concentration of the solution increases with the precipitation of amorphous silica and kaolinite and the dissolution of K-feldspar, resulting in a leftward shift of the solubility line of K-feldspar. This line becomes passing point c at the K^+ concentration defined by equation (9); this means that the solution becomes saturated with K-feldspar and the series of reactions described in the previous section terminates leaving an amorphous silica–kaolinite–K-feldspar assemblage behind.

Application to zoning

Natural weathering and hydrothermal alteration take place by the action of migrating aqueous solutions. Therefore, the successive reactions shown in Figure 2 are expected to occur as a function of migration distance as well as that of time. The only difference from Figure 2 is the absence of stage b in which the dissolution of gibbsite and the precipitation of kaolinite occur simultaneously. When the solution becomes saturated with respect to kaolinite, it moves away from where it precipitated gibbsite; therefore, no gibbsite is available for dissolution. Thus, a zoning, dissolution zone → gibbsite zone → kaolinite zone → kaolinite–silica zone → unaltered zone, may be formed. This state is illustrated in the left part of Figure 3.

Continuous flow of the solution will bring about a shift of zone boundary by the following reason. When a new solution reaches the boundary of the dissolution zone and the gibbsite zone, the Al^{3+} concentration is not yet high enough for precipitating gibbsite, because the amount of K-feldspar has decreased by dissolution in the dissolution zone. Therefore, the boundary of these two zones shifts further from the source with time. Other zone boundaries shift likewise, resulting in the migration of zones as illustrated in Figure 3. Accordingly, the mode of

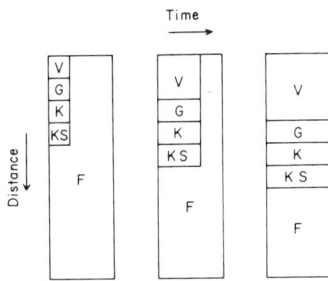

Figure 3. Schematic diagram showing development of zoning pattern under conditions shown in Figure 1. F—feldspar, G—gibbsite, K—kaolinite, S—amorphous silica, V—vacant space.

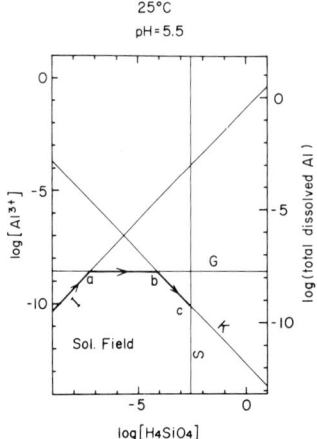

Figure 4. A solubility diagram under weathering conditions. G—gibbsite, K—kaolinite, S—amorphous silica, I—initial path for K-feldspar dissolution.

genesis of a mineral is not uniform; gibbsite, for example, can be formed in either of the following two ways,

(1) K-feldspar → solution → gibbsite,
(2) K-feldspar → solution → kaolinite → solution → gibbsite.

DIAGRAMS FOR VARIOUS CONDITIONS

Diagrams for weathering conditions

For applying zone formation under weathering conditions, the diagrams at 25°C and at a weakly acid pH solution, were prepared. A diagram at 25°C and pH 5.5 is shown in Figure 4. The sequence of precipitation is gibbsite → kaolinite → amorphous silica plus kaolinite. A diagram at 25°C and pH 4 is shown in Figure 5. The sequence is kaolinite → amorphous silica plus kaolinite. A zonal sequence without a gibbsite zone is expected under such low pH conditions.

Diagrams for kaolin and pyrophyllite–diaspore deposits

Interesting zonal sequences similar to hydrothermal deposits, such as kaolin deposits and pyrophyllite–diaspore, were prepared at 200 and 300°C. The minerals concerned are boehmite, kaolinite and amorphous silica at 200°C, and diaspore, pyrophyllite and α-quartz at 300°C.

According to Kennedy (1959), boehmite is formed metastably around 200°C, while diaspore is formed at higher temperatures. Tsuzuki and Mizutani (1969, 1971) showed that kaolinite was transformed into pyrophyllite at 270°C. Mizutani (1966) reported that amorphous silica was formed at a fumarole at 200°C. Examination of solubility diagrams at different pH revealed that there are two different types of sequence of precipitation: (A) sequence from weakly acid solutions and (B) that from strongly acid solutions. Examples of type A are shown on the diagram at 200°C and pH 4 in Figure 6 and that at 300°C and pH 3 in Figure 7. In these diagrams, the initial paths meet the boehmite line or the diaspore line, and the paths turn clockwise similar to the example above. The sequence of precipitation is boehmite → kaolinite → amorphous silica plus kaolinite, or diaspore → pyrophyllite → α-quartz plus pyrophyllite. If $[K^+] = 1.0$ on the diagram in Figure 6, muscovite is precipitated instead of kaolinite. A sericite zone, therefore, may be formed by K^+ rich solution.

If the initial solution is a little more acid, say pH = 2.3, the initial path shifts to the right and first meets the kaolinite line (Figure 8). Boehmite or diaspore zone does not occur in alteration by such solutions.

An example of the type B sequence is shown in the diagram at 200°C and pH 1.8 in Figure 9. In

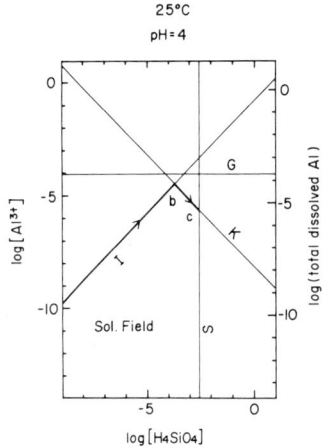

Figure 5. A solubility diagram under weathering conditions. The solution is more acid than in Figure 4. G—gibbsite, K—kaolinite, S—amorphous silica, I—initial path for K-feldspar dissolution.

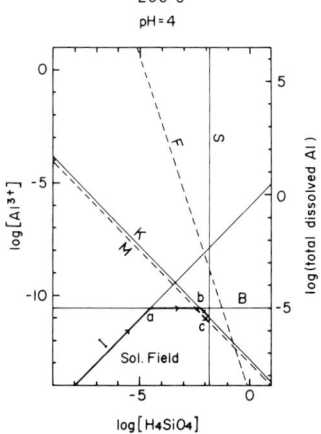

Figure 6. A solubility diagram under hydrothermal conditions. B—boehmite, K—kaolinite, S—amorphous silica, M—muscovite at $[K^+] = 1.0$, F—K-feldspar at $[K^+] = 1.0$, I—initial path for K-feldspar dissolution.

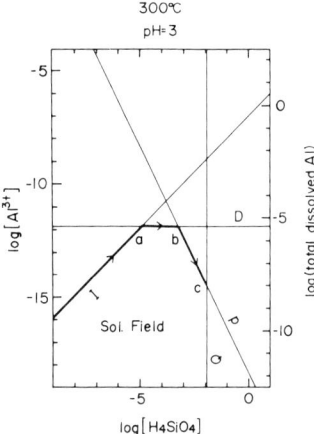

Figure 7. A solubility diagram under hydrothermal conditions. D—diaspore, P—pyrophyllite, Q—α-quartz, I—initial path for K-feldspar dissolution.

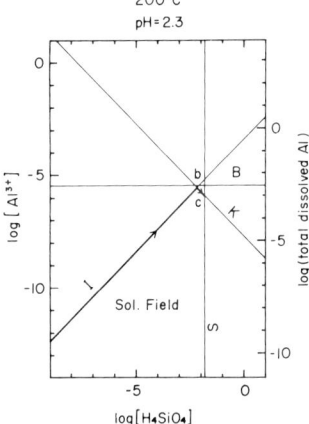

Figure 8. A solubility diagram under hydrothermal conditions. B—boehmite, K—kaolinite, S—amorphous silica, I—initial path for K-feldspar dissolution.

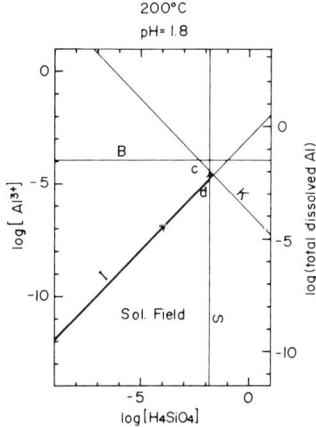

Figure 9. A solubility diagram under hydrothermal conditions. The solution is more acid than in Figure 6. B—boehmite, K—kaolinite, S—amorphous silica, I—initial path for K-feldspar dissolution.

this diagram, the initial path first meets the amorphous silica line at point d, and then the solution changes its composition upward along this line, precipitating amorphous silica. Thus the direction is anticlockwise. When the path reaches point c, kaolinite begins to be precipitated together with amorphous silica. Consequently, the sequence expected is silica → silica plus kaolinite.

Acknowledgements—The author is grateful to Dr. W. H. Huang of University of South Florida for reading the manuscript and making many helpful comments. He is indebted to Dr. K. Nagasawa of Nagoya University for encouraging support, helpful suggestions and critical comments. He also wishes to express his thanks to Dr. N. Nakai and Dr. S. Mizutani of Nagoya University for their constructive suggestions.

REFERENCES

Barin, I. and Knacke, O. (1973) *Thermochemical Properties of Inorganic Substances*: Springer, Berlin.

Correns, C. W. (1961) The experimental chemical weathering of silicates: *Clay Min. Bull.* **4**, 249–265.

Curtis, C. D. and Spears, D. A. (1971) Diagenetic development of kaolinite: *Clays & Clay Minerals* **19**, 219–227.

Frink, C. R. and Sawhney, B. L. (1967) Neutralization of dilute aqueous aluminum salt solutions: *Soil Sci.* **103**, 144–148.

Gardner, L. R. (1970) A chemical model for the origin of gibbsite from kaolinite: *Am. Mineralogist* **55**, 1380–1389.

Garrels, R. M. and Christ, C. L. (1965) *Solutions, Minerals, and Equilibria*: Harper & Row, New York.

Haas, H. (1972) Diaspore–corundum equilibrium determined by epitaxis of diaspore on corundum: *Am. Mineralogist* **57**, 1375–1385.

Haas, H. and Holdaway, M. J. (1973) Equilibria in the system Al_2O_3–SiO_2–H_2O involving the stability limits of pyrophyllite, and thermodynamic data of pyrophyllite: *Am. J. Sci.* **273**, 449–464.

Helgeson, H. C. (1968) Evaluation of irreversible reactions in geochemical processes involving minerals and aqueous solutions—I. Thermodynamic reactions: *Geochim. Cosmochim. Acta* **32**, 853–877.

Helgeson, H. C. (1969) Thermodynamics of hydrothermal systems at elevated temperatures and pressures: *Am. J. Sci.* **267**, 729–804.

Helgeson, H. C., Garrels, R. M. and Mackenzie, F. T. (1969) Evaluation of irreversible reactions in geochemical processes involving minerals and aqueous solutions—II. Applications: *Geochim. Cosmochim. Acta* **33**, 455–481.

Huang, W. H. and Keller, W. D. (1972) Geochemical mechanics for the dissolution, transport, and deposition of aluminum in the zone of weathering: *Clays & Clay Minerals* **20**, 69–74.

Huang, W. H. and Keller, W. D. (1973) Gibbs free energies of formation calculated from dissolution data using specific mineral analyses. III. Clay Minerals: *Am. Mineralogist* **58**, 1023–1028.

Keller, W. D., Hanson, R. F., Huang, W. H. and Cervantes, A. (1971) Sequential active alteration of rhyolitic volcanic rock to endellite and precursor phase of it at a spring in Michoacan, Mexico: *Clays & Clay Minerals* **19**, 121–127.

Kennedy, G. C. (1950) A portion of the system silica–water: *Econ. Geol.* **45**, 629–653.

Kennedy, G. C. (1959) Phase relations in the system Al_2O_3–H_2O at high temperatures and pressures: *Am. J. Sci.* **257**, 563–573.

Kittrick, J. A. (1969) Soil minerals in the Al_2O_3–SiO_2–H_2O system and a theory of their formation: *Clays & Clay Minerals* **17,** 157–167.

Meyer, C. and Hemley, J. (1959) Hydrothermal alteration in some granodiorites: *Clays & Clay Minerals* **6,** 89–100.

Mizutani, S. (1966) Transformation of silica under hydrothermal conditions: *J. Earth Sci.* **14,** 56–88.

Morey, G. W. and Chen, W. T. (1955) The action of hot water on some feldspars: *Am. Mineralogist* **40,** 996–1000.

Nakagawa, Z., Hatahira, S., Urabe, K. and Yamada, H. (1973) Studies on the crystallization process in the system feldspar–NaOH–H_2O at low temperatures (in Japanese): *J. Japan. Assoc. Mineral. Petrol. Econ. Geol.* **68,** 58–69.

O'Neil, J. R. and Taylor, H. P., Jr. (1967) The oxygen isotope and cation exchange chemistry of feldspars: *Am. Mineralogist* **52,** 1414–1437.

Parks, G. A. (1972) Free energies of formation and aqueous solubilities of aluminum hydroxides and oxide hydroxides at 25°C: *Am. Mineralogist* **57,** 1163–1189.

Tsuzuki, Y. and Mizutani, S. (1969) Kinetics of hydrothermal alteration of sericite and its application to the study of alteration zoning: *Proc. Intern. Clay Conf.* 1969 **1,** 513–522.

Tsuzuki, Y. and Mizutani, S. (1971) A study of rock atteration process based on kinetics of hydrothermal experiment: *Contr. Mineral. Petrol.* **30,** 15–33.

Tsuzuki, Y., Mizutani, S., Shimizu, H. and Hayashi, H. (1973) Kinetics of alteration of K-feldspar tp kaolinite and its application to the genesis of kaolin deposits: *Proc. Intern. Clay Conf.* 1972, 313–319.

Tsuzuki, Y., Mizutani, S., Shimizu, H. and Hayashi, H. (1974) Kinetics of alteration of K-feldspar and its application to atteration zoning: *Geochem. J.* **8,** 1–20.

Wilson, M. J., Bain, D. C. and Mchardy, M. J. (1971) Clay mineral formation in a deeply weathered boulder conglomerate in north-east Scotland: *Clays & Clay Minerals* **19,** 345–352.

Wollast, R. (1967) Kinetics of the alteration of K-feldspar in buffered solution at low temperature: *Geochim. Cosmochim. Acta* **31,** 635–648.

AUTHOR CITATION INDEX

A'cquaye, D. K., 196
Aguilera, N. H., 59
Alcover, S. F., 170
Alexander, L. T., 53
Alexander, G. B., 196
Alexander, L. E., 67
Altschuler, Z. S., 252
Anderson, G. M., 210
Aomine, S., 67
Armitage, T. M., 74
Armstrong, F. A. J., 252
Aspinall, J. D., 80

Bailey, S. M., 197, 254
Bailey, S. W., 210
Bain, D. C., 86, 260
Barany, R., 196, 252
Barin, I., 258
Barnes, H. L., 252
Barnes, V. E., 122
Barshad, I., 59, 122, 127
Barton, P. L., 252
Bassett, W. A., 122
Beane, R., 225
Beck, K. C., 254
Beilby, G., 74
Benson, S. W., 254
Berner, R. A., 252, 254
Bethke, P., 252
Bickely, F., 253
Bird, G. W., 210
Block, F. E., 254
Boettcher, A. L., 136
Bohn, H. L., 210
Bohor, B. F., 86
Bolt, G. H., 123, 226
Boltz, D. F., 80
Bondham, J., 86
Bonnet, J. A., 210
Bonnickson, K. R., 187, 253
Bormann, F. H., 163
Boulange, B., 86
Bourbeau, G. A., 196

Bradley, W. F., 81, 123
Bragg, W. L., 26
Bray, R. H., 53
Bricker, O. P., 196, 225, 252, 253
Bridge, J., 196
Brindley, G. W., 59
Brinkman, R., 163
Brown, G., 59, 136
Broyer, T. C., 163
Bruggenwert, M. G. M., 163
Bruynzeel, L. A., 163
Bullwinkel, E. P., 186
Burrough, P. A., 164
Burton, E. F., 74
Buseck, P. R., 170
Byers, 53

Campbell, J. M., 53
Carlson, L., 80
Carmoure, J. P., 163
Carroll, D., 53
Cervantes, A., 258
Chen, W. T., 260
Childs, C. W., 80
Christ, C. L., 136, 196, 210, 225, 226, 252, 258
Chukhrov, F. V., 80
Chute, J. H., 74, 136
Clabaugh, S. E., 122
Clark, J. L., 196
Clarke, F. W., 26, 186, 196
Clelland, D. W., 74
Cline, M. G., 196
Cloos, E., 59
Cobble, J. W., 252
Coleman, N. T., 53, 163
Copeland, C. S., 254
Corey, R. B., 59
Correns, C. W., 252, 258
Cosslett, V. E., 74
Couture, R., 225
Cox, J. D., 225
Crank, J., 122
Crawford, D. V., 196

Author Citation Index

Criss, C. M., 252
Cromack, K., 163
Crosson, L. S., 136
Cumming, W. M., 74
Curtis, C. D., 258

Dana, E. S., 26
Day, J. H., 80
De Bruyn, P. L., 197
De Coninck, F., 86
De Prater, B. L., 252
De Wit, C. T., 163
Deer, W. A., 123, 196
Deffeyes, K. S., 252
Deist, J., 225
Deltombe, E., 187
Delvigne, J., 86
Dempster, P. B., 74
Dijkshoorn, W., 163
Doyne, H. C., 53
Drever, J. I., 225
Dreyer, R. M., 187
Driscoll, C. T., 163
Dwornik, E. J., 252
Dyal, R. S., 128

Eaton, J. S., 163
Eggleton, R. A., 170
Ellis, S. G., 74, 136, 252
Engelhardt, W. N., 74
Ennos, A. E., 74
Ermilova, L. P., 80
Eswaran, H., 7, 86
Eugster, H. P., 187
Evans, L. J., 80, 197, 254
Evans, W. H., 210

Farmer, V. C., 74, 136
Fawcett, J. J., 210
Feder, H. M., 197
Feth, J. H., 175, 252
Fieldes, M., 74
Fischer, W. R., 81, 163
Fisher, D. W., 163
Fisher, J. R., 210, 225
Fitts, 53
Fleming, R. H., 187
Fogel, R., 163
Folk, R. L., 170
Foster, M. D., 210
Fournier, R. O., 196, 197, 254
Fox, C. S., 53
Franzini, M., 123
Frederiksen, R. L., 163, 196
French, R. C., 74
Frick, C. R., 196, 258
Fyfe, W. S., 196

Gardner, L. R., 258
Garrels, R. M., 136, 163, 187, 196, 210, 225, 253, 258
Gatineau, L., 170
Gibb, J. G., 74
Gile, P. L., 26
Gilkes, R. J., 136
Glasstone, S., 123
Goldich, S. S., 187
Goldman, M. I., 187
Good, W. D., 252
Gordon, M., Jr., 187
Gordon, R. L., 74
Gordy, W., 226
Gorshkov, A. I., 80
Greenland, D. J., 136
Grier, C. C., 163
Griffin, O. G., 74
Gunter, W. D., 225

Haas, H., 258
Hallsworth, E. G., 196
Hanlon, 53
Harlow, I., 197, 254
Hanson, R. F., 258
Hanway, J. J., 123
Harris, G. W., 74
Harrison, 53
Hatahira, S., 260
Hayashi, H., 260
Helgeson, H. C., 170, 210, 226, 252, 253, 258
Hem, J. D., 137, 196, 253
Hemingway, B. S., 210
Hemley, J. J., 226, 253, 260
Hemond, H. F., 163
Hendricks, S. B., 59, 128, 163
Henmi, T., 80
Hess, P. C., 253
Heston, W. M., 196
Hietanen, A., 59
Hirota, 74
Hitchen, C. S., 196
Hoagland, D. R., 163
Hoda, S. N., 136
Hofmann, U., 123
Holdaway, M. J., 258
Holdren, C. R., 170
Holland, H. D., 163, 253
Hollander, M. A., 196
Holmes, R. S., 26
Honess, 53
Honjo, Y., 67
Hood, W. C., 136
Hostetler, P. B., 225, 226
Howard, P., 187, 225
Hower, J., 226
Howie, R. A., 196

Author Citation Index

Hsu, P. H., 136
Huang, P. M., 136, 258
Hubbard, W. N., 197
Huber, N. K., 187
Hughes, R. E., 86
Hunziker, R. R., 123

Ihler, R. K., 196
Iijima, S., 170

Jackson, M. L., 53, 59, 67, 80, 128, 136, 137, 196, 197, 210, 226, 227
Jaffe, I., 254
James, W. C., 170, 253
Janitzky, P., 163
Jefferson, M. E., 59, 196
Jeffries, C. D., 53
Jenne, E. A., 67
Jenny, H., 53
Johnson, M. W., 187
Johnson, N. M., 163
Jones, N. K., 53
Jordens, E. R., 164

Kanno, I., 67
Karpourch, R. P., 86
Karpov, I. K., 226
Kashik, S. A., 226
Keay, J., 123
Keller, W. D., 170, 187, 196, 197, 258
Kelly, K. K., 187, 196, 252, 253
Kennedy, G. C., 196, 258
Kerr, P. F., 26
Khanna, P. K., 163, 164
King, E. G., 187, 253, 254
Kitahara, S., 196
Kittrick, J. A., 196, 210, 226, 253, 260
Klotz, I., 136
Klug, H. P., 67
Knacke, O., 258
Koopmans, K., 74
Koutler-Anderson, E., 80, 163
Kramer, H., 252, 253
Krauskopf, K. K., 187, 196
Kuwano, Y., 67

Lacina, J. L., 252
Laidler, K., 123
Langmuir, D., 226
Larsen, E. S., 26
Latimer, W. M., 187, 196, 253
Laudelout, H., 226
Leckie, J. O., 197
Leeflang, K. W. F., 163
Lentze, W., 80
Levine, S., 254
Lewis, D. G., 136

Likens, G. E., 163
Lindheim, M., 163
Lindsay, W. L., 196, 210
Livingstone, D. A., 136, 253
Lotse, E. G., 136, 210
Lovelock, J. E., 163

McCorison, F. M., 163
McCullough, J. P., 252
MacEwan, D. M. C., 8, 59
McHardy, W. J., 74, 86, 260
Mack, G., 170
MacKenzie, F. T., 163, 196, 210, 225, 252, 253, 258
Mackintosh, E. E., 136, 196
Mahon, W. A. J., 252
Margrave, J. L., 197, 254
Marshall, C. E., 59
Martell, A. E., 254
Martin, F. J., 53
Martin, H., 226
Mathieson, A. McL., 123
Mattson, S., 8, 163
Matzner, E., 163
Mayer, R., 164
Mehler, A., 123
Mehlich, A., 53
Mehra, O. P., 80
Meiwes, K. J., 163
Mellon, M. G., 80
Mel'nik, Y. P., 226
Mellvaine, T. C., 197
Meunier, A. F., 170
Meyer, C., 253, 260
Miller, H. G., 163
Miller, J. D., 163
Miller, R. W., 197
Minderman, G., 163
Mitchell, B. D., 74
Mizushima, 74
Mizutani, S., 260
Mohr, E. C. J., 187
Montoya, J. W., 226
Moore, G. W., 253
Morey, G. W., 197, 254, 260
Morgan, J. J., 164
Mortland, M. M., 123, 136
Murad, E., 80

Nagelschmidt, G., 74
Nakagawa, Z., 260
Nesbitt, H. W., 225
Newman, A. C. D., 136
Nixon, R. A., 170
Noake, H., 74
Noggle, J. C., 128, 136, 163
Norrish, K., 136, 137

Author Citation Index

Nriagu, J. O., 210
Nygren, H. D., 253

O'Keefe, M. A., 170
O'Neil, J. R., 260
Orr, R. C., 187, 253
Overbeck, J. T. G., 197

Paepe, P. D., 86
Page, R., 170
Page, A. L., 226
Pankratz, L. B., 254
Parfitt, R. L., 80
Parker, V. B., 197, 210, 254
Parham, W. E., 86, 170
Parks, G. A., 260
Paver, H., 59
Pearson, D., 254
Peech, M., 59, 196
Pennington, R. P., 196
Perry, E. A., Jr., 226
Petrovic, R., 53
Pierce, R. S., 163
Pierre, W. H., 197
Pilcher, G., 225
Pohlman, G. G., 197
Polzer, W. L., 175, 252
Pourbaix, M. J. N., 187
Prenzel, J., 163

Quirk, J. P., 136

Radoslovich, E. W., 123
Raghu Mohan, N. G., 86
Raman, K. V., 136
Rausell-Colom, J. A., 136
Reed, M. G., 123, 136
Reesman, A. L., 197
Reiche, P., 53, 187
Rennie, D. A., 136
Reuss, J. O., 163
Reynders, H. F. R., 164
Rex, R. W., 197
Rich, C. I., 210
Richards, L. A., 128
Richter, D. H., 253
Ridder, T. B., 164
Rieck, G. D., 74
Ritchie, P. D., 74
Roberson, C. E., 67, 175, 196, 252, 253
Robie, R. A., 136, 210, 226, 254
Robinson, W. O., 26
Rosenqvist, I. T., 123, 163
Ross, C. S., 26
Rossini, R. D., 254
Roth, C. B., 136, 210

Routson, R. C., 137, 226
Rowe, J. J., 197, 254
Rowsell, J. G., 80
Russell, J. D., 80
Russell, K. L., 226

Sadig, M., 210
Sand, L. B., 187
Sawhney, B. L., 258
Schiaffino, L., 123
Schofield, R. K., 137, 197
Schulze, D. G., 80
Schumm, R. H., 197, 254
Schwertmann, U., 80, 81, 123
Scott, A. D., 123, 136
Sennett, R. S., 74
Serratosa, J. M., 123
Setlow, L. W., 86, 197
Shannon, E. V., 26
Sharp, J. W., 74
Shawney, B. L., 137
Sherman, G. D., 196, 197
Shimizu, H., 260
Siebert, R. M., 225
Siever, R., 197, 254
Sillen, L. G., 254
Simpson, 53
Slaughter, M., 226
Sollins, P., 163
Spears, D. A., 258
Sridhar, K., 137
Stanford, G., 59
Stephen, I., 59
Stober, W., 197
Stoops, G., 86
Strahler, A. N., 254
Stumm, W., 164, 197
Summer, M. E., 123
Suttner, L. J., 170
Sverdrup, H. J., 187
Sweatman, T. R., 136
Syers, J. K., 136, 137, 210, 226
Sys, C., 86

Tardy, Y., 210
Taylor, A. W., 197
Taylor, H. P., Jr., 260
Taylor, R. M., 81, 137
Thomas, G. W., 163
Thomas, W. J. O., 226
Thompson, M. E., 252
Tinsley, J., 196
Thrush, P. W., 74
Todd, S. S., 187, 253
Toke de Wit, 164
Tollan, A., 164

Toulmin, P., 3rd, 252
Towe, K. M., 81
Tracey, J. I., 187
Truesdell, A. H., 226
Tsuzuki, Y., 260
Tyler, S. A., 196

Uehara, G., 197
Urabe, K., 260
Ulrich, B., 8, 164

Van Baren, F. A., 187
Van Beek, C. G. E. M, 164
Van Bladel, R., 226
Van Breemen, N., 164
Van Dobben, H. F., 164
Van Grinsven, J. J. M., 164
Van der Marel, H. W., 74
Vanden Heuvel, R. C., 59
Van Lier, J. A., 197
Veblen, D. R., 170
Velde, B., 170
Velthorst, E. J., 164
Voight, G. K., 137
Von Englehardt, W., 252

Wagman, D. D., 197, 210, 254
Waldbaum, D. R., 136, 226, 254
Walker, G. F., 123
Walker, G. J., 123, 128
Waring, G. A., 137, 254
Watson, J. H. L., 74, 80
Wayman, C. H., 197
Weaver, R. M., 226

Weiss, A., 123, 197
Weller, W. W., 253, 254
Wells, C. B., 136, 137
Wells, N., 80
Wenk, H. R., 170
Wherry, E. T., 26
White, D. E., 137, 254
Whittemore, D. O., 226
Whittig, L. D., 197, 227
Wicks, C. E., 254
Wielemaker, W. G., 164
Wild, A., 123
Wilde, P., 254
Wildman, W. E., 197, 227
Williamson, K. I., 74
Willis, A. L., 53, 196
Willis, 53
Wilson, M. J., 86, 136, 260
Wise, S. S., 197, 254
Wollast, R., 254, 260
Wong, C. B., 7, 86

Yamada, H., 260
Yeow Yew Heng, 86
Yoder, H. S., 187, 210
Yoshinaga, N., 67
Young, R. C., 136
Young, S. W., 170

Zen, E-an, 210, 225, 227
Zobell, C. E., 210
Zussman, J., 123, 196
Zvyagin, B. B., 80

SUBJECT INDEX

Acidification or alkalization
    arid soils, 162
    reduced soils, 159–161
    well drained soils, 156–159
Acid neutralization capacity, 142
Aluminum
    interlayers, 57
    nonexchangeable, 6
Amorphous material
    on clay minerals, 69
    in feldspar weathering, 82
Amorphous silica
    solubility, 189
    as unstable intermediate, 189

Bauxite, solubility diagrams, 255
Beidellite, chemical composition, 15
Biotite
    effect of plant growth on, 124
    weathering by wheat, 124
Boehmite, free energy of formation, 190

Calcium carbonate, free energy of formation, 177–179
Carbon dioxide, effect on weathering, 239–241
Cerussite, free energy of formation, 185
Chlorite
    free energy of formation, 199
    interstratification, 208
    solid solution components, 209
    solubility, 199
    solution equilibrium, 207
Chrysotile, free energy of formation
Clay mineral
    amorphous coatings, 68
    Beilby layer, 71
    coatings, 7
    crystalline, 5, 26
    optical properties, 10–13
Clinochlore, free energy of formation, 200

Diagenesis, seawater, 244–248
Diaspore, free energy of formation, 191

Dickite
    chemical composition, 15
    X-ray diffraction, 17

Electron microscopy
    clay mineral coatings, 68, 69
    electron beam damage, 73, 166
    high resolution, 165
    holey carbon films, 69, 166
Evaporative concentration, 242–244
Exchangeable cations, energy associated with, 215

Feldspar
    dissolution, 256
    free energy of formation, 183–184
    weathering, 165
        to illite, 165
        to 10 Å lattice planes, 167
        to 10 Å rings, 167
        mass transfer, 229–240
        microsite equilibrium, 169
        montmorillonite, 169
        scanning electron microscope, 82
Ferrihydrite, 7, 75
    color, 76
    Mössbauer spectroscopy, 77
    X-ray diffraction, 77
Free energy of formation. *See also specific minerals*
    estimating from oxide and hydroxide components, 212
    from geologic relations, 176
        check calculations, 177
        layer silicates, 212
    from weathering processes, 180

Gibbsite
    equilibrated with chlorite, 202
    free energy of formation, 190
    from kaolinite, 84
    solubility diagrams, 256

Subject Index

Halloysite
  chemical composition, 15
  from feldspars, 84
  X-ray diffraction, 20
Hematite, equilibrated with chlorite, 202
Hydrothermal rock alteration, 248

Illite, estimation of free energy, 220-224
Immiscible displacement method, 206
Imogolite, 6, 60
  CEC and delta value, 66
  electron micrographs, 61
  infrared analysis, 65
  thermal analysis, 64
  X-ray diffraction, 61-64
Interstratification, random or regular, 56
Ion thinning, 166

Kaolinite
  chemical composition, 15
  from feldspars, 84
  free energy of formation, 181-184, 191, 215
  solubility diagrams, 255
  weathered from feldspar, 231
  X-ray diffraction, 17, 18

Layer silicate, estimating free energy of formation, 212

Malachite, free energy of formation, 185
Mica
  biotite transformed to vermiculite, 124
  chemical analysis, 93, 94
  free energy of formation, 182-184
  potassium replacement, 95, 114-117
    critical K concentration, 106-108, 119
    diffusion, 102-106, 108
    electron probe analysis, 100, 109-111
    optical boundary, 95-97
    X-ray diffraction, 97
  weathering, 91, 119-121, 134
    critical K level, 135
    from feldspars, 231
    fluorine content, 118
Microenvironment equilibria, 173
Minnesotaite, free energy of formation, 215
Montmorillonite
  chemical composition, 15
  energy of exchangeable ions, 218
  estimation of free energy, 220-224
  solubility, 192
  X-ray diffraction, 19
Muscovite
  alteration to clay minerals, 54
  free energy of formation, 215

Ore deposition, 248-252

Paragonite, free energy of formation, 215
Proton
  budget, 152-156
  cycle, 89
  dissociation constants, 140
  energy levels, 141
  sinks, 154
  sources, 154
Proton transfer process
  aqueous phase, 149
  aqueous to atmosphere, 149
  aqueous to solid, 150, 151
  biota to aqueous, 145-149
  cation cycling, 154
Pyrophyllite, free energy of formation, 215
Pyrophyllite-diaspore deposits, solubility diagrams, 255

Quartz
  rate of dissolution and precipitation, 189
  solubility, 189

Rutherfordite, 179

Sepiolite, free energy of formation, 212
Silica-sesquoxide ratio, 25
Soil acidification, 139
  intensity and capacity factors, 140
Soil alkalization, 139
Soil colloid
  chemical analysis, 22
  X-ray, 21

Talc, free energy of formation, 212

Vermiculite
  from biotite, 124
  dioctahedral, 55
  stability diagram, 132
  as unstable intermediate, 89, 129

Weathering
  binary transformation, 30, 32
  broadscale, 1
  chemical changes, 4
  control by solution silica, 195
  definition, 4
  depth function, 42
  desilication, 47
  equilibrium, 52
  fundamental generalization, 33
  irreversible reactions, 228
  laterization, 49
  mass transfer, 229

microenvironments, 1, 7, 90
partial equilibrium model, 228
particle size function, 41
permitted and nonpermitted
 associations, 194
podzolization, 48
reversal, 39
secondary depositions, 36
stability diagram, 193
Weathering rate
 capacity factors, 38
 effect on mineral formation, 196
 intensity factors, 38
 in Sierra Nevada, 241
Weathering stages, 5, 30, 31–38

X-ray diffraction, powder patterns, 14–20, 27–29, 56. See *also specific minerals*

Zoning
 rock alteration, 257
  at elevated temperatures, 258

About the Editor

J. A. KITTRICK obtained the Ph.D. in soil science at the University of Wisconsin in 1955. Since then he has worked in soil mineralogy and soil chemistry at Washington State University. His main interest is mineral weathering. He is a Fellow of the American Society of Agronomy and the Soil Science Society of America. Dr. Kittrick is active in environmental and conservation issues. He led the effort to obtain wilderness classification for the Wenaha-Tucannon Wilderness, and helped to obtain conservation legislation in the 1985 Farm Bill. He is currently chairman of the Program in Environmental Science and Regional Planning at Washington State University.